Plastic Pollution

Malti Goel · Neha G. Tripathi
Editors

Plastic Pollution

Challenges and Green Solutions

Climate Change Research Institute

Springer

Editors
Malti Goel
Climate Change Research Institute
and Ministry of Science and Technology
New Delhi, India

Neha G. Tripathi
School of Planning and Architecture
New Delhi, India

ISBN 978-981-97-5527-1 ISBN 978-981-97-5528-8 (eBook)
https://doi.org/10.1007/978-981-97-5528-8

This Springer imprint is published by the registered company Springer Nature Singapore Pte Ltd.
The registered company address is: 152 Beach Road, #21-01/04 Gateway East, Singapore 189721, Singapore

This book is dedicated to the memory of Smt. Kanta Mittal, a gracious, kind-hearted, and devoted wife of Prof. S. R. Mittal of Birla Institute of Science & Technology, Pilani in Rajsthan, India. Fondly remembered as 'Amma', she was a mother figure to all. Her selfless commitment, cultural values, and dedication inspired us to strive for a better life.

Foreword

In the not-so-distant past, the world functioned without the ubiquitous presence of plastics. We carried our groceries in cloth bags, refilled glass bottles with milk, and stored leftovers in steel containers. Then the synthetic plastic was patented in 1909, and it was followed by a surge in its production and use especially after World War II. The plastics industry grew across the globe, and its use quickly spread to the United States, the United Kingdom, France, and Russia. Plastic, then a marvel of modern science, promised convenience and durability. It seemed like the answer to so many problems.

Hailed as the "material of the future," this versatile material brought convenience, serving as a cost-effective means to transport both solids and liquids. Plastic wares and furniture became symbols of pride, and their uses expanded exponentially, particularly in developing regions. However, as its popularity soared, a harsh reality emerged. The evidence of the collection of plastic waste in the ocean was seen as a threat to living beings on the planet. Soon, it became a "global problem of the future" because plastic showed remarkable resistance to breaking down.

Every challenge comes with solutions, and plastic pollution is no different. The use of plastic to carry hot foods is a huge health hazard. It's crucial that we tackle plastic pollution with comprehensive policies governing its creation, distribution and individual use. Education, no doubt, takes center stage in this battle against plastic pollution. We should find solutions to plastic pollution, control it, and adopt innovative technologies for recycling single-use plastics. We need a collective movement, akin to the 'No Smoking' campaign, and 'No Plastic' to stand against the pervasive use of plastic. Simultaneously, exploring and adopting new materials that are environmentally friendly and versatile is a must. Biodegradable plastic emerges as a promising alternative.

This book, "Plastic Pollution: Challenges and Green Solutions," eds. Dr. Malti Goel and Dr. Neha G. Tripathi, deals with this critical challenge head-on. Covering a vast array of dimensions related to plastic pollution, it underscores the importance of finding green solutions in achieving a number of Sustainable Development Goals. It dives deep into the plastic pollution crisis, exploring its impacts on our environment and ecosystems. I commend the work of each of the eminent scholars and the yeowomen efforts of Dr. Malti, Dr. Neha, and Springer Nature in making this knowledge accessible to a wider audience in India and beyond.

Prof. G. D. Sharma
Former Secretary, University Grants
Commision and Director, Consortium
for Educational Communication
New Delhi, India

Preface

Plastic has become an indispensable material in our modern lifestyles. From packaging and consumer products to medical devices and construction materials, plastics are deeply embedded in every facet of society. However, the surge in global plastic production and consumption over the past decades has given rise to a rapidly escalating plastic pollution crisis that now threatens environmental and public health. Plastic has become an essential material in modern life. It is used in packaging, consumer products, medical devices, and construction materials. However, the increase in global plastic production and consumption has led to a growing plastic pollution crisis that threatens the environment and public health. A recent study by the University of Naples, Italy, found microplastic and nanoplastic particles (MNPs) in 150 out of 257 patients undergoing cardiovascular plaque removal. Polyvinyl chloride (PVC) was found in 31 patients, increasing the risk of heart attack.

The Global Plastic Pollution Partnership (GPAP) is a pivotal alliance, bringing together governments, international organizations, businesses, and communities committed to finding comprehensive solutions. The UN Environment, the Ellen MacArthur Foundation, and the World Economic Forum launched it in 2018. The GPAP has led to national plastic pacts in various countries to drive change through coordinated national strategies and targets. Recently, the G7 and G20 commitments with private sector and civil society initiatives have jointly spotlighted curbing single-use plastic proliferation as an urgent worldwide priority. In an effort to address the crisis, India has implemented various partnerships and policies to tackle plastic pollution. Recently, India has also become a part of the global Clean Seas campaign led by the United Nations Environment Programme, which is dedicated to fighting marine plastic litter.

Plastic pollution solutions would help achieve the Sustainable Development Goals (SDGs), including zero hunger (SDG2), good health and well-being (SDG3), clean water and sanitation (SDG6), requires reducing plastic waste in waterways, oceans, and on land, making responsible consumption choices (SDG 12), addressing climate change (SDG 13), protecting marine and terrestrial ecosystems (SDGs 14 and 15), and building sustainable cities (SDG 11).

This timely new book, 'Plastic Pollution: Challenges and Green Solutions,' lays the foundation and sets the overall context of the issue. The book contains 15 chapters contributed by the experts. The book is in two parts. The first part of 'Plastic Pollution Challenge' discusses land and ocean pollution from plastic waste. In the first chapter on "Understanding Plastic Pollution," Malti Goel appraises the genesis, composition, and structure of plastic materials; the implications of plastic pollution arising on land, ocean, and air; and potential remedies through recycling technologies, sustainable material development, and management strategies - Mission LiFE to reduce plastic pollution on land and at sea. The second chapter on "The Global Plastic Pollution Crisis—How Norway Deals with the Challenge" is by Beate Kværnes Langset. She sheds light on Norwegian strategies and government policies to deal with the plastic crisis and encourages solutions. Since 2019, several projects have been started under the India-Norway Marine Pollution Initiative, focusing on the benefits of the circular economy. The cooperation between Norwegian and Indian stakeholders, policymakers, and researchers aims to work for a cleaner and healthier ocean.

In the third chapter, "Plastics and Human's Health", Priyanka Chaudhury examines the environmental and health risks associated with plastic molecules. Many properties of plastics depend on the additives used as plasticizers, and exposure to these toxic chemicals can lead to irreversible health effects, underscoring the urgent need for action. Chapter fourth by Shyamala K. Mani is on "Plastic Pollution and Health Impacts". She covers case studies of impacts on environmental and health hazards associated with the plastic manufacturing, recycling, and incineration processes. She describes the need for aligning with global trends towards a circular plastic economy, and a case study on the impact of COVID-19 on plastic pollution.

The fifth chapter "Plastic Debris, Microplastics, and Marine Pollution" by Satish Sinha addresses the global crisis of plastic waste, aiming at its impact on the marine ecosystem and human health and emphasizing its role as a vector for toxic chemicals. The study highlights the growing threat of marine plastic pollution and microplastic contamination of the ocean surrounding India, emphasizing the need to preserve coral reefs through policy interventions. Chapter sixth on "Current status of marine litter and microplastic contamination across India's coastal boundaries: Challenges and Management" by Pravakar Mishra et al. further highlights the sources of marine litter. It tracks the research papers that have been written on plastic trash, marine litter, and microplastics (MLMP) along the different coasts of India, suggesting there are big gaps in understanding how these things affect the environment and people's health. The need for a national marine litter policy is highlighted. Tapti Sengupta and Abhijit Mitra in chapter seventh on "Abundance of microplastics in the mangrove dominated Indian Sundarbans: A mini-review" present a quantitative case study approach converging on higher levels of microplastic pollution in coastal sediments of the mangrove-dominated ecosystem in the Indian Sundarbans. The study relates it to harm it may cause to marine environment due to anthropogenic activities, suggesting that an effective policy roadmap to retard plastic pollution growth is needed.

The second part of the book is on 'Green Solutions and Technologies'. Chapter eighth by Gautam Sen on "Overview of Plastics, Usage, Pollution, and Mitigation" is an overview of the exponential growth of the global plastics industry, which has resulted in high plastic waste generation and pollution. Ongoing scientific solutions using bacteria, worms, enzymes, and fungi to mitigate the menace of plastic pollution; and deploying new technologies such as artificial intelligence to address this pressing environmental challenge are highlighted. Chapter ninth on "Role of Digitization in Achieving Sustainable Waste Management of Plastic and Other Materials" by Sourabh Manuja addresses the city challenges of solid waste management and global concerns about marine litter. A digital portal called 'Sansaadhan' is deployed and tested to facilitate visualization of waste flow patterns in cities and enable informed decision-making to accelerate progress towards sustainable plastic waste management. Neha G. Tripathi, in chapter tenth on "Plastic Waste Crisis: How Smart Cities Can Lead the Change", presents key findings from case studies of smart cities. Indore and Panaji cities in India have successfully integrated smart waste management practices leveraging technology and implementing innovative solutions, including advanced segregation systems, IoT-enabled tracking, and localized recycling ecosystems, to tackle plastic pollution and promote sustainability.

The next three chapters provide a practical approach to plastic waste management in India. Chapter eleven by S. Nand and M. Goswami on "Plastic Waste Management in Fertilizer Sector—A Case Study" highlights the extensive use of high-strength plastic bags needed for packaging and distribution of fertilizers to millions of farmers across the country in the fertilizers sector. It showcases compliance with the Plastic Waste Management Rules in India and actions taken for recycling the waste as per EPR guidelines through registered waste management agencies and recyclers. Chapter twelfth of "Plastics and the Sustainability Paradox" by Swapan Ray considers the relevance of plastic materials to Sustainable Development Goals and suggests aggressively pursuing the path of the circular economy. Pathways to mitigating plastic pollution include promoting changes in consumer behavior to eliminate indiscriminate littering, the viability of reducing the overall demand for such products and reducing the waste burden. R. Venkatesan et al., in chapter thirteenth on "Plastics in the Indian Ocean: Survey, Transport, and Removal" present key initiatives taken in India to address the issue of marine plastic pollution. The impact of monsoon circulation on the transport and classification of plastic pollution cleanup technologies for marine litter and potential implications such as the need for comprehensive monitoring systems, intercepting plastic waste before it harms the ocean using autonomous robots, and using biodegradable plastics to effectively are described to address this pressing transboundary issue.

Chapter fourteenth by Shamim Ahmad on "Microbial Biodegradation of Plastics and Microplastics: Recent Development" provides a review giving insights into the mode of action and potential enzymatic mechanisms. It underscores the necessity for optimizing processes to achieve more effective and thorough results, as well as the requirement for further research. Improved efficiency can be achieved by reducing plastic pollution using pre-treatment methods and identifying suitable microbial strains. Chapter fifteenth, "Plastic Waste to Value: Desirable Technology Interventions" by Rakesh K. Khandal and his team provides an overview of the full plastic lifecycle, from design and production of plastics to waste management encompassing environmental, health, business, policymaking, and technological dimensions of the issue. The need for devising strategies and techniques depending on the nature of the constituent waste material is emphasized through technological interventions for control of plastic pollution. Vital recommendations for value addition in the control of pollution are made.

The book "Plastic Pollution: Challenges and Green Solutions" offers a comprehensive understanding of local and global challenges, spanning the scale of the problems arising from plastics and microplastics, their drivers, effects, and emerging solutions. The book stands as a testament to the collective efforts of its contributors, commendable work of authors, and the Springer Nature, is meant to be a vital resource for those interested in working to address this critical environmental challenge of our time. This includes environmental managers, sustainability managers, industry leaders, policymakers, scientists, students, reporters, and society at large. Certainly, the book will earn a place in the libraries of academic institutions, professionals working in the plastics industry, medical doctors, and policymakers.

New Delhi, India
March 2024

Dr. (Mrs.) Malti Goel
Dr. Neha G. Tripathi

Acknowledgements

The book "*Plastic Pollution: Challenges and Green Solutions*" would not have been possible without the invaluable contributions and unwavering support of numerous individuals and organizations. We would like to extend our deepest appreciation to the esteemed authors who generously shared their expertise and insights. We also express our gratitude to Prof G. D. Sharma for providing the foreword to the book and the members of the review panel comprising Prof. V.S. Verma, Mr. Gautam Sen, Prof. Shyamala Mani, and Prof S. Ahmad for their valuable suggestions. We sincerely thank Prof. D. P. Agrawal, Chairman of the Governing Council, CCRI for his encouragement.

We express our special gratitude to the Climate Change Research Institute (CCRI) for organizing the discussions and funding pivotal awareness workshops titled "Cities and Plastic Pollution: Challenges & Solutions" and "Implementing Sustainable Development Goals in India: Beating Plastic Pollution". These workshops played a crucial role in shaping the book, with thought-provoking deliberations and policy papers. The book is a reflection of our shared commitment for creating a sustainable future.

We are grateful to Springer Nature for providing us with the platform to disseminate this crucial research to a broader audience. Our sincere appreciation goes to Ms. Priya Vyas, Senior Editor, and Ms. Swati Meherishi, Editorial Director, Applied Sciences & Engineering, Springer Nature, and their team for their dynamic support and commitment.

It is our hope that this volume will increase awareness about how plastic pollution harms us and stimulate future strategies to tackle these challenges.

About This Book

In 2024, an important global treaty on plastic pollution control is scheduled to take place. The United Nations Environmental Assembly (UNEA) established the Intergovernmental Negotiating Committee (INC) in March 2022 due to concerns about the severe impacts of plastic pollution on the environment and health. The INC is expected to complete its negotiations by the end of 2024 towards a global legally binding instrument aimed at ending plastic pollution, including marine pollution, by 2040.

The book "Plastic Pollution: Challenges and Green Solutions" by the Climate Change Research Institute lays the foundation and sets the overall context of the issue. The book consists of 15 chapters contributed by experts and specialists. It provides an overview of the current challenges in managing plastic waste and offers a multi-dimensional perspective on plastic pollution control while advocating for technology and green solutions. The book describes local challenges, such as plastic waste mismanagement and lack of investment in waste recycling, and global challenges, such as marine pollution and the impact of microplastics on health. It suggests the need for comprehensive science-based solutions and the use of AI technologies.

Contents

Editors and Contributors

About the Editors

Dr. Malti Goel has a Doctorate in Physics and D.I.I.T. in Solid State Physics from Indian Institute of Technology, Delhi. She did her Masters in Physics from Birla Institute of Technology & Science, Pilani, with the first rank receiving a Gold Medal in 1967. She had distinguished service in the Ministry of Science & Technology, Government of India. An emeritus scientist and a renowned climate change expert with over 40 years' experience in carrying out scientific research and policy planning, she has strong track record of multi-partnership and multi-institutional projects' implementation with industry, academia and government participation in science & technology areas of national importance. Dr. Malti Goel is recipient of many prestigious Awards such as Er Avinash Chandra Medal 2006 for excellence in environment education; Bharat Jyoti Award 2012 for outstanding contribution to society; and Build India (Corporate) Award 2015. She became Fellow of National Environment Science Academy in 2008 and of International Conference on Ecology & Environment in 2015. She was awarded the 'Lifetime Achievement Award' in 2016 by the Pearl Foundation Madurai for her exceptional contributions to climate change education and research. She is the recipient of 'Outstanding Contributions to National Development (OCND '23)' Award from IITDAA in Academics and Social Impact category. Her original research has been cited in the handbooks and patents. She is a member of several prestigious professional organizations and is currently leading the Climate Change Research Institute with science and technology solutions for a sustainable energy future.

Dr. Neha G. Tripathi is a distinguished climate and sustainability educationist with experience spanning over 20 years in city planning, architecture, and teaching. She obtained her Doctorate in 2013 and Master of Planning (Environmental Planning) degree from the School of Planning and Architecture, New Delhi, where she received the Gold Medal for overall performance. Dr. Tripathi served as Assistant Professor of Environmental Planning at the School of Planning and Architecture in New Delhi from 2013 to 2022. She has led planning projects for the Eco-sensitive Area Mount Abu, Indian Institute of Management Indore, Central Drug Research Institute Lucknow, among others. Dr. Tripathi has published 5 chapters in technical books by Springer and 15 international papers on topics spanning urban planning, energy, climate change, and smart cities. Her areas of expertise include low carbon cities, smart cities, climate change, and solar energy. She co-authored the book "Solar Energy Made Simple for a Sustainable Future" published by Springer in 2022. Dr. Tripathi is an Associate Member of the Institute of Town Planners, Member of the Council of Architecture, Indian Building Council, and National Environmental Science Academy. Her extensive research, publications, leadership roles, and awards demonstrate her deep knowledge and commitment to the field of environmental planning and architecture. Currently engaged in pioneering efforts, Dr. Tripathi is innovating the climate change curriculum for K-12 students in the United States.

Contributors

Atiya Ahmad Brentwood Tower, Eros Garden, Charmwood Village, Faridabad, Haryana, India

Faisal Ahmad Brentwood Tower, Eros Garden, Charmwood Village, Faridabad, Haryana, India

Shamim Ahmad Brentwood Tower, Eros Garden, Charmwood Village, Faridabad, Haryana, India

Priyanka Chaudhary Department of Oral and Maxillofacial, Trauma Plastic Reconstructive Surgery, All India Institute Of Medical Sciences (AIIMS), New Delhi, India;
HOD Chemistry (Teaching and Research), Brilliants' Convent Sr. Sec School, Pitam Pura Delhi, India

Pranshu Chhabra Self Employed Scientist, Ahmedabad, India

K. Dhineka National Centre for Coastal Research, Ministry of Earth Sciences (MoES), Government of India, NIOT Campus, Pallikarani, Chennai, India

Malti Goel Climate Change Research Institute and Ministry of Science and Technology, New Delhi, India

Manish Goswami The Fertiliser Association of India, FAI House, New Delhi, India

Ruchi Gupta Chemcred, Delhi, India

T. Kaviarasan National Centre for Coastal Research, Ministry of Earth Sciences (MoES), Government of India, NIOT Campus, Pallikarani, Chennai, India

R. Keerthivasan National Institute of Ocean Technology, Chennai, India

Rakesh K. Khandal SDGI Global University, Ghaziabad, U.P., India

Beate Kværnes Langset Counsellor Climate and Environment 2022–2024, Royal Norwegian Embassy, New Delhi, India

Shyamala K. Mani Centre for Chronic Disease Control, New Delhi, India;
NIUA, New Delhi, India

Sourabh Manuja CCP-ME Project, Deutsche Gesellschaft für Internationale Zusammenarbeit (GIZ) GmbH, New Delhi, India

Pravakar Mishra National Centre for Coastal Research, Ministry of Earth Sciences (MoES), Government of India, NIOT Campus, Pallikarani, Chennai, India

Abhijit Mitra Department of Marine Science, Calcutta University, Kolkata, India

S. Nand The Fertiliser Association of India, FAI House, New Delhi, India

Ayesha Nazeer Brentwood Tower, Eros Garden, Charmwood Village, Faridabad, Haryana, India

K. Nidhi CFRA, NIFTEM-K, Sonipat, Haryana, India

S. Ramasundaram National Institute of Ocean Technology, Chennai, India

Swapan Ray Indian Center for Plastics in the Environment (ICPE), Mumbai, India;
Somaiya Vidyavihar University, Vidyavihar, Mumbai, India

Sunanda Sahu National Centre for Coastal Research, Ministry of Earth Sciences (MoES), Government of India, NIOT Campus, Pallikarani, Chennai, India

M. Sambandam National Centre for Coastal Research, Ministry of Earth Sciences (MoES), Government of India, NIOT Campus, Pallikarani, Chennai, India

Gautam Sen Ex, RIL, Mumbai, India;
Ex, ONGC, New Delhi, India

Tapti Sengupta Department of Microbiology, West Bengal State University, Barasat, Kolkata, India

Satish Sinha Toxics Link, New Delhi, India

Yazhini Sivaram Former Intern, University of Liverpool, Liverpool, UK

Meenu Talwar Material Science Division, Shriram Institute for Industrial Research, Delhi, India

Neha G. Tripathi School of Planning and Architecture, New Delhi, India

R. Venkatesan Institute of Ocean Management, Anna University, Chennai, India

Plastic Pollution Challenge

Understanding Plastic Pollution

Malti Goel

Abbreviations

ASEAN	Association of Southeast Asian Nations
ABS	Acrylonitrile Butadiene Styrene
CPCB	Central Pollution Control Board
CSP	Center for Sustainable Polymers
EPR	Extended Producer Responsibility
EPS	Expanded Polystyrene
GSM	Gram Per Square Meter
GPAP	Global Plastic Action Partnership
HDPE	High-Density Polyethylene
INC	Intergovernmental Negotiating Committee
LDPE	Low-Density Polyethylene
LiFE	Life Style for Environment
PCC	Pollution Control Committee
PC	Polycarbonate
PDK	Poly-Di-Ketoenamine
PE	Polyethylene
PET	Polyethylene Terephthalate
POPs	Persistent Organic Pollutants
PP	Polypropylene
PS	Polystyrene
PVC	Polyvinyl Chloride
SPCB	State Pollution Control Board
SUP	Single-Use Plastic

M. Goel (✉)
Climate Change Research Institute and Ministry of Science and Technology, New Delhi, India
e-mail: maltigoel2008@gmail.com

M. Goel and N. G. Tripathi (eds.), *Plastic Pollution*,
https://doi.org/10.1007/978-981-97-5528-8_1

UNFCCC United Nations Framework Convention on Climate Change
UNEA United Nations Environment Assembly

1 Introduction

Plastic is very much on the menu

—King Charles III in 'Our Ocean' conference held in Malta in 2017 [2].

King Charles III, King of the United Kingdom (then Prince Charles), said that we are very close to reaching the point when whatever wild-caught fish you eat will contain plastic and called for decisive action to stop overfishing [2]. Horrified by the plastic littering of the seas, in another marine conference held earlier, he had observed that we absolutely cannot ignore an increasing quantity of plastic waste in the marine environment. The marine plastic pollution problem has further magnified since then, and efforts are being made to find technological solutions.

Plastics such as, rubber, horn, and shellac, date back to ancient civilizations [3]. These were derived from natural products. Other forms of plastic, namely celluloid and cellophane, became known in the nineteenth century [4]. Leo Beakeland, a Belgian chemist, created the first synthetic plastic, named Bakelite, in 1907 by reacting two chemicals, phenol and formaldehyde in a laboratory. He described it as 'plastic material'. Other plastic materials were synthesized subsequently. In the 1930s, plastics were produced from crude oil. Wallace Carothers and his team at DuPont developed nylon, one of the first synthetic fibers made of a polymeric material [5]. Synthetic rubber 'neoprene' was produced by Julius Nieuwland while working for DuPont. Plastics continued to evolve with the development of new plastics with improved properties. The commercial production of plastics from petrochemicals has advanced significantly with these developments. Production of polyethylene (PE) began in the 1930s. Polypropylene (PP), which was synthesized in the 1950s, became commercially available in the 1960s. Common types of plastics include polyethylene (PE), polypropylene (PP), polyvinyl chloride (PVC), polystyrene (PS), and many more, each with its own unique characteristics and applications. In due course, plastics like high-density polyethylene (HDPE), low-density polyethylene (LDPE), polycarbonate (PC), polyethylene terephthalate (PET), and polytetraphthalate (PTFE-Teflon) were produced in large quantities and became integral to consumer goods, automotive components, medical devices, packaging, and many more. Plastics have emerged as versatile, moldable, durable, cost-effective, and lightweight materials. Cumulative plastic production from 1950 to 2017 was estimated at 8.3 BT [6]. It is projected to reach 34 billion metric tons (BT) by 2050, and 12 BT will be plastic litter on land and sea.

With this background in mind, the structure of the chapter is as follows: Sect. 1 introduces the history of plastics and the seriousness of their impact. Section 2 deals with plastic materials. Section 3 delves into pollution from plastic waste and its impacts on the ecosystem. Section 4 is about technological and strategic solutions and regulations to manage plastic waste. Finally, Sect. 5 makes a concluding recommendation.

2 Plastic Materials: The Awakening of a New Era in Materials Science

Plastics are essentially made of macromolecules, known as polymers. The basic building block of a polymer is a 'monomer', a single molecular entity that combines in large numbers with identical or similar molecules through chemical arrangements and forms a long chain. The process is known as polymerization. The polymers are therefore long chains of identical small molecules linked by 'weak' or 'strong' chemical bonds. Most polymers in large-scale use contain carbon atom chains with hydrogen and other atoms such as oxygen, chlorine, fluorine, nitrogen, silicon, phosphorus, sulfur, and others. A polymer is named according to the structure of the monomer.

Polymers can be engineered into plastics to have a wide range of properties, making them suitable for various applications. Plastics are lightweight and can be molded into various shapes and sizes using different processes like injection molding, extrusion, blow molding, etc. This makes the material ideal for creating complex and custom-designed components. Depending on the chemical composition and structure, some can be rigid and strong, while others are flexible and elastic. Plastics are, therefore, versatile materials. Due to their cost-effectiveness, they are adopted in numerous applications [7]. Different plastics have been the subject of intense research and innovation. Their use grew in building construction, paints, clothing, packaging, automobiles, and in many other functions of everyday life (Fig. 1). With advancements in science and technology, polymers are becoming vital for advanced research into hydrogen energy storage materials, smart technologies, robotics, artificial intelligence, etc. Soon we will be in a situation where it is hard to find things not made of plastic materials.

Plastics come in various types, each with its own set of properties and applications. They are classified based on their chemical composition and properties. Two broad categories according to how they behave when they're heated are; (1) Thermoplastics and (2) Thermosets.

(1) Thermoplastics: In a thermoplastic, the long polymer molecules are joined to one another by very weak bonds in a linear chain (Fig. 2a), which easily break apart when heated and quickly reform again when the heat is taken away. They soften when heated, are easy to melt down and recycle. However, when mechanical force is applied, they show ***creeping*** behavior as they do not come back to their previous shape. Some common examples are polyethylene (plastic bottles and sheets), polystyrene (crumbly white packaging material), polypropylene (plastic

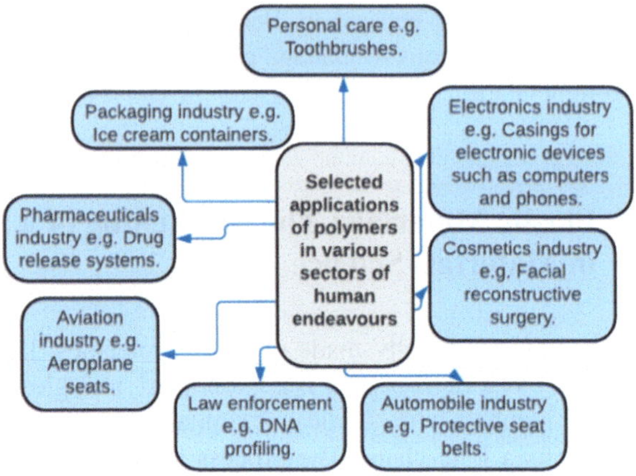

Fig. 1 Plastics are one of man's greatest feats in the field of science and technology [7]

ropes), polyvinylchloride/PVC (toys and credit cards), etc. Thermoplastics have a nearly 85% share of total plastic consumption.

(2) Thermosets: Thermosetting plastics are usually made in cross-linked bonds from much bigger polymer chains than thermoplastics (Fig. 2b). They do not soften when heat is applied. They are molded into shapes during the manufacturing process by the application of heat or pressure to form a dense, hard structure with strong cross-links binding to their neighbors. This process is known as ***curing***. Thermosets, once they're set or cured during manufacture, cannot be remolded. Some examples of common uses are polyurethane (insulating material in buildings), polytetrafluoroethylene/PTFE (nonstick coatings on cooking pans), computer casings, and epoxy resin (as adhesives).

Additives: In plastics, to get desired chemical and physical properties, extra ingredients such as colorants (to add color to plastics), plasticizers (to make it more flexible, viscous, and easier to shape), stabilizers (to stop plastics from breaking apart), fillers (as low-cost minerals to reduce the cost), and reinforcements (to make them tougher) are added during processing. These are known as additives or plasticizers. The additives affect the chemical composition, chemical, and mechanical properties of the plastics and also influence their cost while making them more versatile.

3 Plastic Waste Pollution

Knowing polymers is vital to understanding plastic pollution. The plastic materials offer numerous benefits in use; they were considered inert, lightweight and durable, but their environmental and health impacts were not considered. Their durability meant that they did not decompose for years after use. Having a long life, they are not recyclable easily and contribute to pollution; effective environmental management requires a proper understanding of both. It is estimated that nearly 8 billion metric

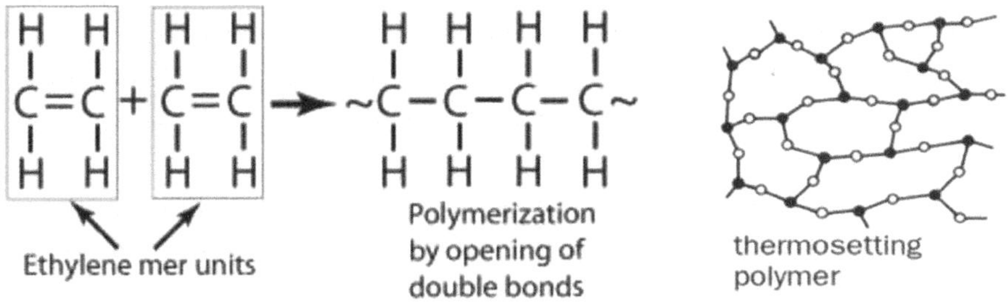

a Thermoplastics linear chain b Thermosets cross-links

Fig. 2 **a** Thermoplastics linear chain. **b** Thermosets cross-links

tons of plastic waste exist on the planet. Most plastics are derived from crude oil refinery waste, produced from fractional distillation. As the plastic industry had put crude waste from the petroleum industry into a usable form, Marc Hillmyer, director of the Center for Sustainable Polymers (CSP) at the University of Minnesota, once observed about plastics, '*... although they're durable in use, they're also durable in waste* [8].' This aptly describes the reasons for high plastic pollution and the need for its proper management. The question therefore arises: is the pollution solely due to plastic materials, or is it caused by a lack of waste management?

3.1 Sources of Plastic Pollution

The accumulation of plastic waste in the environment from its use in various sectors of the economy is giving rise to large-scale global concern. Notably, plastic pollution occurs when plastics are littered after use and not managed well. Major sectors contributing to plastic waste can be identified as households, industries, agriculture farms, and medical centers. Household waste arises from food wrappers and containers, shopping bags, bottles, and caps, and electrical equipment like computers, mobiles, etc. It is directly related to plastic consumption per capita in households. In advanced nations like the USA, plastic consumption could be as much as 100 kg per capita or more, whereas in most developing countries, including India, it averages 20 kg per capita [9], producing waste in proportion.

Industrial waste is mostly manufacturing-based; it arises from the automotive and construction industries, small and medium enterprises, and the electronics industry. For agriculture sector industries are major consumers of plastics in the form of packaging materials for seeds, fertilizer bags, irrigation pipes, and nets. Plastic covers used for fertilizer transportation, storage, and pest control are large sources of plastic waste. In medical centers, plastic waste includes used hand gloves, syringes, medical cartridges, blood bags, face masks, etc. During COVID-19, demand for single-use plastics grew enormously. It was seen that health centers' plastic waste contribution in India grew as much as six times [10]. The medical litter is non-recyclable, highly toxic, hazardous, and harms the ecosystem. The pollution is dependent on the composition of the plastic molecules. Common types of plastics, their chemical formulae, uses, and pollution caused by them are depicted in Table 1.

Plastic waste spreads in landfills, agricultural fields, the ocean, coastal regions, coral reefs, roadways, urban water runoff, freshwater lakes, and streams. Long-term plastic pollution threatening the ecosystem originates primarily from (1) single-use plastic (SUP) and (2) microplastic wastes. Plastic grocery bags were first introduced in the USA in 1979. The SUPs soon became favorites for attractive and cost-effective packaging and for items like straws, coffee cups, cutlery, bottles, and plastic bags. These are meant to be disposed of after use. About 343 MT of plastic waste [11] is being produced every year, and nearly half of it is single-use plastics (SUPs). Among the various uses of plastics in the construction industry, transport industry, and consumer goods packaging alone has a share of nearly 40%. Of plastic packaging, 95% is SUPs.

Microplastics and Nanoplastics are contributed by synthetic textiles such as polyester and nylon shedding fibers when washed, which add to microplastic pollution generated from marine debris in the oceans. Microparticles from personal health care and beauty products end up in water bodies. They accumulate in the environment and in water, posing further ecological and health risks.

3.2 *Impacts of Plastic Pollution*

The impacts of plastic waste are becoming a threat to the planet's ecosystem, affecting soil, water, and air, which in turn causes adverse impacts on human health and global climate change.

3.2.1 Soil Pollution from Open Dumping

The SUP waste dumped in open spaces gets collected in landfills. It can interact with multiple soil properties, change soil porosity, and hinder the growth of earthworms and other useful microorganisms, which consequently leads to the infertility of the soil. The chlorinated plastics release harmful chemicals that can seep into groundwater from the soil. The microplastic gets transferred between soil areas via physical, chemical, and biological processes and climatic factors such as wind and heavy precipitation. The presence of microplastics as pollutants has detrimental effects on the soil as well as contaminates surface and underground water resources; thus, such pollution threatens human health and food security.

In agricultural fields, microplastics can stimulate soil enzymes, change soil aggregation, and increase water absorption in the soil. The chlorinated plastic releases harmful chemicals upon breaking, which can seep into groundwater from the soil. By changing soil properties microplastics affect plant growth, flora, and fauna in the soil environment. Microplastic foams were found in agricultural soils from wastewater

Table 1 Common types of plastics, their chemical formulae uses and pollution caused

Sr. no	Name	Molecular structure	Uses	Pollution
1	Polyethylene (PE) **Types**: High-density polyethylene (HDPE) – Low-density polyethylene (LDPE) $\|C_2H_4\|_n$	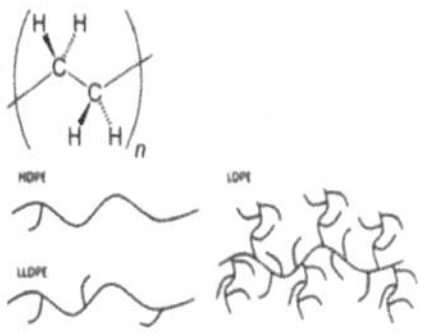	Carry bags (LDPE) Packaging materials (e.g., plastic bags), milk bottles (HDPE), shampoo containers, squeeze bottles, buckets, toys	Improperly discarded plastic bags can clog storm water drains and harm aquatic life when they enter water bodies. Plastic bottles take a long time to decompose
2	Polypropylene (PP) $\|C_3H_6\|_n$	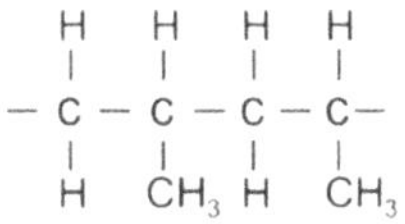	Food containers, bottle caps, MW trays, vehicle upholstery, medical devices, fibres	Like other plastics, improper disposal of polypropylene items can lead to litter and contribute to visual pollution
3	Polyvinyl chloride (PVC) $\|CH_2–CHCl\|_n$	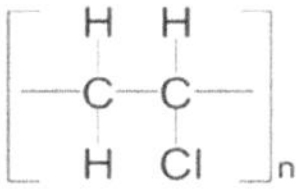	Pipes, electrical cable insulation, vinyl flooring, toys, credit cards, medical tubing, clothing	Releases toxic chemicals in air when burned. Disposal landfills can lead to leaching of harmful chemicals into the soil and groundwater
4	Polystyrene (PS) **Types**: Expanded polystyrene (EPS, commonly known as Styrofoam) $\|C_8H_8\|_n$	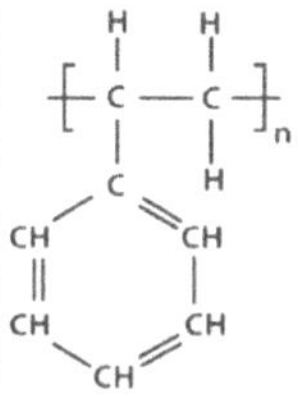	Food containers, disposable cutlery, cups, protective packaging material, insulation	Breaks into small, lightweight pieces that pose a threat to wildlife when ingested. Polystyrene and Styrofoam are banned from recycling due to hazardless waste creation
5	Polyethylene terephthalate (PET) $\|C_{10}H_8O_4\|_n$	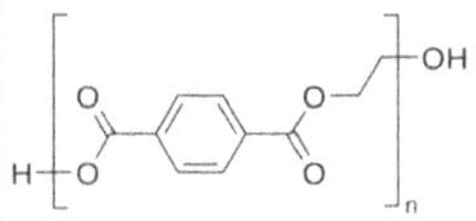	Beverage bottles, food containers, fabrics, ropes	Bottles and food containers are a source of litter. They can persist in the environment if not recycled

(continued)

Table 1 (continued)

Sr. no	Name	Molecular structure	Uses	Pollution
6	Polycarbonate (PC) $\|C_{15}H_{16}O_{2}\|_n$	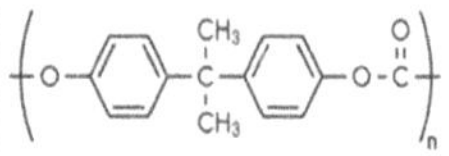	Optical lenses, eyeglasses, automobiles, heat resistant uses, car windows, dielectrics in capacitors	Contribute to plastic pollution if they are not properly disposed of or recycled
7	Acrylonitrile butadiene styrene (ABS) $\|C_8H_{8x}C_4H_{6x}C_3H_3N\|_n$	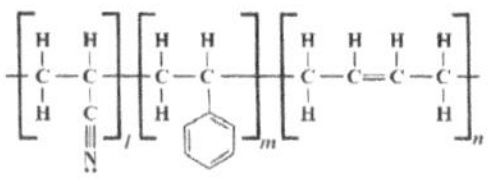	Automotive parts, toys, consumer electronics, computer key boards, construction industry	Contributes to litter and visual pollution, incineration causes air pollution
8	Polymethamethacrylate (PMMA) $\|C_5H_8O_2\|_n$	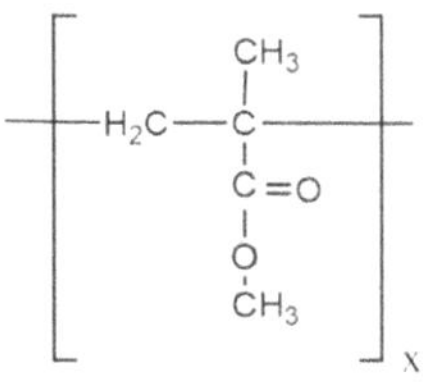	Shatter-resistant windows, aquariums and hockey rinks, dentures, bone implants	Breaks into microparticles, which are taken up into the gills and digestive gland of the mussel causing toxicity
9	Poly urethane (PU) $\|C_{27}H_{36}N_2O_{10}\|_n$	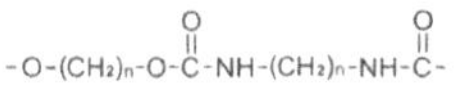	Foam insulation, furniture, automotive industry, building interiors, footwear	Releases toxic fumes when burned, contributing to air pollution

and sewage sludge [12]. Recent studies on the sources and transportation routes of plastics to agricultural soils suggest that plastic micro-contaminants in the soil environment reach cosmetic and cleaning products through wastewater. Contaminants also arise from the use of plastic foam materials for storage and fruit protection. Sa'adu and Farsang [12] estimated a microplastics abundance of 210–2000 pieces kg^{-1} in different agricultural soils of different farming types. According to a study by the UN Food and Agriculture Organization, the earth's soil could be more saturated by plastic pollution than the oceans [13]. 'Plasticulture' has been promoted for increasing crop productivity and to achieve food security.

Fig. 3 Plastic pollution in the coastal region

3.2.2 Marine Pollution by Ocean Dumping

From land, plastic travels to the ocean, continues beyond the EEZ, and becomes transboundary pollution. The first evidence of marine pollution was seen in 1969, when plastic items were found in the stomachs of seabirds. Jenna et al. estimated that out of 275 MT of plastic waste generated in 192 coastal countries in 2010, 4.8–12.7 MT entered the ocean [14]. The UNEP predicts that 10–20 MT [15] of plastic are being added to oceans every year, in addition to the 150 MT of plastic litter that already exists. The Ellen MacArthur Foundation anticipated that by 2050, oceans are expected to contain more plastic than fish [16]. In partnership with World Economic Forum and McKinsey & company it came out with a circular economy vision, in which plastic never becomes a waste. Plastics in marine litter at coasts enter the oceans and continues to accumulate in the marine environment (Fig. 3).

The low thermal conductivity of plastics when distributed in aquatic systems causes global warming by barricading CO_2 absorption. Planktons in the ocean grow by absorbing CO_2 from the atmosphere in the presence of sunlight and releasing oxygen. Plankton growth helps the ocean become an important carbon sink. By grazing on microplastics, these plankton could reduce the ocean's capacity to act as a CO_2 sink and accelerate the loss of ocean oxygen. Controlling marine pollution is therefore crucial to not only protecting the health of marine ecosystems, aquatic life, and the well-being of communities dependent on ocean resources but also to mitigating climate change. Currently, there could be about 50–75 trillion pieces of plastic and microplastic in the ocean [17].

3.2.3 Air Pollution from Plastic Waste

Three major sources of air pollution from the open dumping of plastics are (i) greenhouse gas emissions from incineration and burning of plastics; other toxic gases are also emitted; (ii) methane emissions from landfill sites and oceans; and (iii) microplastic pollution in the environment. During the burning of plastic waste, greenhouse gases are released along with other toxic gases. In the case of chlorinated plastic, incineration leads to emissions of carbon dioxide, nitrogen oxide, sulfur oxide, etc. It is also known to generate volatile organic compounds such as dioxins, black carbon, pyrene, and polychlorinated furans. The SUPs dumped in landfill sites give rise to emissions; 50% of this gas is methane, which is a greenhouse gas. Microplastics are discharged into the air from many sources. Kole et al. [18] estimated the global average emission from the wear and tear of automobile tires of microplastics to be 0.81 kg/year. About 3–7% of the plastic particles in the air we breathe are generated from tires. In addition, 5–10% of plastic pollution in aquatic systems is derived from tires.

3.3 Impact on Human Health

Pure plastics are generally insoluble in water and nontoxic. They may not have a direct impact on health, except when the contents are hot. But many of the additives in plastics are toxic and may leach into the environment even at room temperature; once ingested, they can cause severe harm. The burning of plastic litter produces toxic fumes, which are harmful for human health. Burning releases highly toxic gases like dioxins, furans, mercury, and polychlorinated biphenyls into the atmosphere. Once inhaled, the risk of heart disease increases, aggravates respiratory ailments, causes nausea or headaches, and damages the nervous system. Polystyrene emissions are found to be very harmful to the central nervous system.

Microplastics can originate in the air, water, and dust. Inhalation of polluted air and dust causes adverse health effects. They enter the circulatory system and affect the respiratory system. Their presence has been reported in the lungs. Microplastics and nanoplastics are also added from the textile and cosmetic industries. Indigestion of microplastics in human bodies from dust, water in plastic bottles, and seafood has been reported [19]. Plastic microparticles mix with the environment and get accumulate in the body through food, water, and air (Fig. 4).

Microplastics are known to impact diversity of microbial communities in the human gut, which in turn impact endocrine, immune, and nervous system [20]. According to recent research published in the Proceedings of National Academy of Science, one liter water bottle may contain nearly a quarter of a million pieces of nanoplastics on average [21].

3.4 Plastic Pollution and Climate Change

Plastic pollution in the ecosystem is a challenge, becoming a global climate change threat. It can be more damaging than climate change, and its control requires minimizing solid waste and reducing greenhouse gas emissions. According to a study by the Center for International Environment Law [22] in 2019, plastics generated 1.8

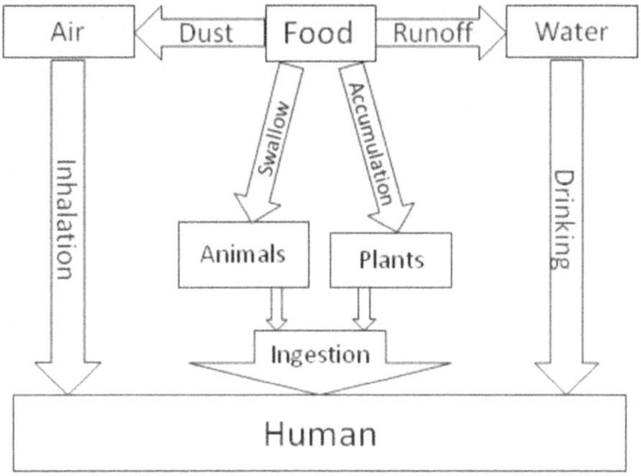

Fig. 4 Human exposure to microplastic through different routes

billion metric tons of greenhouse gas emissions, with 90% of these coming during the plastic supply chain, which includes plastic production and conversion from fossil fuels. The refinery alone emits approximately 200 million metric tons of greenhouse gases each year. Plastics, which emerged as versatile, moldable, durable, and cost-effective materials and a solution for climate change by saving energy and precious metal resources (when plastic is used in place of metals), are seen to contribute to greenhouse gas emissions from each stage of the plastic life cycle, i.e., fossil fuel extraction from the earth, transport, refining into plastics, and management of waste. The incineration of plastic waste results in major greenhouse gas emissions of CO_2. Landfills account for more than 15% of methane emissions.

Polyvinyl chloride produces hazardous halogens, and the impact of pollution in the air is climate change. Studies conducted in the USA and Australia have suggested that, from production to consumption and disposal, plastic is one of the most carbon-intensive industries. 1 kg of plastic releases 2–3 kg of CO_2, which is similar to that of steel production from ore. Throughout their life cycle, plastics have carbon footprints and contribute to 3.4% of global GHG emissions. According to the CIEL study, the CO_2 eq. GHG emissions released by the plastic life cycle in 2030 may equal 295 coal-based power plants of 500 MW capacity, and in 2050 they may equal 615 such plants [23]. The plastic life cycle emissions from 'Çradle to Grave' are projected to grow up to 56 Giga tons of CO_2 equivalent emissions by 2050 and have a share of 10–13% of the entire remaining carbon budget.

Microplastic fragments (~5 μ) found in the oceans interfere with the largest carbon sink capacity of the ocean to absorb and sequester carbon dioxide from the atmosphere. Plastic waste on the coastlines, riverbanks, and landscapes releases more methane. Unmanaged plastic in the environment therefore continues to impact climate change and contribute to climate change [24]. Conservation of resources through responsible production, consumption, reuse, and recovery of materials through recycling can only reduce emissions.

4 Remedies for Controlling Plastic Pollution

Launched by the Ellen MacArthur Foundation in October 2018, the 'New Plastics Economy Vision' [25] aligns with the principles of the circular economy, an industrial system that is restorative and regenerative by design. The circularity concept involves optimum design, waste minimization, energy recovery, education, awareness, employment, resilience, growth, and skills. A comprehensive circular economy approach can address the root causes and contribute to global efforts to combat the plastic pollution, climate change, and biodiversity crises while delivering economic, environmental, and social benefits. The 6 'Rs' of pollution management are reuse, recycle, reduce, refuse, rethink, and repair. They are also called the 6Rs of sustainability.

Reuse: Keeping in view the entire life cycle of items, from source to manufacturing to use, bulk plastics can be reused by recycling and remoulding of the plastic products.

Recycle: Waste plastic can be reused by making usable products by physical, mechanical, chemical or bio-recycling.

Reduce: The consumption of SUPs in plastic packaging products should be reduced to be stopped eventually.

Refuse: Single-use plastic products should be refused. More sustainable options like reusable steel, glass bottles and non-disposable utensils should replace them.

Rethink: Policies and regulations banning SUPs and other regulatory measures for introducing guidelines to manufacturers should be in place.

Repair: Give more emphasis to using a broken item after repair, extending its life thereby reducing burden on resources.

Plastics being moldable, lightweight, durable, innovative, and cost-effective, the production has grown exponentially since the 1950s, and the forecast is that from the current global production of ~400 million metric tons, it will triple by 2060. Plastics have a long life, so waste is leading to critical disposal issues by collecting in landfills, causing marine debris, and becoming a major human health hazard. Two key remedies for controlling plastic pollution in the short and medium term are:

(1) Recycling of Plastic Waste (2Rs—Reuse and Recycle) and
(2) Management of Plastic Waste (4Rs—Reduce, Refuse, Rethink and Repair).

4.1 Plastic Waste Recycling—Technology Approach

In view of the special characteristics of plastics and their polymeric nature, many tangible and intangible benefits can be accrued by using plastic recycling technologies. Conversion of plastic waste for energy recovery or material recovery is a promising solution to beat plastic pollution. As hydrocarbon materials, plastics can be converted into oil or high-calorie gaseous fuels.

Fig. 5 Recreation hut made of plastic waste in Myanmar (photo-2017)

Plastics-to-fuel technologies offer an opportunity to reduce greenhouse gas emissions by up to 60–70% over new forms of crude oil extraction [26]. The obvious advantages are the production of green fuels, reduction in pressure on landfills, reduced collection of marine debris, etc. Evidently, the vast experience of research institutions needs to be translated to commercial plastic (waste)-to-fuel plants, and the technology should be recognized as an important part of plastic waste management. An overview of various recycling processes is presented below.

4.1.1 Physical Recycling

Recycling is the process of converting used plastic into another usable product. Physical recycling is reuse in a similar or modified form by recovering waste. Physical recycling helps reduce the amount of plastic waste ending up in landfills or oceans by reusing the material to manufacture new products without changing its chemical structure. It has a relatively low cost and extends the lifespan of plastic by physically transforming it into functional products. A wide variety of products, from jewelry to exhibition halls, can be created. An example of physical cycling is depicted in Fig. 5.

4.1.2 Mechanical Recycling

Mechanical recycling is similar to physical recycling, but in this case the plastic waste is sorted, cleaned, shredded, melted, molded, and reprocessed into pellets or flakes, which can then be used to manufacture new plastic products. This process does not involve any chemical reactions that change the polymer structure but instead relies on physical processes to transform the plastic waste into reusable materials. Another option is to apply some force, like mechanical means, to shred plastic waste and then convert it into products. An example is its use in road resurfacing by adding bitumen. The mechanically recycled product has comparatively lower performance demand than the original.

4.1.3 Thermal Recycling

Incineration is a commonly adopted method of thermal recycling for producing energy from waste. Incineration reduces waste volume and can generate energy in the form of heat or electricity. Plastic incineration, however, gives rise to pollutants and ash residues, adding toxic chemicals like dioxin and furan that may pose environmental and health risks if not properly controlled. Incineration is therefore not proving to be a viable solution for most plastics.

4.1.4 Advanced Thermal Recycling

Pyrolysis is an advanced method of thermal recycling. In pyrolysis, waste is heated in the absence of oxygen at a high temperature of around 300–900 °C. Under these conditions, the long-chain polymer molecules of plastics break down into smaller hydrocarbon molecules, which can be further refined into fuels or chemicals. Depending on the structure of the polymer, the pyrolysis process is selected. As of now, pyrolysis continues to be the most accepted technology for plastic recycling.

Plasma pyrolysis is carried out in an oxygen-starved environment at plasma temperatures, which helps break bonds even in thermosetting polymers and is found to destroy medical waste to get value-added products in an efficient and eco-friendly manner [27, 28].

Catalytic pyrolysis is a two-step process of thermal degradation that takes place in the absence of oxygen and in the presence of a catalyst [29]. The catalyst breaks long polymer chains at lower temperatures than those of plasma pyrolysis. By breaking bonds, polymers get converted into monomer molecules, which are then condensed and fractionated to get fuels like diesel or gasoline. The process is proving a cost-effective conversion method for solving the acute environmental problem of disposing of plastic waste.

Hydrothermal recycling is a process, also known as hydrocracking or hydrothermal liquefaction, carried out in the presence of hydrogen and a catalyst to convert waste into valuable products such as biofuels and biochemicals [30]. To depolymerize polyethylene (PE) and polypropylene (PP), their derivatives and mixtures require supercritical water conditions at a temperature >374 °C and a pressure >23 MPa. This process is expensive and needs to be commercially scaled up.

4.1.5 Gasification

Gasification is the partial combustion of plastics carried out to produce syngas (CO, H_2, CH_4, and CO_2) and char at high temperatures (650–1600 °C) in the first stage. A gasification system comprises three fundamental elements: the gasifier—to produce the syn gas; the gas clean-up system—required to remove harmful compounds from the combustible gas; and the energy recovery system [31]. Subsequent reduction of the product gases is done into CO and H_2 using steam reforming.

In *plasma gasification*, a wide range of feedstock plastic waste in the temperature range of 1200–12,000 °C can be converted into high-calorific fuel gases and inert solid slag. Plasma arc gasification is a process that uses an electric arc. It is suitable for converting entire waste, including plastics (without separation), into fuel of high calorific value. Inert solid slag is generated in the process. The flue gas is cleaned up before it is released into the atmosphere. Considering the fact that raw materials are MSW, which comprises all types of waste, including plastics, cost-effectiveness can be improved as the cumbersome step of plastic segregation is avoided. Already, such systems are in operation in European countries and the USA.

4.1.6 Chemical Recycling

Chemical recycling is a solvent-based treatment of plastic waste and can also be applied to all plastics, including thermosets. It breaks down polymer chains to varying degrees, partly by means of reactants, potentially aided by catalysts. Solvents such as supercritical fluids or certain chemicals facilitate the depolymerization process and are selected according to the waste plastic material [32]. Depending on the solvent used, polymers are cracked back into hydrocarbons. It can either be to extract the polymer back from the waste or to depolymerize the plastic back into its constituent monomer, which can then be used to produce new plastics or other chemicals. It offers a step toward reducing greenhouse gas emissions. Chemical recycling has advantages over mechanical recycling in a circular economy. Nonetheless, many challenges remain, and industrial-scale plants for chemical or solvent-based plastic recycling are still lacking [33].

Recent advancements suggest that combining thermo-mechanical, thermo-chemical, and bio-chemical processes of waste conversion could reap further benefits in conversion of waste to recovery of fuels.

4.1.7 Recycling of Marine Debris

The plastic pollution issue got aggravated when light-weight plastic drifted into the ocean and caused marine pollution. According to the UN Environment, single-use plastics account for almost 40% of total production and have become a threat to marine ecology. In 2014, oceans had 1 kg of plastic per 5 kg of fish; in 2050, they may have more plastic by weight than fish [34]. Marine debris is difficult to manage.

In the recycling of marine debris, several difficulties related to the degradation of materials, the distribution of residues, the size of plastic waste, etc., are faced [35]. Various chemical, thermal, and biological processes have been applied to recycle marine debris. Hydrothermal carbonization using sea water as a solvent appears feasible for converting marine plastic debris into fuel [36]. Raising the temperature to 400 °C increases hydrochar formation and emission of fuel gases.

4.1.8 Bio-decomposition of Plastic Waste

Biodegradable plastics are made from materials other than petrochemicals. They can undergo biological conversion and decompose in the presence of selected enzymes and other microorganisms to degrade plastic waste into simpler organic compounds, which can then be metabolized or further processed into useful products. Many fungus species are capable of bioremediating plastics. Biodegradation of such plastics can occur with selected enzymes in the ocean environment. Fermentation helps in converting polymers into simpler products. Pestalotiopsis, a genus of fungi, degraded polyurethane and converted it into organic matter in both solid and liquid suspensions [37]. In anaerobic conditions, bio-decomposition makes it ideal for use in landfills. In aerobic (oxygenated) environments, they are capable of removing plastic from the ocean. Such plastic-eating mushrooms are emerging as an organic solution for decomposing plastic in the ocean while still creating an edible mushroom. With an estimated 2.2 and 3.8 million species of fungi on Earth, more than 100 types of fungi are documented to have plastic-degrading capabilities.

4.2 Development of Alternate Sustainable Materials

Bio-based polymers as options for achieving a circular economy for plastics, have the ability to break down into natural compounds, such as carbon dioxide, water, and biomass, through the action of microorganisms under specific environmental conditions. Bio-based plastics are derived from renewable resources and biomass such as corn starch, sugarcane, cactus, or cellulose. It can be biodegradable or safe, reducing the environmental impact compared to conventional plastics. Edible films and coatings made from materials like seaweed, starches, or proteins can be used to package food products. Safe additives, such as natural fibers of bamboo, hemp, or flax, can be used as fillers or reinforcements, reducing the reliance on synthetic additives and improving the environmental footprint of the plastic product. They can be added to conventional plastics to enhance their biodegradability without compromising their mechanical properties and facilitate microbial degradation after use.

For finding alternatives that are more sustainable and recyclable, new ways of making plastics from sources other than fossil fuels are developing. Unlike biodegradable polymers, which provide an end-of-life option that reduces the accumulation of plastic waste in landfills or the ocean, alternative plastics made by using molecular assembly make it easier to break into monomers and recycle. Poly-di-ketamine (PDK) is a custom-designed material designed for efficient and indefinite recycling. It is fully recyclable and provides a low-carbon manufacturing solution for plastic products that no longer have to go to landfills [38]. PDK is seen as a wonder material and provides for a circular economy.

The choice of material depends on factors such as the desired end-of-life options, mechanical properties, environmental impact, and specific application requirements. Advanced technologies of chemical recycling help in depolymerization and enable the breakdown of plastics into their constituent monomers, which can then be used to produce new plastics without degradation in quality. Research at the Indian Institute of Technology, Madras, demonstrated success in degrading fluoropolymers, which are used for missiles and are very stable, into tiny fragments under certain chemical and mechanical treatments.

4.3 Management of Plastic Waste

Mass production of SUPs has led to the destruction of ecosystems. Management of non-recyclable plastic waste has gained momentum through policies and regulations, banning SUPs and other regulatory measures for introducing guidelines to manufacturers and consumers. There are many challenges hindering reduction of plastic pollution.

4.3.1 Ban on SUPs

Many countries and regions have taken steps to regulate or ban single-use plastics in various capacities due to environmental concerns. However, the specifics of these bans, their enforcement, and their effectiveness can vary significantly. Some countries have introduced regulations aimed at promoting alternatives and encouraging recycling, whereas others have implemented comprehensive bans on certain types of single-use plastics, such as plastic bags, straws, or cutlery. Some others have focused on imposing taxes or fees on their use. In Europe, Denmark and Germany were the first countries to ban plastic bags in 1991 and began charging a tax on them in 1994. In South Asia, India and Bangladesh were the first to ban thinner plastic bags in 1999 and 2002, respectively. In Rwanda, the plastic ban has been rigorously enforced. It is known for having one of the strictest bans on single-use plastics, introduced in 2008. In 2014, the Netherlands became the first country to ban the use of microbeads in cosmetics. The governments across the world have taken steps to ban plastic bags and minimize SUP waste and marine debris. Kenya banned plastic bags in 2017, making it the most recent one out of more than two dozen countries that have sought to reduce plastic bag use.

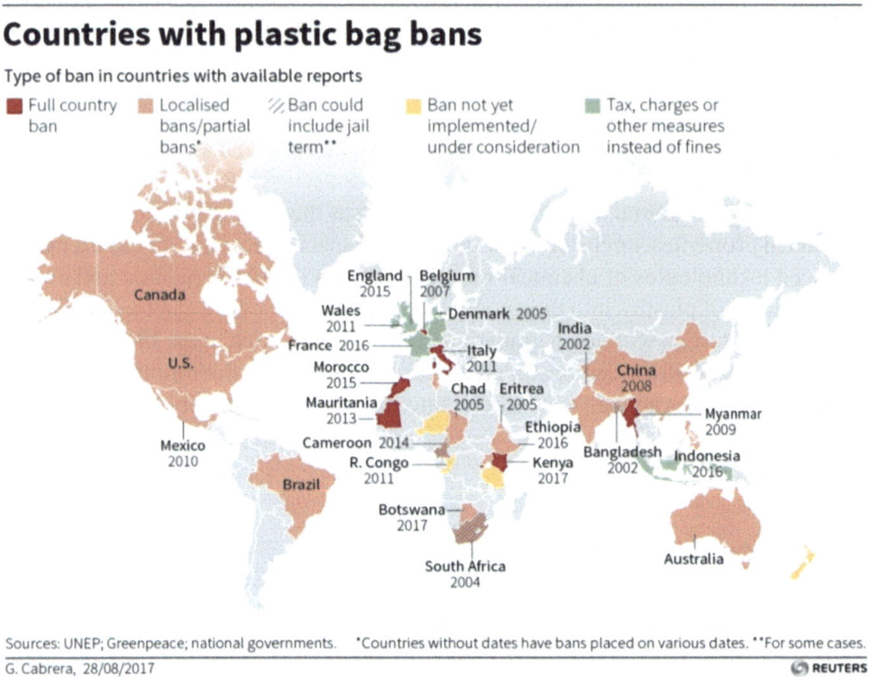

Fig. 6 Countries having introduced restrictions on use of plastic bags (*Source* ABC News Australia [39])

Banning plastic bag initiatives, introduced in different countries across the globe are depicted in Fig. 6.

Yet global efforts to curb plastic pollution by banning SUPs continue to be a challenge. The alternatives are more expensive for the customer and consume more energy. Implementation of biodegradable or bioplastics faces several challenges, including scalability, cost-effectiveness, and environmental impacts associated with their production and disposal. At the same time, small and medium-scale industries involved in plastic processing are affected, impacting the economy and jobs at a place. A behavioral change in plastic use is therefore required to achieve success.

4.3.2 Awareness Campaign

To implement bans on single-use plastics, countries need to promote behavior change and raise awareness to educate the public about the adverse impacts of plastic pollution. Along with a cleanliness drive for SUPs, the elimination of the use of plastic and the disposal of plastic waste have to gain momentum throughout the region. Public institutions, business houses, and other public institutions can play a major role in creating awareness by participating in the campaign mode to pick up any plastic waste in their surroundings and bring it to a central location where it is collected and taken for disposal. Various strategies to engage and inform the public could be: education—developing curriculum materials and activities focused on plastic pollution control for schools; workshops and campaigns—organizing events through the participation of society; religious bodies—making announcements during large public gatherings about the importance of reducing single-use plastics. In addition, celebrities can take the lead through slogans and social media.

Awareness campaigns for limiting the increase of SUPs as waste have become an urgent worldwide priority. In the Philippines, the city of Manila has created a garden using discarded plastic bottles and turned them into colorful tulips, capturing tourists' attention and building awareness about plastic recycling [40]. At the same time, local bodies and municipalities also need to be trained to dispose of collected plastic waste and to develop business models in consultation with the industry and recyclers.

4.3.3 Toward a Global Treaty

Starting with the 'Stockholm Convention on Persistent Organic Pollutants', which came into force in 2004, to restrict the production and use of persistent organic pollutants (POPs), a number of international efforts have been made aimed at addressing various aspects of plastic waste management. The Basel Convention on the Control of Transboundary Movements of Hazardous Wastes and Their Disposal, which included plastic waste, was evoked in 2011. In 2015, the EU created a Circular Economy Action Plan and the European Union's Single-Use Plastics Directive to strengthen the regulations on recycling plastic products and provide more of a science-based approach to legislation [41]. The Association of Southeast Asian Nations (ASEAN) Framework of Action on Marine Debris was developed in 2017 to address trans-border issues in plastic pollution. In May 2019, the Amendments to the Basel Convention on the Control of Transboundary Movements of Hazardous Wastes and Their Disposal aimed to regulate the movement of plastic waste more effectively. These amendments aim to reduce the generation of plastic waste and minimize marine debris.

The World Economic Forum, the Ellen MacArthur Foundation, and the UN Environment Global Plastic Pollution Partnership (GPAP) are committed to finding comprehensive solutions [42]. Launched in 2018, the aim is to drive change through coordinated national strategies and targets. On March 2, 2022, the resumed session of the Fifth Session of the United Nations Environment Assembly (UNEA-5.2) adopted a resolution to initiate an international law-making process to combat global plastic pollution, including marine plastic pollution, by 2024. An Intergovernmental Negotiating Committee (INC) was created to develop a global binding treaty on plastic pollution, including in the marine environment. It held its first session at Punta del Este, Uruguay, November 28–December 2, 2022. The High Ambition Coalition, which has 133 countries as its members and is chaired by Norway and Rwanda, has set the goal of ending plastic pollution by 2040. A treaty was agreed to address plastic pollution during its life cycle, from feedstock production to end-of-life [43]. The coalition is expected to provide information on science-based objectives in the international negotiations.

4.3.4 UNEP India–Norway Bilateral Cooperation Initiative

Launched in 2019, the India-Norway Marine Pollution Initiative seeks to contribute to the Sustainable Development Goal 14 and specifically SDG14.1. The overall objective of this initiative is to improve the management of marine pollution in India to actively engage globally on this issue. The initiative is complementary to the UNEA Resolution 4 (Sub-resolutions 7 and 10 on marine plastic litter and microplastics, and pollution by single-use plastic products, respectively) and the Global Programme of Action for the Protection of the Marine Environment from Land-based Activities (GPA). Launched under the India-Norway Task Force on Blue Economy for Sustainable Development, it is aligned with the Indian Government's Action Plan for the Clean Sea's Campaign, and other major policies and missions such as the 'Swachh Bharat Mission' and 'Clean Ganga Mission'. The initiative has active projects being implemented by the institutions in India and Norway.

4.3.5 G20 New Delhi Leaders' Declaration

The G20 Session held in New Delhi with India's Presidency commended the resolution UNEP/EA.5/Res.14 which established an Intergovernmental Negotiating Committee (INC) to develop an international legally binding instrument on plastic pollution, including the marine environment. The G20 Marine Litter Action Plan noted the ERP guidelines for e-waste and plastic waste. The leaders pledged to build on the G20 Marine Litter Action Plan as also elucidated in the Osaka Blue Ocean Vision. On July 1, 2022, the Indian government has implemented a ban on single-use plastics, a step which will help curb plastic pollution because single-use plastics account for 43% of the plastic waste generated in India.

4.3.6 Plastic Waste Management Regulations in India

India's plastics industry began in the 1960s and is fairly big. According to the Central Pollution Control Board (CPCB), 5939 registered units are engaged in plastic processing in 29 states and union territories in India. In addition, 615 unregistered plastic recycling units exist in five states or UTs [44]. Almost all sectors of the economy are dependent on plastics, in one form or another. Plastic consumption in India is estimated to be 22 million metric tons per year. Nearly 85% of all plastic packaging is SUP, resulting in waste generation on land and more so in the ocean. The gradual increase in plastic waste has been a growing concern for Indian cities. The growth rate of the plastics market has been one of the highest, and it is higher than the country's GDP growth. The estimated waste generation in India is 4 million metric tons of plastic annually.

To address the subject of plastic pollution, the Government of India has regulations dating back to 1999. These regulations gave thrust to the recycling of plastic bags in their different modes, such as municipal solid waste, waste from petroleum refineries, or non-recyclable plastic waste. The Plastic Waste Management and Handling Rules of 2011 were notified to prohibit the manufacture of small plastic bags of a thickness less than 49 microns. The New Plastic Waste Management Rules, 2016 and amendments issues in 2018 attempted to resolve and sought to extend the responsibility of the plastic producers and generators to create an effective waste management system and included energy recovery as well as material recovery as the preferred solution for end-of-life management use of non-recyclable multilayer plastic [45].

India has rich experience in introducing large-scale behavioral change programs for eco-friendly development. In 2014, 'Swachh Bharat Mission' was launched, and as part of this, there was no less challenge to tackle the issue of plastic waste, which had been growing by leaps and bounds in the past 50 years and has become a threat to the planet. To increase recycling rates, the use of plastic waste was made mandatory in road construction in 2015. Several Indian cities—Pune, Indore, Surat, and Chennai—with some pilot demonstrations in Delhi—have constructed roads using the plastic-bitumen mix [46]. These roads were found to be longer lasting, did not develop cracks and potholes, and there was a savings in the cost of bitumen when plastic waste was used.

The SUPs phase-out plan officially began on October 2, 2019, the 150th birth anniversary of Indian revered leader Mahatma Gandhi, called the father of the nation. The 'Swachhta hi Sewa' 2019 mission, a plastic waste-free campaign, gained popularity in schools, colleges, and universities [47]. India was designated the host country by the United Nations Campaign on World Environment Day 2018 with the theme Break Plastic Pollution and resolved to phase out SUPs by 2022. Plastic waste management rules have been amended in 2021, 2022, and 2023 mainly to curb pollution due to SUP items and strengthen Extended Producer Responsibility (EPR) compliance. Many civil bodies are collecting tens of thousands of tons of plastic waste from the coastal areas and seashores in India. 'Swachh Sagar Surakshi Sagar Campaign' was launched in the 75th year of India's independence to remove waste from 75 beaches in 75 days. In November 2023, WWF-India teamed up with students from Veraval Fisheries College and picked up 550 kg of waste along a two-kilometer stretch—from the Veraval Chowpati beach in Gujarat. Plastic constituted about 350 kg of it, comprising food-packaging material, including plastic sachets and single-use plastic cups, bags, wrappers, etc. [48].

A glossary of plastic rules in India is compiled in Annexure I.

4.3.7 Mission LiFE for Plastic Pollution

Mission LiFE (Life Style for Environment) is an India-led global mass movement aimed at promoting sustainable lifestyles and addressing environmental challenges through environmentally friendly culture and traditional practices. The *mantra* LIFE: Lifestyle for Environment to Combat Climate Change was shared at the 26th session of the Conference of the Parties (COP26) to the United Nations Framework Convention on Climate Change (UNFCCC) held in Glasgow, United Kingdom. It was stated that Lifestyle for the Environment has to be taken forward as a campaign to make it a mass movement of environmentally conscious lifestyles. The world needs mindful and deliberate utilization instead of mindless and destructive consumption.

India has announced LiFE as the theme for World Environment Day 2023 in the context of plastic pollution control and launched the Meri LiFE App for public participation. India has implemented policies and introduced strategies to control and manage plastic waste. These initiatives aim to reduce plastic usage, encourage recycling, and address the environmental impact of plastics. In 2022–23, Mission LiFE focuses on 75 individual LiFE actions across the 7 categories [49]. In the SUPs category, it is targeted to reduce the use of plastics in various applications, from shopping bags to packaging bags used for milk and buttermilk, and to encourage eco-friendly choices like the use of bamboo toothbrushes and neem combs. Lifestyle strategies to reduce plastic pollution on land and sea, behavioral change including adopting a lifestyle to minimize plastic usage; are as follows:

i. *Reducing Single-Use Plastics*: say no to plastic straws, utensils, and disposable cutlery.
ii. *Managing Plastic Waste Disposal*: recycle wherever local guidelines exist.
iii. *Choosing Sustainable Packaging*: packaging made from sustainable recyclable materials.
iv. *Replace Disposable Items*: like razors, cleaning pads, and sponges with reusable alternatives.
v. *Use Plastic Alternatives*: choosing products made from eco-friendly alternative materials like glass, metal, bamboo, or paper.
vi. *Awareness and Education*: creating awareness among the community about the environmental impact of plastics and the latest developments in plastic pollution reduction.
vii. *Avoid Use of Microplastics*: using natural, biodegradable personal care products free from microbeads.
viii. *Careful disposal of SUPs*: prohibiting flow of single-use plastics in rivers and oceans.
ix. *Minimize Food Waste*: avoiding use of excess packaging waste.

5 Concluding Remarks

Plastics, made of polymers, are moldable, lightweight, durable, innovative, and cost-effective materials. Because of their versatile properties, they have uses in all sectors of the economy: infrastructure, construction, agriculture, automobiles, consumer goods, telecommunications, packaging, and scientific and medical equipment, among others. Since the 1960s, there has been exponential growth in the use of plastics. Plastic products help reduce greenhouse emissions during the operation of a device by consuming less energy. Soon it was realized that plastics have a long life of 400–500 years, and plastic waste accumulates in the environment, causing long-term pollution. Plastic debris in the ocean is a huge threat to human health and marine life, resulting in the ingestion of microplastic particles by mammals, leading to food chain. Plastic pollution is posing many environmental, health, and social challenges. Evidently, long-chain plastic materials are source of pollution, and poor waste management is to blame.

This chapter provides a comprehensive outlook to achieve an understanding of plastic pollution, with an emphasis on ecosystem impacts. The focus is on two immediate concerns: (i) managing single-use plastic on land and in the ocean, and (ii) the state-of-the-art processes for recycling waste through waste-to-fuel and waste-to-product technologies. Control of pollution would require running campaigns for public awareness by educational institutions, religious bodies, and celebrities on social media. Research laboratories, universities, and IITs, in collaboration with the plastics industry, could intensify research and development for scientific solutions of plastic recycling technology. Global and local environmental movements such as the Mission LiFE and many others have begun to address this insurmountable challenge and to go plastic-free. Plastics are important to us and the solutions cannot be assessed in isolation and must be considered in terms of the whole plastic value chain. Strategies for managing plastic pollution can contribute to achieving Sustainable Development Goals (SDGs). These would relate to taking actions for reducing the production of waste; promoting reuse and recycling and managing marine debris through regulations, respectively.

To minimize plastic waste, plastic-to-fuel technologies should be recognized as complementary to recycling in a policy framework and an important part of a community's integrated solid waste management. At the same time ways to develop new bio-based plastics and use of biodegradable plastics to create substances that are more environmentally friendly are essential. Plastic pollution continues to be a challenge, and to resolve the paradox of plastic use it calls upon industries, scientists, local bodies, individuals, and governments to come together and explore sustainable alternatives and green solutions for reducing the production and excessive use of SUPs. Each and everyone in the society has to take responsibility to solve the crisis of plastic pollution.

Annexure I

Plastic Waste Management Rules in India since 1999 (Adopted from Policy Paper [45]).

Regulation	Specific provisions	Intent
Recycled Plastics Manufacture and Usage Rules, 1999	Stipulated both size and thickness restrictions on plastic bags, requiring them to be at least 8 × 12 inches in size and 20 microns in thickness The use of recycled and virgin-colored polybags for non-food applications was allowed but discouraged for packaging food items	To regulate the manufacture, sale, use and recycling of plastic bags
The Recycled Plastics Manufacture and Usage (Amendment) Rules, 2003	Amendment issued in 2003, prohibited manufacture, stocking, distribution, or selling of carry bags made of virgin or recycled plastic less than 20 × 30 cm in size and 20 microns in thickness. It also banned the use of recycled plastic bags and containers for storing, carrying, dispensing or packaging of food items	To regulate methods of collection segregation, treatment and disposal of plastic waste
The Plastics (Manufacture, Usage and Waste Management) Rules, 2009	Prohibited the use of carry bags or containers made of recycled or non-biodegradable plastics for storing, carrying, dispensing or packaging of food stuffs. Adherence to new guidelines for manufacturing and recycling was regulated	To regulate the use of plastics in its various purposes and of plastic waste
Plastic Waste (Management and Handling) Rules, 2011	Prohibition of the manufacture, stocking, distribution, or selling of carry bags made of virgin or recycled plastic less than 20 × 30 cm in size and 40 microns in thickness Units manufacturing plastic bags are required to register with the respective state pollution control board (SPCB) or pollution control committee (PCC) before starting production	To regulate the use, collection, segregation, transportation and disposal of plastic waste

(continued)

(continued)

Regulation	Specific provisions	Intent
Plastic Waste Management Rules, 2016	Producers are required to develop a waste collection system based on extended producer responsibility (EPR) Primary responsibility for collection of used multi-layered plastic sachets or pouches or packaging is that of producers, importers or brand owners who introduce them in the market identified Manufacture and use of non-recyclable multilayer plastic should be phased out within two years	To extend the responsibility of the plastic producers and generators to create an effective waste management system To expand its jurisdiction to rural areas, to Gram Panchayats, because plastics has reached rural areas as well
Plastic Waste Management (Amendment) Rules, 2018	The amendment to 2016 rules is to establish a centralized registration system by mandating brand owners and producers operating in more than two states to register with the CPCB	To include energy recovery as well as material recovery as the preferred solution for end of life management
Plastic Waste Management Amendment Rules, 2021	Prohibited identified single-use plastic items by 2022. Non-woven plastic carry bag shall not be less than 60 gram per square meter (GSM) i.e., thickness of plastic carry bags will increase from 50 to 75 microns from 30th September, 2021 onwards and to 120 microns with effect from 31st December, 2022	To curb pollution due to single-use plastic items, it provides guidelines for extended producer responsibility given legal force
Plastic Waste Management (Amendment) Rules, 2022	Notified the instructions on extended producer responsibility (EPR) for plastic packaging. stipulate mandatory targets on EPR, recycling of plastic packaging waste, reuse of rigid plastic packaging and use of recycled plastic content	To reduce plastic foot prints from plastic packaging To provide assistance under the Swachh Bharat Mission for plastic waste management, in urban and rural areas of the country
Plastic Waste Management (Amendment) Rules, 2023	Notified on April 2023 for giving insights to industry into management of plastic waste, particularly related with EPR, compostable plastic and biodegradable plastic technologies	To make EPR compliance mandatory and include biodegradable polymers as separate category for special products

(continued)

(continued)

Regulation	Specific provisions	Intent
Plastic Waste Management (Second Amendment) Rules, 2023	The second amendment to 2016 rules notified in October 2023 is to strengthen the regulation of plastic waste management, particularly in the areas of compostable and biodegradable plastics, waste processing, labeling, and extended producer responsibility	To make implementation of the rules effective for managing plastic waste and promoting environmental sustainability Biodegradable plastics use in carry bags is permitted

References

1. Borrelle, S. B., Ringma, J., Law, K. L., Monnahan, C. C., Lebreton, L., McGivern, A., Murphy, E., Jambeck, J., Leonard, G. H., Hilleary, M. A., Eriksen, M., Possingham, H. P., De Frond, H., Gerber, L. R., Polidoro, B., Tahir, A., Bernard, M., Mallos, N., Barnes, M., & Rochman, C. M. (2020). Predicted growth in plastic waste exceeds efforts to mitigate plastic pollution. *Science, 369*(6510), 1515–1518. https://doi.org/10.1126/science.aba3656
2. https://phys.org/news/2017-10-prince-charles-plastic-menu-world.html
3. https://www.sciencemuseum.org.uk/objects-and-stories/chemistry/age-plastic-parkesine-pollution
4. https://www.sciencehistory.org/stories/magazine/celluloid-the-eternal-substitute/
5. http://scihi.org/wallace-hume-carothers-nylon/
6. https://www.statista.com/statistics/1019758/plastics-production-volume-worldwide/
7. Austine, I., Chinomso, O., Sinha, R. S., Vuyelwa, M., Carlos, B. J., & Paulo, S. J. (2021). ACS Omega 2021, 19343–19355. http://pubs.acs.org/journal/acsodf. Plastic pollution: A perspective on matters arising: Challenges and opportunities. https://pubs.acs.org/doi/full/https://doi.org/10.1021/acsomega.1c02760
8. https://csp.umn.edu/2016/08/17/interview-with-director-hillmyer-featured-in-nature-news/
9. Md. Golam, K., Imtiaz, M. N., Rafat, S., Quoc, N. H., & Monjur, M. (2023). Plastic waste: Challenges and opportunities to mitigate pollution and effective management. *International Journal of Environmental Research, 17*, 20. https://doi.org/10.1007/s41742-023-00507-z
10. Shams, M., Alam, I., & Mahbub, Md. S. (2021). Plastic pollution during COVID-19: Plastic waste directives and its long-term impact on the environment. *Environmental Advances, 5*. https://doi.org/10.1016/j.envadv.2021.100119
11. Geyer, R., Jambeck, J. R., & Law, K. L. (2017). Production, use, and fate of all plastics ever made. *Science Advances, 3*, e1700782.
12. Sa'adu, I., & Farsang, A. (2023). Plastic contamination in agricultural soils: A review. *Environmental Sciences Europe, 35*, 13. https://doi.org/10.1186/s12302-023-00720-9
13. Muhammad, S., et al. (2022). Microplastics in the soil environment: A critical review. *Environmental Technology & Innovation, 27*, 102408.
14. Jambeck, J. R. (2015). Plastic waste inputs from land into the ocean. *Science, 347*, 768. https://doi.org/10.1126/science.1260352
15. Gemma, J. (2019). Principles for responsible investment, the plastics landscape: The challenges and possible solutions. https://www.unpri.org/plastics/plastics-the-challenges-and-possible-solutions/4773.article

16. World Economic Forum. (2016). Ellen MacArthur Foundation and McKinsey & Company. *The new plastics economy: Rethinking the future of plastics*. https://www3.weforum.org/docs/WEF_The_New_Plastics_Economy.pdf
17. World Ocean Day: How much plastic is in our oceans? https://www.aljazeera.com/news/2023/6/8/
18. Kole, P. J., Lohr, A. J., Van Belleghem, F., & Ragas, A. (2017). Wear and tear of tyres: A stealthy source of microplastics in the environment. *International Journal of Environmental Research and Public Health, 14* (1265).
19. Guo, J.-J., Huang, X.-P., Xiang, L., Wang, Y.-Z., Li, Y.-W., Li, H., Cai, Q.-Y., Mo, C.-H., & Wong, M.-H. (2020). Source, migration and toxicology of microplastics in soil. *Environment International, 137*, 105263.
20. Teles, M., Balasch, J. C., Oliveira, M., Sardans, J., & Penuelas, J. (2020). Insights into nanoplastics effects on human health. *Science Bulletin, 65*, 1966–1969.
21. https://www.nbcnews.com/science/science-news/scientists-find-quarter-million-invisible-nanoplastic-particles-liter-rcna133021
22. CIEL. (2019). Plastic and climate: The hidden costs of a plastic planet. https://www.ciel.org/plasticandclimate/
23. Climate change and plastics pollution, Synergies between two crucial environmental challenges, Policy highlights, OCED. https://www.oecd.org/environment/plastics
24. Eleni Kemene. (2023). World Economic Forum. https://www.weforum.org/agenda/2023/06/plastic-pollution-and-climate-change-perspectives-on-the-path-forward
25. https://www.unep.org/new-plastics-economy-global-commitment
26. https://waste-management-world.com/
27. Nema, S. K., & Ganeshprasad, K. S. (2002). Plasma pyrolysis of medical waste. *Current Science, 83*(3), 271–278.
28. https://wedocs.unep.org/bitstream/handle/20.500.11822/41263/Plastic_Science_E.pdf
29. Punčochář, M., Ruj, B., & Chatterjee, P. K. (2012). Development of process for disposal of plastic waste using plasma pyrolysis technology and option for energy recovery. *SciVerse ScienceDirect*. https://doi.org/10.1016/j.proeng.2012.07.433, pp. 1877–7058 © 2012 Published by Elsevier Ltd. Open access under CC BY-NC-ND license.
30. Laredo, G. C., Reza, J., Ruiz, E. M., Laredo, G. C., Reza, J., Ruiz, E. M. (2023). Hydrothermal liquefaction processes for plastics recycling: A review. *Cleaner Chemical Engineering, 5*, 100094. ISSN 2772-7823. https://doi.org/10.1016/j.clce.2023.100094
31. Dave Pragnesh, N., & Kumar, J. A. (2010). Plasma pyrolysis and gasification of plastics waste: A review. *Journal of Scientific and Industrial Research, 69*(3), 177–179.
32. Vora, N., Christensen, P. R., Demarteau, J., Baral, N. R., Keasling, J. D., Helms, B. A., & Scown, C. D. (2021). Leveling the cost and carbon footprint of circular polymers that are chemically recycled to monomer. *Science Advances, 7*, eabf0187.
33. Klotz, M., Oberschelp, C., Salah, C., Subal, L., & Hellweg, S. (2024). The role of chemical and solvent-based recycling within a sustainable circular economy for plastics. *Science of the Total Environment, 906*, 167586. ISSN 0048-9697. https://doi.org/10.1016/j.scitotenv.2023.167586
34. Pandey, S., & Manuja, S. (2020). *Plastic waste management-turning challenges into opportunities*. The Energy and Resources Institute, New Delhi.
35. Peña-Rodriguez, C., Mondragon, G., Mendoza, A., Mendiburu-Valor, E., Eceiza, A., & Kortaberria, G. (2021). Recycling of marine plastic debris. In: Parameswaranpillai, J., Mavinkere Rangappa, S., Gulihonnehalli Rajkumar, A., Siengchin, S. (Eds.), *Recent developments in plastic recycling*. Composites Science and Technology. Springer, Singapore. https://doi.org/10.1007/978-981-16-3627-1_6
36. Iñiguez, M. E., Conesa, J. A., & Fullana, A. (2019). Hydrothermal carbonization (HTC) of marine plastic debris. *Fuel, 257*, 116033. https://doi.org/10.1016/j.fuel.2019.116033
37. Vasarhelyi, K. (2021). *Plastic eating mushrooms: A viable solution to plastic decomposition.*
38. Bruce Adams. (2022). *New circular polymer may solve mixed-use plastics recycling dilemma*, July 21. https://www.plasticstoday.com/materials-research/new-circular-polymer-may-solve-mixed-use-plastics-recycling-dilemma

39. ABC News Australia. https://www.greenecoservices.com/countries-with-plastic-bag-restrictions/
40. https://www.reuters.com/article/us-asia-waste-philippines-tulips/philippine-city-turns-plastic-trash-into-tulips-to-fight-waste-idUKKBN1WA0BB/
41. https://repak.ie/driving-change/circular-economy-eu-legislation/eu-plastics-strategy-and-sup-directive/
42. https://www.unep.org/
43. https://plasticseurope.org/changingplasticsforgood/global-plastics-treaty/
44. https://recykal.com/blog/the-rising-problem-of-plastic-waste-in-india/
45. https://www.researchgate.net/publication/333356328_Strategies_for_controlling_Plastic_Pollution_in_India_Policy_Paper
46. Plastic Pollution and Climate Change, Climate SAR Bulletin. http://ccri.in/bulletin.htm
47. Goel, M., & Chatterjee, S. (2019). Beat plastic pollution—Towards Swachch Bharat Abhiyan. *University News, 57*, 44.
48. https://www.wwfindia.org/news_facts/feature_stories/paving_the_way_for_a_plastic-free_ocean/
49. https://www.niti.gov.in/sites/default/files/2023-06/Thinking-For-Our-Planet-75-Ideas-to-Promote-LiFE.pdf

The Global Plastic Pollution Crisis—How Norway Deals with the Challenge

Beate Kværnes Langset

Abbreviations

CSE	Centre for Science and Environment
EEA	European Economic Area Agreement
EU	European Union
HIC	High Ambition Coalition to end plastic pollution
INC	Intergovernmental Negotiating Committee
INOPOL	India–Norway cooperation project on capacity building for reducing plastic and chemical pollution in India
NORCE	Norwegian Research Centre AS
NIVA	Norwegian Institute of Water Research
OPTOCE	Ocean Plastic Turned into an Opportunity in Circular Economy
POPs	Persistent Organic Pollutants
UNEA	United Nations Environmental Assembly

1 Introduction

As a counsellor for climate and environment at the Norwegian Embassy in New Delhi, I have enjoyed travelling north, east, west and south in India, both through work and in private. I always look out of the window from a car, from a train or a rickshaw. I have noticed that waste management is managed differently around the country. It is interesting to see how different cities tackle waste management—and in particular, plastic pollution. Poor waste management causes air pollution, water

B. K. Langset (✉)
Counsellor Climate and Environment 2022–2024, Royal Norwegian Embassy, New Delhi, India
e-mail: beatelangset@live.no

M. Goel and N. G. Tripathi (eds.), *Plastic Pollution*,
https://doi.org/10.1007/978-981-97-5528-8_2

and soil contamination. Open and unsanitary landfills contribute to contamination of drinking water and can cause infection and transmit diseases. Cities have a key role in fighting the challenges and developing solutions regarding plastic pollution. In my opinion, two factors must be in place to succeed: systems and commitment.

I visited Indore in March 2023, India's cleanest city, and I was impressed in how they had holistic thinking and a genuine circular economy. The waste management sector employs 75,000 people, and involve all levels, from the street swipers to the recyclers, managers and politicians. It was not only the technical and digital solutions that impressed me, but also how the citizens had changed their behaviour, and how proud they were to live in Indore. Many factors have led to Indore's success; however, I think that systems and commitment are some of the key factors.

2 The Global Plastic Problem

> Plastic is a remarkably useful material, primarily because of its durability, but it is this durability that makes plastic a serious environmental concern when it is discarded or ends up in the environment. Plastic litter and plastic pollution are one of the fastest growing global environmental problems of our time, with major consequences for oceans and seas, nature, wildlife, and human well-being [1].

The plastic pollution crisis is global. According to OECD plastic outlook, 8–10 million tonnes of plastic are released into the ocean every year. The amount of plastic in the ocean is expected to double in the next 15 years.

The Bouvet Island is a heartbreaking example that marine plastic pollution is a global challenge. Bouvet Island is a Norwegian volcanic island in the Southern Ocean in Antarctica. It is the world's most remote islands, as far away from cities and rivers, and waste sources as you can get, and yet, you will find plastic waste. Researchers from the Norwegian Polar Institute visited the island in 2018 and discovered large amounts of plastic waste, like bottles and fishing gear. Four seals were painfully entangled in fishing net pieces and needed assistance from the researchers to get free from the plastic waste.

Researchers all over the world are engaged in the global plastic crisis and hundreds of reports have been written on topics like microplastics, chemicals in plastics, recycling, single use plastics, extended producer responsibility, the informal sector, circular economy, etc. This chapter will highlight some of the recently published reports written by Norwegian researchers and Norway's partners in India.

The recently published report «*Hvordan plast og rydding av plast påvirker økosystemet på.*

'*Lisle Lyngøy*'[1] (September 2023) is one of the more encouraging reports. Beach clean-ups have of many been looked at as an activity with a strong visual effect that is not making a big difference in the long run. While preventing the release of plastics

[1] "How plastic and clean-up of plastic affects the ecosystem on Little Lyngøy".

into the environment is the obvious priority, this report shows that the local effort done by beach cleaners actually can make a difference [2].

'Lisle Lyngøy' has been selected for a case study done by NORCE (Norwegian Research Centre AS). Lisle Lyngøy is a small island located close to Bergen, Norway's second biggest city. The island is facing the North Sea and is representing a so-called hot spot for the accumulation of marine waste. Marine plastic pollution has led to significant landscape changes, because of undisturbed accumulation of marine waste over a period of 50–60 years.

Surface clearance of plastic and other waste was carried out on the island in August 2021, followed by a maintenance clean-up in April 2022. The clean-up led to a significant decrease in the total mass of microplastics in the fresh water and ocean at the beach zone as early as the following year. The rapid decline in microplastic mass shows how important cleaning up macroplastic is for limiting microplastic pollution. Comparison of the size distribution of microplastics before and after cleaning indicated that plastic quickly breaks down in freshwater on the island. The project results emphasise how important it is to clean plastic from the environment and regularly carry out cleaning. This is a good argument for cleaning up beaches and other places that are heavily polluted by plastic. Clean ups must furthermore be balanced against the possible negative effects of the risk of increased spread of microplastics [3].

The Nordic Council of Ministers is the official body for inter-governmental cooperation in the Nordic Region (Norway, Sweden, Denmark, Finland and Iceland). The council recently launched the report 'Towards Ending Plastic Pollution by 2040'. The report shows that an international treaty with global rules for plastics can reduce plastic pollution by 90% by 2040. The report shows that there is a need for a holistic approach to plastics throughout the entire life cycle in order to stop plastic pollution. The report contributes to the policy debate by defining 15 far-reaching policy interventions across the plastic lifecycle, and estimating its impact on plastic stocks and flows, GHG emissions, costs and jobs. The focus is on minimising the negative impacts of mismanaged plastics and plastic releases into the environment. The model presents five pillars and two alternative scenarios of how the plastic system could evolve by 2040: 'The Business-as-Usual Scenario' and 'The Global Rules Scenario' [4].

The first pillar in the report is about reducing virgin plastic production and consumption. Targets to reduce virgin plastic volumes can be set, and virgin plastic fees can be used to fund solutions across the plastic lifecycle. The second pillar is addressing elimination of avoidable and problematic plastics and chemicals. Bans and reuse targets for avoidable single use plastics can be developed, in addition to phase-out criteria for problematic plastics, polymer applications and chemicals of concern.

The third pillar focuses on safe circularity. This can be done by developing design rules for safe reuse, repair, durability and cost-effective recycling, combined with targets for collection and recycling and EPR schemes. The fourth pillar focuses on ensuring controlled disposal of waste that can be followed by restrictions on export of plastic waste and mitigation and removal programmes for legacy plastics in the environment. The fifth pillar describes which steps can be taken to preventing the use of microplastics and how to reduce microplastics releases into the environment.

According to the report, the Global Rules Scenario would provide important savings in public expenditure compared to the Business-as-Usual Scenario. Cumulative public spending from 2025 to 2040 in the Global Rules Scenario will amount US$1.5 trillion, compared to US$1.7 trillion in the Business-as-Usual Scenario. "*The savings would mainly accrue from reductions in plastic volumes, resulting in less plastic to collect and manage. However, this would primarily apply to regions with well-developed infrastructure; other regions would still need to invest more in expanding their waste management systems.*" [4, p. 19].

The reduction in virgin plastic volumes that could be achieved in the Global Rules Scenario would vary significantly by geography. The largest reductions would be required in regions that are predominantly composed of high-income economies and are forecasted to have high relative economic and consumption per capita growth. Low and middle-income countries in regions such as India, South and Southeast Asia, Africa and the Middle East face greater challenges in scaling reduction levers.

The report concludes that it is crucial to acknowledge that plastic pollution is a broader problem. Critical issues such as health risks, chemicals of concern and negative impacts on biodiversity are not addressed in the report but must be tackled. The Global Rules Scenario outlined in the report, is intended merely as a starting point for systems change in the global plastics system, rather than as a comprehensive solution [4].

The Norway–India Marine Pollution Initiative was established between Indian and Norwegian authorities in 2019. The initiative has resulted in several publications, including a report from the Centre for Science and Environment (CSE) on 'Managing plastic waste in India', and two UNEP publications: a flyer on 'Best practice on plastic waste' and a report on 'Status of research capacities and information on marine litter in India'. All these reports are available at UNEPs website: indianorwaymarinepollutioninitiative.in.

'The India–Norway cooperation project on capacity building for reducing plastic and chemical pollution in India (INOPOL) is managed by NIVA (Norwegian Institute of Water Research) and is supported by the Norway–India Marine Pollution Initiative. The project has published several reports and articles regarding plastic waste, that combines Norwegian and Indian research on plastic and marine litter. INOPOL phase 1 (2019–2022) was carried out in Gujarat, in the catchment areas of Tapi and Daman Ganga River. The project published a Plastic Waste Management Strategy for Gujarat and an action plan for reducing persistent pollutants in Gujarat.

The project has already published several interesting publications regarding plastic and plastic pollution in phase 2 (2022–2025). In addition, some materials are behind academic paywalls, others are waiting to be published.

One of the articles gives an overview of the main interlinkages between persistent organic pollutants (POPs) and plastic in the waste management system of India. In recent years, the focus on policy, regulation and management of plastic has increased both internationally and in India. The article concludes that in collaboration with the international community and guided by the recommendations developed under the

Basel and Stockholm Conventions, the best available technologies should be adapted to the Indian situation, preventing serious contamination and impacts on human health and the environment [5]. This article may aid policy decisions to combat the release of POPs from different waste types and processes in India.

Another publication from the INOPOL2 project addresses the socio-economic and environmental complexities regarding the Indian plastic bans. The Indian ban on a selected group of single-use plastic products entered into force in July 2022 and provided a change towards a more progressive plastic waste legislation. This article concludes among other things that bans as a policy tool can be "*considered a resolute and decisive way to limit inputs of plastic pollution into the environment.*" At the same time, bans only address a few aspects of a very complex challenge. On the positive side, bans can generate public awareness, but on the other side, bans can have harmful social and economic impacts [6].

The article also points out challenges regarding biodegradable and compostable plastics. These products can create additional waste streams and contaminate existing recycling value chains. Many reports published in the last couple of years conclude that there is a need for more research on this topic.

November 2022 The European Union released its EU policy framework on biobased, biodegradable and compostable plastics. The framework concludes that biobased, biodegradable and compostable plastics could possibly have some beneficial effects if it is designed for circularity, and produced safely and sustainable. However, these plastics present challenges and require substantial research before the environmental benefit can be established. It is important to ensure that they contribute to the circular economy, which aims to keep the value of resources, materials and products in the economy for as long as possible and to avoid waste. In 2023, the INOPOL-project published an article regarding methods, harmonisation and quality control when measuring riverine macroplastic. The article presents a '*critical review of methods used to measure macroplastic flows in rivers, with a specific focus on opportunities for methods testing, harmonisation, and quality assurance and quality control*" [7, p. 1].

River systems are a key environmental recipient of macroplastic pollution, but there is no single method representing a standard approach. Several studies have revealed important findings, but there is a disparity in the reporting of methodologies and data. This article concludes that implementing quality assurance and quality control protocols in sampling, and reaching agreements internationally to set standards for harmonisation are urgently needed. An established consensus on harmonised monitoring standards for riverine macroplastic could strengthen the legally binding global agreement on plastic pollution that will emerge by 2024. And it can provide essential information to help framing local, national and international solutions [7, p. 1].

3 Norway's Plastic Pollution and Circular Economy Strategies and Policy

In March 2022, the United Nations Environmental Assembly (UNEA) decided to work towards an international legally binding instrument to end plastic pollution. Norway's Minister for Climate and the Environment, Espen Barth Eide, who also was the president of UNEA-5 said: '*Plastic pollution knows no national borders. That is why binding international cooperation is so important. The world took an important and necessary step closer to a global plastics agreement during the international negotiations in Paris last week, but there are still many demanding questions that need to be resolved. Norway is impatient and wants an international agreement against plastic pollution both for nature's sake and to protect people's health*'.

To follow up the resolution, an Intergovernmental Negotiating Committee (INC) was established to develop an international legally binding instrument on plastic pollution. The INC began its work during the second half of 2022, with the ambition to complete the negotiations by the end of 2024.

Norway has a history of taking initiatives to end plastic pollution. And at UNEA-5 a group of like-minded countries took the initiative to form a coalition of ambitious countries following the adoption of the resolution. The High Ambition Coalition to end plastic pollution (HIC) was announced during UNEA 5.2 and is co-chaired by Norway and Rwanda. HIC is committed to develop an ambitious international legally binding instrument based on a comprehensive and circular approach that ensures urgent action and effective interventions along the full lifecycle of plastics. The High Ambition Coalition now has 66 members, but India has not yet joined the coalition.

The Norwegian Government released a Norwegian Plastics Strategy in August 2021. The Government's vision is a more sustainable plastics value chain globally, regionally and nationally. The strategy brings together the Norwegian Government's plastics policy and describes most of the measures that have been implemented or are being implemented to reduce plastic litter and plastic pollution [1].

On a global level, it is vital to get in place a new legally binding global agreement on plastic pollution to realise this vision. Key elements in a global agreement must be that all countries take effective measures to reduce the production and consumption of primary plastic polymers, to avoid chemicals of concern, to increase the safe circularity of plastics in the economy, to manage plastic waste in an environmentally sound and safe manner, and eliminate the release of plastics, including microplastics, to air, water and land.

As a signatory of the Basel Convention on the Control of Transboundary Movements of Hazardous Wastes and their Disposal, Norway has made strides in strengthening control of the global trade in plastic waste. In 2019, following a Norwegian proposal, the Basel Convention adopted new and stricter rules for the global trade in plastic waste. To engage businesses and other actors in efforts addressing plastic waste, Norway has taken the initiative for, and is leading, a global partnership on plastic waste under the Basel Convention. The main goal of this partnership is to improve plastics management on a global level, both in the design of goods and services and in waste management.

At the regional level, it is especially important to support the EU's efforts for an improved product policy framework and its implementation in Norway. As a European Economic Area Agreement (EEA), country, and as an integral part of the EU's Single Market, most of Norway's plastic policy must be developed in collaboration with the EU.

In Norway, at national level, it is important to promote more sustainable plastic products and sustainable consumption, to reduce the quantity of plastic waste that cannot be reused and recycled, and to prevent plastic waste from being discarded in nature. The Norwegian Plastics Strategy includes measures for more sustainable plastic value chains and measures to ensure that hazardous substances are removed from the cycle, which will increase the recycling of plastic waste and ensure safe new products. Various chemical substances are added to plastics to give the desired properties. Some are substances of very high concern and others are hazardous substances that may cause harm to human health and the environment. A more circular plastics economy must be non-toxic, in order to prevent the spreading of hazardous substances that can cause harm to human health and to the environment. Norway's plastics strategy assumes that the same limit value must be set for hazardous substances in secondary plastic raw materials and the products that are made from these, as required for products produced by primary raw materials.

The Norwegian Plastics Strategy follows established principles in Norway's environment policy. This includes the principle that those responsible for pollution must bear the costs of measures to prevent and reduce plastic pollution. It is often difficult to identify the origin of discarded plastic waste. To bring about lasting changes, measures must be put in place early in the lifecycle of plastic products and at the same time ensure that plastic waste is more properly managed and not discarded in nature. These challenges cannot be solved by a single country alone; international cooperation is needed to effect the necessary changes.

Targeted measures are needed for various value chains and sectors to reduce plastic pollution from both land-based and sea-based sources. Single-use plastic products are a particularly large source of littering, as well as being a poor use of resources.

The plastics strategy also addresses the question of alternatives to plastic made of primary raw materials, such as bio-based plastic, and the concept of biodegradable plastic. This is a complex area, and more knowledge is needed. It is important that bio-based plastic replacements and new alternative plastics should reduce the overall environmental impact of plastic, and not create new problems.

Plastic packaging is the largest category of both new plastics put on the market and of the plastic waste generated in Norway and Europe. Reducing the environmental impact of plastic packaging and making use of the resources in this plastic waste is a high priority for both Norway and EU. The targets for recycling in the EU waste regulations, which also apply in Norway through the EEA agreement, are already high, and will continue to increase incrementally until 2030. Even if Norway is not a part of the European Union, as an EEA country, Norway is committed to the EU targets on recycling, which states that 50% of all plastic packaging shall be recycled in 2025, and 55% in 2030.

Microplastics are found in all environmental compartments, in the air, lakes, oceans, ice, soil, and in living organisms. Consequences for ecosystems and food security caused by microplastics are uncertain but can be serious. Measures are needed to prevent and reduce microplastic emissions from a range of different sources. Significant sources in Norway include wear and tear of tyres, artificial turf pitches, paint, textiles and intentionally added microplastics to products. Microplastics are also formed from the decomposition of larger pieces of plastic, so measures against littering are also important in reducing microplastic pollution [1].

The Norwegian Plastics Strategy must be seen as an important part of Norway's circular economy strategy. In 2021 the Ministry of Climate and Environment, Ministry of Local Government and Modernisation, Ministry of Agriculture and Food, Ministry of Trade, Industry and Fisheries delivered a strategy for developing a green, circular economy [8].

The strategy has identified construction and buildings, process industries, wholesale and retail trade and the bio-based sectors (agriculture, forestry, aquaculture and fisheries) as sectors with the greatest potential for value creation in a more circular economy. The strategy identifies barriers and policy instruments that can be used to promote the transition to a more circular economy.

Implementation of the *Circular Economy Action Plan* by the European Union (EU) is a strong driver of the shift to a circular economy in Norway. As a party to the EEA Agreement and a participant in the internal market, Norway has close links to the EU, and the initiatives that are being taken under the action plan therefore have major implications for Norway as well.

The Government shall facilitate efforts to enhance circularity in the Norwegian 'key products value chains' that are pointed out in EU's Circular Economy Action Plan: construction and buildings, batteries and vehicles, packaging, plastics, textiles, food, water and nutrients, electronics and Information and Communications Technology.

To promote technology development and make Norwegian industry more competitive, the Government cooperates closely with the EU on the development of waste and product legislations, and access to shared funding instruments.

The greatest barrier to the transition to a circular economy is the challenge to make circular business models profitable. The Norwegian Government supports the development of expertise and innovative business models in retail and wholesale trade and other service industries [8].

The financial sector can play a key role by channelling capital to sustainable investments. The aims of the EU action plan on financing sustainable growth are to reorient private capital towards more sustainable investments, manage climate-and environment-related risks and foster transparency and long-termism in financial and economic activity. The Norwegian Government supports these aims [8].

Norway has strategies and policies that cover both plastics and circular economy. The strategies involve local, regional and global topics. Norway's commitment in the global work against marine pollution and the cooperation with India on the same topic is important for the work towards an international legally binding instrument to end plastic pollution, and for the follow-up after the agreement is concluded.

4 Norway–India Collaboration

In most ways, Norway and India cannot be compared. Although India is much larger than Norway, both in inhabitants and land area, we have similar size of ocean areas to manage: two million square kilometers each. But while 5.5 million people are living in Norway, India is the world's most populous nation with more than 1.4 billion people.

Despite the many differences, Norway and India cooperate on numerous areas. Ties between Norway and India have deep historical roots. As early as the 1600s, a Danish-Norwegian trading station was established in Tranquebar, and Norway was one of the first countries to recognise India's independence in 1947.

In 2018, the Norwegian Government launched a strategy for cooperation with India.

There are many reasons why India and Norway find common ground, and the strategy points out four priority areas for cooperation: Democracy and a rules-based world order, The Oceans, Energy and Climate and Environment [9].

Mutually beneficial Norway–India cooperation on Oceans and Blue Economy goes back to the 1950s but has been substantially strengthened since the establishment of a formal *Ocean Dialogue* and a *Joint Task Force on Blue Economy for Sustainable Development* in 2019.

The Ocean Dialogue was constituted through an MoU signed by the Indian Minister of External Affairs and the Norwegian Minister of Foreign Affairs. The MoU states that Norway and India will hold regular interactions and cooperate in areas of mutual interest, such as shipping, fisheries, aquaculture, ocean energy and technology, ocean research and combating marine plastic litter.

The creation of a Task Force on Blue Economy for Sustainable Development was proposed by Prime Minister Narendra Modi in a meeting with Norwegian Prime Minister Erna Solberg in 2018 and established during Solberg's visit to India in 2019. Both Prime Ministers agreed that Blue Economy can deliver economic growth, new jobs, improved nutrition, and increased food security, if managed sustainably. The Blue Economy cooperation is therefore among one of the main pillars of the Norway–India relationship. Since the Task Force on Blue Economy was started in 2019, it has catalysed increased cooperation on marine pollution and more—to the benefit of both countries.

Combating plastic pollution and marine litter is a key priority of both Norway and India. Since 2019, several projects have been started under the India–Norway Marine Pollution Initiative, focusing on enhancing the circular economy, waste management, co-processing of plastic waste in coastal areas, testing and monitoring of chemical and plastic wastes in rivers and encouraging voluntary citizen participation in beach clean-up. Through Project Steering Committee meetings and other fora, Norway and

India have an extensive dialogue on policies for enhancing the circular economy and reducing pollution of rivers and oceans.

In 2018, the Norwegian government launched a new development programme to combat marine litter and microplastics. The programme is intended to contribute to Sustainable Development Goal 14.1 which states that by 2025, the world should prevent and significantly reduce marine pollution of all kinds. Norway has committed to spending NOK 1.6 billion (approx. 200 mill. USD) in the period 2018–2024 combating marine litter and microplastics in the world's oceans.

The main objective of this Norwegian development program is to prevent and greatly reduce the extent of marine litter from large sources in developing countries.

To achieve this, funding is set to focus on four outcomes:

- Management of plastic waste in partner countries is improved.
- Selected coastal areas and rivers are cleared of waste and the waste is sustainably managed.
- Private sector performance regarding sustainable production and use, and responsible waste management, is improved.
- Global commitments and national and regional instruments to prevent marine litter are strengthened.

The focus is mostly on populous and economically fast-growing countries in Asia with long coastlines.

Contributing to the creation and development of land-based waste management systems is highly prioritised as sound waste management is the most important measure to reduce the supply of litter to the ocean. Funding is also provided for research which will give knowledge that can contribute to finding solutions to the problem. Furthermore, it is important to support initiatives that contribute to increased awareness about marine litter. Actors with the required technical expertise contribute with training, knowledge exchange and technology transfer. Some initiatives focus on innovation, development of new technology and sustainable products. Other actors receive support to influence and advise on government policies, regulations and action plans. Norway also supports processes to achieve stronger international commitments and agreements to prevent marine litter. Beach clean-ups and other clean-up activities are supported in some cases, and particularly where they can help contribute to increased awareness raising.

In India, the marine pollution initiative has five projects funded by the Norwegian Development Program to Combat Marine Litter and Microplastics:

1. UNEP India is coordinating and providing a common platform for the India–Norway Marine Pollution Initiative in India. UNEP is contributing to the management of marine pollution in India, and supports India to actively engage globally on this issue. They provide evidence-based inputs on key policy and governance aspects related to marine plastics pollution. They facilitate international knowledge exchange on the issues of EPR, SUP product bans, and municipal level plastics waste management. And they are generating awareness through regional

and national level programs of the government to support national initiatives on implementation of SUP product bans.
2. OPTOCE (Ocean Plastic Turned into an Opportunity in Circular Economy) is led by SINTEF, one of Europe's largest independent research organisations. The project investigates involvement of energy intensive industries (cement, manufacturing) to increase the treatment capacity for non-recyclable plastic waste in China, India, Myanmar, Thailand and Vietnam. The outcome is that plastic recovery shall contribute to improved resource efficiency and circular economy, and that private sector's involvement and performance in environmentally sound waste management is improved. In India, the project focuses on using non-recyclable plastic waste in cement production (co-processing) that saves CO2 emissions and substitutes the use of coal.
3. INOPOL is an India–Norway cooperation project on capacity building for reducing plastic and chemical pollution in India. The first phase of the INOPOL project (2019–2022) focused on 'Developing Coherent Systems for Data Collection and Analysis' through developing river monitoring capacity on plastics and POPs pollution in specific project areas of Vapi and Surat in the state of Gujarat. INOPOL 2 (2022–2025) has taken this work further, applying a multidisciplinary, cross-sectorial, and integrated approach to collect data and develop a coherent system for data collection in Tamil Nadu and Uttarakhand. The project will scale up by establishing baselines in the selected states, develop monitoring and data collection capacity, support the implementation of current policies, assess local gaps and hurdles, identify opportunities associated with implementation, and promote science-based advice to local and national government bodies. INOPOL is a collaborative project between NIVA (Norwegian Institute for Water and the Environment) and four Indian research institutions (Mu Gamma Consultants, SRM Institute of Science and Technology, CIPET and Toxics Link).
4. Center for Science and Environment (CSE) coordinates 'Mainstreaming Circular Economy in the Waste Management Sector in India'. CSE has a forum of 100 cities that are strengthened to catalyse cross learning, innovation and sharing of best practices, and works on waste management in Agra and Gurugram. CSE has 'Zero Waste' school campus and they are doing trainings, capacity building, exposure visits and gives technical support. They also give input on city government on waste bylaws.
5. Afroz Shah Foundation manages 'Marine Litter and mindset change in Mumbai'. This project aims to create scalable future model plastic producers while engaging the local communities, municipalities, and students to beat the plastic pollution. The project is specifically looking into creating sustainable models at the ground level so that individuals easily can adopt a sustainable lifestyle. They carry out beach- and mangrove cleanup demonstrations as a preventive action to stop plastic from entering the ocean, and thereby reducing marine litter in the ocean. They involve in communities to reduce garbage and learn how to handle garbage properly by involving the locals to protect and preserve the environment.

Information about the projects, including the publications and contact information, are available on www.indianorwaymarinepollutioninitiative.in.

The differences between India and Norway are massive in almost every area, and waste management is no exception. Norwegians segregate their waste at home, in paper, plastic, metal, food waste, hazardous waste, etc. This system works, but it is dependent on individual dedication. It is a vulnerable system, so in addition more and more cities are investing in innovative technological sorting solutions. In India, waste management differs tremendously from city to city. Many good, local initiatives, that also among other initiatives, include education and training for schoolchildren and local bodies.

To stop plastic from ending up in the oceans, a proper waste management must be in place, in addition to reduce the plastic production and consumption. According to OECD, the world is producing twice as much plastic waste as two decades ago, with the bulk of it ending up in landfill, incinerated or leaking into the environment, and only 9% successfully recycled, according to an OECD report. OECD has compared the countries share of plastics treated by waste management category (2019). Almost half of all plastic waste is generated in OECD countries. Plastic waste generated annually per person varies from 221 kg in the United States and 114 kg in European OECD countries to 69 kg, on average, for Japan and Korea. Most plastic pollution comes from inadequate collection and disposal of larger plastic debris known as macroplastic, but leakage of microplastics from things like industrial plastic pellets, synthetic textiles, road markings and tyre wear are also a serious concern [10].

According to the OECD report, 46% of India's plastic waste is mismanaged or not collected. 36% end in landfills and 4% get incinerated and 13% is recycled. Norway is not mentioned in the report, but according to a national report, Norwegians are using 900,000 tons of plastic, and produce 540,000 tons of plastic waste each year [11]. According to the report, most of the plastic is used for packaging, followed by tyres, textiles and construction. Norwegian plastic manufacturers use only 9% recycled plastic in its products. Textiles made of plastic make up around 9% of annual Norwegian plastic waste, almost 10 kg per head per year, in third place after packaging and car tyres. Most plastic is used for packaging, which accounts for 40% of all plastic waste in Norway. Only 24% of plastic waste is recycled. Around 35% of the plastic waste goes to incineration. Smaller amounts end up in landfill or littering and 80% of plastic waste for recycling is exported [11].

5 Summary

Looking at the future, I see challenges and hope. More than three billion people depend on the ocean as a source of food, work, and energy and taking care of the ocean is crucial to combating poverty and reversing the climate crisis [12].

Cooperation between Norwegian and Indian stakeholders, policymakers and researchers can play an important role in the work for a cleaner and healthier ocean and contribute to common global rules and harmonised action, coordination and collaboration. By developing new knowledge, technology and business models, waste can even become a valuable resource and a source for new industries and local employment.

The global plastic crisis can only be dealt with if we work together on the whole plastic lifecycle. But individual engagement also matters. Let me end my chapter with a small example. In 2022, we used 722 million plastic bags in Norway, or to put it another way; each of us used 132 plastic bags. If only one of five say no to a plastic shopping bag once a month, we save 199 tons of plastic, every single year, only in Norway. So small steps also matter when you sum up.

References

1. Norwegian Ministry of Climate and Environment. (2022). *Norwegian plastics strategy* Norwegian Plastics Strategy (regjeringen.no)
2. Bødtker, G., Haave, M., Velle, G., Andersen, G. L., Gomiero, A., Gaasø, R., Bruvik, K., & Bastesen, E. (2023). Hvordan plast og rydding av plast påvirker økosystemet på Lisle Lyngøy. NORCE Report nr. 6 (Bergen, BC: NORCE Research). Rapport-6-2023-NORCE-Klima-og-miljø_HMF11432.pdf (norceresearch.no)
3. Bastesen, E., Haave, M., Andersen, G. L., Velle, G., Bødtker, G., & Krafft, C. G. (2021). Rapid landscape changes in plastic bays along the Norwegian coastline. *Frontiers in Marine Science | Marine Pollution, 8*. Frontiers | Rapid Landscape Changes in Plastic Bays Along the Norwegian Coastline (frontiersin.org)
4. Nordic Council of Ministers/Systemiq. (2023). *Towards ending plastic pollution by 2040*. Towards_Ending_Plastic_Pollution_by_2040 (diva-portal.org)
5. Chakraborty, P., Chandra, S., Dimmen, M. V., Hurley, R., Mohanty, S., Bharat, G. K., Steindal, E. H., Olsen, M., & Nizzetto, L. (2022). Interlinkage between persistent organic pollutants and plastic in the waste management system of India: An overview. *Bulletin of Environmental Contamination and Toxicology*, 1–10. https://doi.org/10.1007/s00128-022-03466-x. Interlinkage Between Persistent Organic Pollutants and Plastic in the Waste Management System of India: An Overview - PubMed (nih.gov)
6. Nøklebye, E., Adam, H. N., Roy-Basu, A., Bharat, G. K., & Steindal, E. H. (2023). Plastic bans in India – Addressing the socio-economic and environmental complexities. *Environmental Science and Policy, 139* (2023). Plastic bans in India – Addressing the socio-economic and environmental complexities - ScienceDirect
7. Hurley, R., Braaten, H. F. V., Nizzetto, L., Steindal, E. H., Lin, Y., Clayer, F., van Emmerik, T., Buenaventura, N. T., Eidsvoll, D. P., Økelsrud, A., Norling, M., Adam, H. N., & Olsen, M. (2023). Measuring riverine macroplastic: Methods, harmonisation, and quality control. *Water Research, 235*.Measuring riverine macroplastic: Methods, harmonisation, and quality control (sciencedirectassets.com)
8. Norwegian Ministry of Climate and Environment, Ministry of Local Government and Modernisation, Ministry of Agriculture and Food, Ministry of Trade, Industry and Fisheries (2021) *Norway's strategy for developing a green, circular economy*. Norway's strategy for developing a green, circular economy - regjeringen.no
9. Norwegian Ministry of Foreign Affairs. (2018). *Norway – India 2030 The Norwegian Government's strategy for cooperation with India*. Norway - India 2030 (regjeringen.no)
10. OECD. (2022). Plastic pollution is growing relentlessly as waste management and recycling fall short, says OECD. Retrieved from: www.oecd.org/environment/plastic-pollution-is-growing-relentlessy-as-waste-pollution-is-growing-relentlessly-as-waste-management-and-recycling-fall-short.htm

11. Syversen, F., Sundt, P., Kirkevaag, K., & Briedis, R. (2020). Materialstrømmen til plast i Norge – hva vet vi? Report by Handelens Miljøfond. Materialstrømmen-til-plast-i-Norge-Hva-vet-vi_-1.pdf (dl8y9d78cbd9m.cloudfront.net)
12. Norad. (2022). Ocean Report 2022 Making Waves. Ocean report 2022: Making Waves Norway's support for a sustainable ocean (norad.no)

Plastics and Human's Health

Priyanka Chaudhary

Abbreviations

EDCs	Endocrine Disrupting Chemicals
MNPs	Microplastics and nano plastics
PAHS	Polycyclic Aromatic Hydrocarbons
PBDEs	Polybrominated Diphenyl Ethers
PBTs	Persistent Bio accumulative and Toxic pollutants
PE	Polyethylene
PET	Polyethylene Terephthalate
PFAS	Polyfluoroalkyls Substances
PFOA	Perfluorooctanoic Acid
POPs	Persistent Organic Pollutants
PP	Polypropylene
PS	Polystyrene
PUR	Polyurethane
PVC	Polyvinyl Chloride
SCCPs	Short Chain Chlorinated Paraffins
VOCs	Volatile Organic Compounds

P. Chaudhary (✉)
Department of Oral and Maxillofacial, Trauma Plastic Reconstructive Surgery, All India Institute Of Medical Sciences (AIIMS), New Delhi, India
e-mail: drpriyankachaudhary701@gmail.com

HOD Chemistry (Teaching and Research), Brilliants' Convent Sr. Sec School, Pitam Pura Delhi, India

M. Goel and N. G. Tripathi (eds.), *Plastic Pollution*,
https://doi.org/10.1007/978-981-97-5528-8_3

1 Introduction

Plastics are the most pervasive materials according to its nature, origins, and impacts. In other words, plastics are synthetic organic polymers-arrays. Plastic products with different chemical compositions are derived from fossil fuels. The utility of plastics, ubiquity, and durability allow plastic resins to be pressed, rolled, stretched, and extruded into every desired shape. Plastics play a major role in everything from plastic bottles, bags, food packaging, clothing, vehicle parts, prosthetics, and construction materials. The health effects of plastics and its additives, such as benzene, phthalates, and bisphenol A, have increased in recent time period. While plastics being cheap and abundant,concern continues to coincide with escalating plastic production and uses. Concern over the impacts of plastics and human health has been denied, and downplayed. The scale and urgency of the plastics pollution crisis have prompted, legal, scientific, and advocacy initiatives around the world to address it, including legal bans on plastic use. It is important to understand the full range scale of health crisis posed by plastic supply chain, and its manufacture, use disposal, and presence in the environment.

2 Life Cycle of Plastic at Every Stage—A Serious Threat to Human Health

Extraction and Transport

As the extraction methods and locations have expanded the release of toxic chemicals into water, air, and food has led the massive public health impacts. Extraction process will cause oil and gas to flow out of the drilled well and in addition to hydrogen sulfide and various hydrocarbons—benzene, toluene, ethylbenzene, and xylene. Oil and gas extraction.

Health Impacts Due to Extraction and Transport of Precursors of Plastic Production

During drilling, fracking, and flaring there is release of toxic chemicals into the environmental elements. BTEX can lead to cardiovascular disease and respiratory conditions such as breathlessness, pulmonary inflammation, cardiac arrhythmia and lung aggravation. Oil and gas production emits millions of tons of methane and other pollutants such as volatile organic compounds and it can spread from the site of production. Chronic exposure will lead to lung function impairment and chronic obstructive lung disease. Projections indicate that by 2025 in the US summer pulmonary attacks in children under the age of 18 years will be 750,000. Asthma-related emergencies are due to ozone smog from oil and gas production.

Communities living nearby production sites will be exposed to chemicals via various routes such as inhalation, ingestion, skin and eye contact and ears for noise exposure. It can impact the respiratory, circulatory, reproductive, immune, neurological, and digestive systems severity depending upon the period of exposure.

Extraction mines also affect mental health and human rights because of loud noise and sleep disturbances that induce stress, increase blood pressure, diabetes, heart disease, depression, and learning difficulties in children.

Risk Zone Factors to Children, Infants, and Pregnant Women (Clinical Trials)

The toxic chemicals used in oil and gas development including benzene, toluene, ethylbenzene, and xylenes are potential endocrine disruptors and cause negative developmental impact on reproductive, neurological, and immune systems. Studies have shown that there is a severe increased risk of cancer, obesity, diabetes, metabolic toxicity, infertility, and increased risk potential of prenatal and early infant developmental period. Studies show evidence that exposure to toxic chemicals will lead to increased risk of preterm birth, poor infant health, and low birth weight baby. There will be higher incidence of birth defects of the brain, spinal cord and spine, congenital heart defects, and higher rates of childhood Leukaemia.

Plastic pollutants used in fracking can enter drinking resources easily through wastewater spills, injection of wastewater into ground water, and dumping of improperly treated wastewater. The United States Environmental Protection (UNEP) agency has identified around 173 chemicals that cause major deadly health impacts such as carcinogenicity, neurotoxicity, immune failure, toxic defects in body weight and blood chemistry, liver defects, renal toxicity, and reproductive toxicity.

Hidden Risk of Pipeline Infrastructure

Pipelines are needed to transport natural gas from the site of extraction to refineries ports and consumers. Pipelines emit methane, ethane, benzene, toluene, xylene, carbon monoxide, ozone, and other pollutants. Thus the damaged or old pipelines increase the risk of fugitive emissions and incidents. Before the plastic reaches consumers it reaches the environment and impacts human health at their initial stages of lifecycle [1].

3 Health Impacts from Chemicals Involved in Refining and Manufacture Process

Major pollutants emitted in plastic productions are affecting human health because of toxicity produced by benzene sulphonates, nitrates, benzene styrene and toluene and 1,3 butadiene [2].

Toxic Effects of 1,3 Butadiene

Flammable, colourless, and mild gasoline odour used as a carbocation toxic intermediate in the manufacture of plastic polymers. Short-term exposure can cause eye inflammations, severe allergies of throat and major cause of upper respiratory infections, headaches, fatigue, low blood pressure and pulse, and central nervous system failure. Long-term exposure leads to Leukaemia prominently lymphatic Leukaemia and Acute Myeloid Leukaemia.

Benzol

Benzene is colourless, flammable liquid with aromatic odour used as a chemical solvent to form monomers. Short-term exposure causes symptomatic disorientation in body like headache, tremors, drowsiness, and dizziness. Severe toxic exposure leads to organ system failure (death) within several minutes or hours. Long-term exposure leads to conditions like anaemia and leukaemia specifically carcinogenicity and Hodgkin lymphomas. Irregularity in menstrual cycles has been studied in women due to exposure and underdeveloped ovaries and egg cycle.

Ethenyl Benzene (Styrene)

Styrene is highly explosive colourless liquid used in the production of polystyrene plastic and resins. It can be released into the air and food chain can cause inflammation of lungs, and sensory organs. High exposure will lead to decreased vision, slower reflexes, and cancerous proliferations in the body.

Phenylmethane (Toluene)

Toluene is colourless liquid with aromatic odour used in the production of toxic polymers such as polyethylene terephthalate (key component in plastic bottles and nylon). Low to moderate short duration exposure levels of toluene will lead to symptoms like fatigue, dehydration memory loss, nausea, and loss of appetite. Long-term toxic direct exposure can cause irritation of retina, lungs, nausea, headache, and dizziness. It also hampers development in the nervous and reproductive system in children or in embryonic intra-uterine stage resulting in developmental defeats in children.

Bimethyl (Ethane)

Ethane is a toxic by-product of natural gas extract used in harmful plastic production through its intermediate dimethyl carbocation conversions to ethylene resulting in excessive production of toxic components like sulphur dioxide, and nitrogen oxides which combine to create hole in ozone in the presence of ultraviolet radiations. These will alter health impacts including retinal and throat inflammation, nausea, headache, nasal bleeds, renal and liver damages, central nervous system failure, and carcinogenicity.

Methyl Ethylene and Propylene Oxide

Methyl ethylene is colourless gas with slight petroleum-like odour—a chemical toxic intermediate used in plastic production (carpet fibres). Toxic chemicals are produced and exposure leads to symptoms like dizziness, drowsiness, and unconsciousness.

Propylene oxide-flammable, colourless liquid has been classified as severe human carcinogen.

Polycyclic Aromatic Hydrocarbons (PAHS)

Environmental major toxicants linked to plastic toxins formulation including anthracene, phenanthrene, and pyrene are enlisted as human carcinogens. Persistent, bioaccumulative, and toxic pollutants (PBTs) are highly toxic, long lasting substances that can be harmful to food chains. PBTs related to plastic production are hexabromobiphenyl ethers, Penta oligothiophene, phenol, propylated phosphate, and 2,4,6 tris phenol, benzo, perylene, lead, mercury and tetra Bromo bisphenol. These pollutants have wide range of health defects such as failure to nervous and reproductive systems and other developmental carcinogenetic defects.

Impacts on Workers in Plastic Industry

They are exposed to several toxic chemicals including carcinogens and endocrine disrupting chemicals (EDCs). These chemical toxins are popular to cause breast cancer, and reproductive system disorientations. Chemical releases such as vinyl chloride, styrene, acrylonitrile, bisphenol A, and formaldehyde are tagged as carcinogen and exposure can lead to mammary gland tumours, liver damage, lung cancer, ovarian cysts, endometriosis, and breast cancer. Further, additives including plasticizers, flame retardants, and metals used in plastic production have carcinogenic and endocrine disruption health impacts.

Plastics Usage and Health Effects

Plastics cracks into three categories: Polymer resins, Synthetic plastic fibres, and Additives. The most prevalent plastics resins are manufactured from Polyethylene (PE), Polypropylene (PP), Polystyrene (PS), Polyvinyl chloride (PVC), Polyethylene Tere-phthalate (PET), and Polyurethane (PUR) resins [2].

Plasticizer Toxins Intervening in Human Health

Plastic refers to numerous chemical forms of polymers which are synthesized from hydrocarbon synthetic chain that are polymerized to form macromolecular chains from monomer units.

A wide array of chemicals and additives may be used in the manipulation processes of plastic with different functional monomer units including initiators, catalysts, and solvents greatly intervene in protein synthesis in RNA. Oleochemical additives are used to provide structural and chemical manipulation of polymerization in DNA toxic replicants that include numerous agents like stabilizers, plasticizers, flame retardants, pigments, and fillers.

As plastic particles continue to degrade human health with chemical additives such as pathulates and BPA severely cause reproductive toxicity and teratogenicity. Vinyl chloride and butadiene are carcinogens. Benzene and phenol are mutagenic (they change the genetic material—through DNA manipulation) Other chemicals used as softeners or brominated flame retardants cause birth defects, cancer and hormonal problems particularly in women. Once the additives have been released, including through incineration of plastic waste, they can persist in the environment, building up in the food chain.

Digestive Toxicity with Food Packaging Chemicals

Plastic chemicals circulate from packaging into food and beverages and is known as main source of degradation of human health with exposure to contaminants associated with plastics. Some polymers undergo catalytic degradation when come in contact with acidic or alkaline elements of food in the presence of UV lights and thermal reactivity against it.

Persistent organic pollutants (POPs) are a class of highly hazardous chemical pollutants that are recognized as a serious global threat to stable human health with pace in evolution and survival of various organisms in ecosystems.

- Complex structural chain of chlorinated formulated paraffins (SCCPs)
- Polybrominated diphenyl ethers (PBDEs)
- Nonyl phenols
- Octyl phenols
- Per and Poly fluoroalkyl substances (PFAS) are recognized as plosive plastic additives by international community as POPs.

Accumulation of POP's in living organisms and other traces of toxicity found at higher concentrations at higher level organisms (consumers) in the food chain. These are linked to cancerous proliferation of various systems and reproductive deterioration and other acute genetic diseases [https://issuelab.org/resources/34349/34349.pdf] [3]

- Cancers and tumours such as tissue sarcoma, non-Hodgkin lymphoma, breast cancer, pancreatic cancer and adult-onset leukaemia.
- Neurological disorders (attention deficit disorder, behavioural problems such as impaired memory, aggression, delinquency, learning disabilities, schizophrenia).
- Mitotic—Meiotic cycle disorders—(abnormal spermiation, miscarriages, preterm delivery, low birth weight, altered sex ratios in offspring, shortened period of lactation in nursing mothers and severe menstrual disorders like Pcos, heavy uterine shedding and prolapsed.

Chemical toxins migrating from food packaging and other food materials can be harmful at very slight doses.

Chemicals of Concern:

Several chemicals commonly found in food packaging have been identified as potential hazards to human health. These include bisphenol A (BPA), phthalates, perfluoroalkyl substances (PFAS), and various types of plasticizers and additives. These chemicals are used in the production of plastics, coatings, and other packaging materials to improve their performance and durability.

- Nonyl-phenol, a crackling product of the antioxidant and thermal stabilizer tris phosphates and polyvinyl chloride food wraps.

- BPA, is found in poly-carbonated plastics, epoxy resin liners of metal cans and paper receipts. Phthalates (di-isononyl phthalate and di-thienyl phthalate, non-phthalate plasticizer) are a potential carcinogen used in meat wraps and straps in automobile industry.
- Styrene, is the monomer used to manufacture polystyrene and polystyrene foam [3].
- Per and polyfluoroalkyl substances (PFAS) such as perfluorooctanoic acid (PFOA), perfluoro octane sulfonate and perfluoroalkyl acids to provide grease proof barrier to paper wraps and paper and fibber containers used in food packing.
- Perchlorate is used in various formulations for food packaging, gasket closures, and antistatic agents in dry food packaging.
- These pollutants toxins of plastics are famous for causing hazardous- (endocrine disruption, carcinogens reproductive toxins and mutagens [3].

Mechanism Involved

When food comes into contact with packaging materials containing these chemicals, there is a risk of migration, where the chemicals leach into the food matrix. Once ingested, these chemicals can interfere with normal digestive processes and have systemic effects on the body. For example, BPA, a known endocrine disruptor, can mimic the action of hormones in the body and disrupt the endocrine system, leading to adverse health effects.

Health Implications

Exposure to food packaging chemicals has been associated with various health issues, including reproductive disorders, developmental abnormalities, metabolic disorders, and cancer. Ingestion of these chemicals can also lead to inflammation of the digestive tract, disruption of gut microbiota, and impaired nutrient absorption, compromising overall digestive health.

Regulatory Measures

In response to growing concerns about the safety of food packaging chemicals, regulatory agencies around the world have implemented measures to limit their use and protect public health. For example, BPA has been banned or restricted in certain applications, and alternative materials and manufacturing processes are being explored to reduce reliance on hazardous chemicals. To minimize the risk of digestive toxicity associated with food packaging chemicals, several mitigation strategies can be employed. These include using safer alternatives to hazardous chemicals, improving packaging design to reduce leaching, enhancing monitoring and testing of packaging materials, and educating consumers about potential risks and safe handling practices. In Conclusion, digestive toxicity associated with food packaging chemicals is a significant public health concern that requires attention from regulators, manufacturers, and consumers alike. By implementing stricter regulations, adopting safer alternatives, and promoting awareness and education, we can mitigate the risks posed by these chemicals and safeguard the health and well-being of consumers.

Types of plastics, their uses and health impacts are summarized in Table 1.

Table 1 Types of plastics and their health impacts

Type of plastic/toxic additives	Uses	Health effects
Polyvinyl chloride (PVC), polycarbonate (PC)/bisphenol	Plasticizers, can liners	Mimics estrogen, ovarian disorders
Polystyrene (PS), PVC/ phthalates	Plasticizers, artificial fragrances	Interference with testosterone, affects sperm motility
Persistent organic pollutants (POPs)	Weedicides, flame fillers, stabilizers	Neurological and reproductive damage
Dioxins	Processed/thermosetting plastics	Carcinogen, testosterone interference
Polycyclic hydrocarbons	Pesticides	Reproductive misfunctioning
Polychlorinated biphenyls	Dielectrics	Thyroid disorders
Polystyrene	Crackled product	Carcinogen, DNA adducts
PVC/nonyl phenol	Anti-static, anti-fog, surfactant	Mimics oestrogen

It is recommended to not use any of the fluorinated alternatives to PFOA and PFOS due to their persistency and mobility as well as potential degradation of health in organisms.

Impacts on Health from Size of Plastics (Accumulation Criteria)

Plastics come in variable sizes and chemical composition from macroplastics to microplastics to nanoplastics, and the health impacts and exposure pathways of interrupting human health vary.

Toxicity of (Macro/Micro/Nano) Plastics

Macroplastics toxicity is termed as processed plastic larger than 5 mm accumulation in living cells and microplastics toxicity is synthetic polymers accumulation less than 5 mm respectively. They occur in infinite shapes for example spheres, fragments, granules, pellets, flakes, breads, filaments, or fibres. Microplastics can be detected in organic sampling down to 1 micron in size. Nanoplastic toxicity is specifically defined as severe accumulation of particles less than 1 micron in the range of 1–100 nm.

Micro and macroplastics migrate from marine organisms as original consumer migration chain in the form of accumulated results of plastics in living cells and morphological disorientations in marine organisms.

Microplastics enter the biosphere either from primary or secondary microplastics aggregation and accumulation in living cells. Primary microplastic toxicity is referred to as microplastics produced in micro sizes. Primary microplastics substances start accumulating in biome in the form of powders and pellets (<5 mm in size). These microplastics intervening in life processes from primary microplastics sources include microbeads used in personal care products, such as hand cleaners, facial scrubs, and toothpastes. Primary microplastics are also aggregated through a variety of industrial applications, specifically in body fluids during oil and gas drilling and other types of abrasives metal extraction, air blasting while paint removal from boat hulls, and engines and metal surface cleaning.

Secondary microplastics are degradants waste of macroplastics of larger consumable items. This type of macroplastic products are found in shoreline and litter. Synthetic textile fibres and particle from car tyre abrasion are the two main sources of secondary microplastics accumulation in the ocean.

Nano plastics are increasingly used in products such as paints, adhesives, pharmaceuticals, and electronics and in 3D painting entering the living cell and systems of organisms through such gateway. Similarly to microplastics, secondary nanoplastics also result from further human health degradation.

Burden of Plastic Additives in Human Body

Chemicals used in food packagingare used in a wide array of consumer products. People are exposed with such toxicity without their knowledge because chemicals in plastics and packaging do not appear on ingredient lists. Health impacts of these toxic additives are given in Table 2 (Fig. 1).

Biomonitoring of Human Body Burden with Plastics

Biomonitoring the human body with plastics involves examining the presence and impact of plastic particles or chemicals within the body. Plastics are ubiquitous in modern life, and they can enter the body through various routes, including ingestion, inhalation, and dermal absorption. Here's how biomonitoring with plastics works:

Table 2 Toxic chemical additives and their health impacts

S. no	Toxic chemical additives	Products	Health impact
1	Acrylonitrile oxides	Textile clothing /automotive appliances/3-d printers	Polycystic syndrome, dermal carcinogenicity
2	Plastic retardants	Lensed glasses/thermal papers	Breast and prostate cancer
3	Lead	Stabilizer/colourant	Endometrium, renal cancer
4	Bisphenol-A	Electronic appliances	Cardiac arrhythmia
5	Cadmium metal	Metal refining straps	Inflammation of skin, throat lung toxicity
6	Per fluorinated substances (PFAS)	Cookware coating/repellent	Endocrine disruptors/ Hormonal imbalance
7	SCCP	Toys/glassware	Respiratory inflammations
8	Vinyl chloride	Wall packaging/covering	Gastro intestinal cancers
9	Styrene	Packaging covering	Skin disorders
10	Phthalates	Pvc toxicity	Nervous damage

Source Adopted from https://issuelab.org/resources/34349/34349.pdf [3]

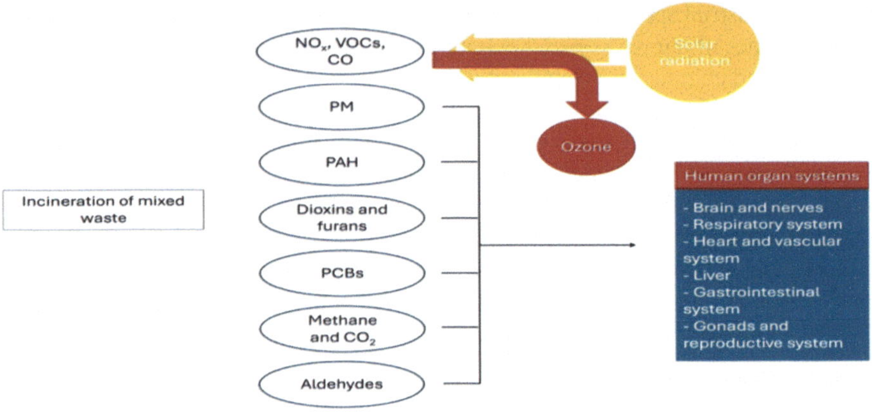

Fig. 1 Incineration of mixed waste and pollutants and human organs affected

1. Sample Collection: Biomonitoring typically involves collecting samples from bodily fluids, tissues, or excretions. Blood, urine, saliva, and breast milk are commonly analyzed, as they can provide valuable insights into the presence of plastics or plastic-related chemicals.
2. Analysis Techniques: Once samples are collected, scientists employ various analytical techniques to detect and quantify plastics or plastic-related compounds. These techniques may include microscopy, spectroscopy, chromatography, and mass spectrometry.
3. Identification of Plastic Particles: Microplastics, which are tiny plastic particles smaller than 5 mm, can be identified and quantified in biological samples. These particles can originate from sources such as microbeads in personal care products, synthetic clothing fibres, or larger plastic debris that has degraded over time.
4. Assessment of Chemicals: Plastics can also release chemicals known as plasticizers, flame retardants, and monomers. These chemicals may leach from plastic products and enter the body, where they can potentially have adverse health effects. Biomonitoring aims to identify and quantify these chemicals in biological samples.
5. Health Implications: Research on the health effects of plastic exposure is ongoing. Some studies suggest that microplastics and plastic-related chemicals may have harmful effects on human health, including endocrine disruption, inflammation, and oxidative stress. Biomonitoring helps researchers understand the extent of human exposure to plastics and assess any associated health risks.
6. Policy and Regulation: Biomonitoring data can inform policymakers and regulators about the need for measures to reduce plastic pollution and mitigate human exposure to plastics. This may include policies to ban certain plastic additives or restrict the use of plastics in food packaging and consumer products.

Overall, biomonitoring the human body with plastics is a critical aspect of understanding the impact of plastic pollution on human health and informing strategies to address this global environmental and public health challenge.

Biomonitoring is considered one of the best methods to determine actual levels of plastics toxicity and exposure. It measures the chemicals, their metabolites, or specific substrate reaction and its products in urine or blood are found in traces or in biochemical increases toxic level. Many of the chemicals used in different areas of industries are used in a wide array of other consumer products or but other industries. Most people are exposed without their knowledge or consent because of chemicals in plastic. The Centres for Disease Control and Prevention Nutrition Examination Survey, is one of the most comprehensive surveys of a population's chemical exposure in 2009–2010. It found BPA in 92% of the urine samples from children (at least six years old) and adults in the United States. Ten of the 15 phthalates were detected in virtually all of the samples, as were PFOA and perchlorate, and 4-nonylphenol was found in 51% of people tested. Other studies have proved that BPA in human blood and other tissues is a serious growing health hazard. Many of the toxic chemicals used in food packaging are also used in a wide array of other consumer products. Biomonitoring data does reveal minimum to maximum presence of a specific chemical in the human body and its prevention conducted by National Health and the result of exposure to plastic or packaging, does confirm that human populations have significant and increasing body burdens as a result of exposure to many toxic plastic chemicals [4].

4 Clinical Trials and Results

4.1 Vascular Toxicity in Human by Nano and Micro Plastics

The pervasive presence of plastic pollution in our environment has raised concerns about its impact on human health. Of particular concern is the potential vascular toxicity caused by nano and microplastics, tiny particles that can easily penetrate biological barriers. This essay explores the mechanisms by which nano and microplastics induce vascular toxicity and the implications for human health.

Understanding Nano and Microplastics:

Nano and microplastics are plastic particles with sizes ranging from nanometers to micrometers, making them virtually invisible to the naked eye. They originate from various sources, including the breakdown of larger plastic debris, industrial processes, and the fragmentation of synthetic fibers. Due to their small size, nano and microplastics can easily enter the environment, including water bodies, air, and soil, where they can be ingested by organisms and ultimately reach human populations through the food chain and air inhalation [5].

Mechanisms of Vascular Toxicity:

Once inside the body, nano and microplastics can translocate across biological barriers, including the gastrointestinal tract, respiratory system, and skin, gaining access to systemic circulation. Within the bloodstream, these particles can interact with endothelial cells lining blood vessels, triggering inflammatory responses and oxidative stress. Additionally, nano and microplastics can serve as carriers for toxic chemicals and pathogens, exacerbating their vascular toxicity.

Impacts on Human Health: The vascular system plays a crucial role in maintaining physiological homeostasis, regulating blood flow, and delivering nutrients and oxygen to tissues. Vascular toxicity induced by nano and microplastics can disrupt these functions, leading to various adverse health outcomes. Chronic exposure to these particles has been associated with endothelial dysfunction, atherosclerosis, hypertension, and thrombosis, increasing the risk of cardiovascular diseases such as heart attacks and strokes. Furthermore, vascular inflammation and oxidative stress induced by nano and microplastics have been linked to systemic effects, including neurodegenerative diseases and metabolic disorders. Despite growing evidence of the vascular toxicity of nano and microplastics, several knowledge gaps and challenges remain. The complex interactions between plastic particles, biological systems, and other environmental stressors require interdisciplinary approaches combining toxicology, epidemiology, and environmental science. Furthermore, standardized methods for the detection and characterization of nano and microplastics in biological samples are needed to accurately assess human exposure levels and health risks [5]. Addressing these challenges will be crucial for developing effective strategies to mitigate the vascular toxicity of nano and microplastics and protect human health. In conclusion, nano and microplastics represent a significant environmental and public health concern due to their potential to induce vascular toxicity. Understanding the mechanisms underlying their adverse effects on the vascular system is essential for mitigating their impact on human health. Concerted efforts from scientists, policymakers, and the public are needed to address this pressing issue and safeguard the health of current and future generations.

4.2 *Education, Health, and Plastics*

Education, health, and plastics are interconnected in various ways that impact individuals, communities, and the environment.

1. Education: Education plays a crucial role in raising awareness about the health and environmental impacts of plastics. By incorporating lessons on plastic pollution, recycling, and sustainable practices into school curriculums, students can learn about the importance of reducing plastic waste and making environmentally conscious choices. Education also empowers individuals to advocate for policies that promote plastic alternatives and waste reduction initiatives.

2. Health: The use of plastics in various products, packaging, and medical devices can have implications for human health. Certain chemicals found in plastics, such as BPA and phthalates, have been linked to adverse health effects, including hormonal disruptions and potential carcinogenic properties. Additionally, plastic pollution in the environment can contaminate water sources and food supplies, leading to health risks for individuals who consume contaminated products. Promoting the use of safer, non-toxic alternatives to plastics can help mitigate these health risks and protect public health.
3. Plastics: The widespread use of plastics has led to a global environmental crisis, with plastic pollution posing significant threats to ecosystems, wildlife, and human health. Single-use plastics, in particular, contribute to the accumulation of plastic waste in oceans, rivers, and landfills, leading to environmental degradation and harm to marine life. Addressing the issue of plastic pollution requires a multi-faceted approach that includes reducing plastic consumption, promoting recycling and waste management practices, and transitioning to biodegradable or compostable alternatives.

5 Conclusions

A disproportionate impact of plastic production and usage on vulnerable populations, particularly children, pregnant women, and communities living near production sites, is taking place. Medically, exposure to the toxic chemicals associated with plastics can lead to irreversible health effects, underscoring the urgent need for action. Moreover, the growing concern about microplastics and nanoplastics in the environment and their potential to enter the human body through various routes, such as food, water, and air, is a stark reminder of the far-reaching consequences of our reliance on plastics.

The evidence discussed in this chapter highlights the importance of adopting a precautionary approach and implementing a global legal framework to address the plastic pollution crisis. By understanding the full scale of the health crisis posed by the plastic supply chain and its presence in the environment, we can work towards solutions that prioritize human well-being and ecological balance.

In conclusion, addressing the interconnected issues of education, health, and plastics is essential for promoting sustainable practices, protecting public health, and mitigating the environmental impacts of plastic pollution. By raising awareness, implementing effective policies, and fostering a culture of environmental stewardship, we can work towards a healthier, more sustainable future for all.

References

1. Report: Plastic Threatens Human Health at a Global Scale (2019) https://www.plasticpollutioncoalition.org/blog/2019/2/20/report-plastic-threatens-human-health-at-a-global-scale; Accessed 11 March 2024 (https://www.ciel.org/wp-content/uploads/2019/02/Plastic-and-Health-The-Hidden-Costs-of-a-Plastic-Planet-February-2019.pdf)
2. Halden, R. U. (2010). Plastics and health risks. *Annual Review of Public Health, 31*, 179–194. https://doi.org/10.1146/annurev.publhealth.012809.103714
3. https://issuelab.org/resources/34349/34349.pdf
4. https://www.cdc.gov/nchs/nhanes/index.htm
5. https://www.nejm.org/doi/full/10.1056/NEJMoa2309822

Plastic Pollution and Health Impacts

Shyamala K. Mani

1 Introduction

1.1 Background

Plastics were invented in the 1950s and quickly became the ubiquitous material to be used everywhere. Today, there is hardly a sector in which plastics are not used. This is mainly due to their durability, flexibility, lightweight and relatively low cost of production [1]. Plastics are materials comprised of repeating units of macromolecules, which could be of natural organic, synthetic or semisynthetic origin. Plastics, so named because of their plasticity, ductility and other malleable and flexible qualities can be used for fulfilling different requirements in the domestic, commercial, and industrial sectors. Plastics, especially synthetic polymers are manufactured from the byproducts of the petroleum sector, namely naphtha and others. Primary plastics are those from which products can be manufactured, e.g. resin pellets, powders, plates, sheets, or strips. Synthetic fibres, foils, pipes, gooses, parts of other manufactured products, are parts of other products where they appear as accessories; Toys, clothes, and furniture are manufactured products from synthetic polymers while bottles, bags, and boxes are packaging materials. Post consumer plastics are often scrapped as waste especially if they are not reusable including mixed plastics. Recycled plastics whether granules or lumps are secondary plastic waste materials [2].

S. K. Mani (✉)
Centre for Chronic Disease Control, New Delhi, India
e-mail: shyamala.mani@gmail.com

NIUA, New Delhi, India

M. Goel and N. G. Tripathi (eds.), *Plastic Pollution*,
https://doi.org/10.1007/978-981-97-5528-8_4

2 The Indian Plastic Industry

The Plastic Industry in India made a beginning in 1957. From manufacturing polystyrene, the industry expanded and is currently manufacturing all categories of plastics and their products. The diversification of the industry has also led to the development of multiple processing units in different parts of the country. It has over 2,000 exporters. Over 4 million people are employed in this sector which primarily comprises small- and medium-sized enterprises [3]. Many consumer goods, including sanitary fittings, electrical accessories, laboratory/medical surgical ware, tarpaulins, laminates, fishnets, travel ware, and others are exported as also raw material [4].

The country, having among the world's largest petroleum refinery at Hazira, Gujarat, produces raw materials, including polypropylene, high-density polyethylene, low-density polyethylene, and PVC. Export of plastic material was US$ 7.045 billion in the financial year 2020, most of it from the export of plastic raw materials which was US$ 2.91 billion [4].

As per CPCB reports, 5.6 million tonnes of plastic waste per annum (TPA) generated in the country contributes to 8% of the total solid waste in India. With plastic consumption on the rise, it is not surprising that it will soon reach 20 kg per capita and most of it will be single-use plastic products (SUPP) which will be discarded and found in our environment. Delhi produces the maximum quantity of plastic waste followed by Kolkata and Ahmedabad. It is well-known that in India, packaging plastics make up around 43% of the plastics manufactured and these mostly go for making SUPPs whose life varies from a few minutes to a day or few days with few takers even if they are collected because of their flimsiness and poor quality [5] (Fig. 1).

The sector-wise demand of plastics in India is 18,307 kilotons in commodity sector, which includes packaging and other areas where thermoplastics are used including construction, agriculture, etc. The demand for thermosets is 1628 Kilotons, engineering plastics, 960 kilotons and high-performance plastics only 2.6 kilotons. The commodity plastic usage, which includes packaging, containers, pipes and others, is 19 times larger than engineering plastics and 11 times that of thermosets.

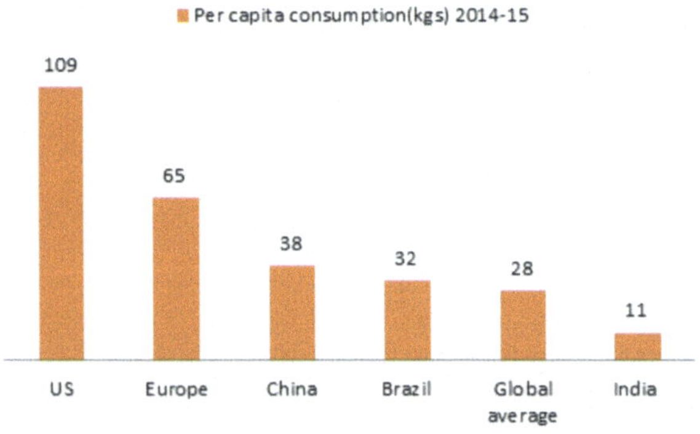

Fig. 1 Average per capita consumption of plastic in India in 2014–15 was about 11 kgs. (*Source* [5])

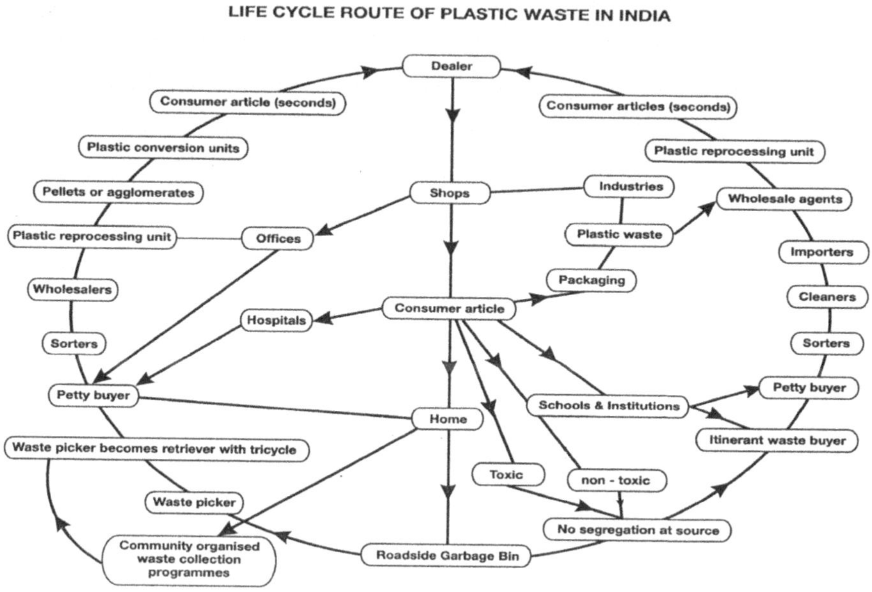

Fig. 2 Life cycle route of plastic waste in India. (*Source* Life Cycle Study of Plastics, International Symposium on Plastics, CIPET Chennai, 1997 [7])

The estimated growth in the packaging sector is 7% and in appliances and consumer electronics it is 7.5%. Healthcare and medical devices saw a CAGR of 15% in the last three years, and 3 D printing is likely to see a 30% increase. The estimated potential use of plastics in government initiatives like Swachh Bharat Mission (SBM), Jal Jeevan Mission, City Gas Distribution (CGD), PMAY and auto components is 5.4 million tons [6]. As per an Economic Times report, India generates 3.5 million tons of plastic waste annually while recycling only 30% of the plastic waste generated in the country (Fig. 2).

Plastics made from natural organic substances, such as cellulose and other biomaterials, account for a small proportion of the total plastics produced and manufactured [8].

The construction industry is among the biggest consumers of plastics which are used for packaging, containers, bags, and parts in building such as plastic pipes and vinyl siding. With as much as 75% of all plastic becoming waste, the cost of managing plastic waste plus that associated with greenhouse gas emissions from plastic production is US$40 billion annually UNEP [9].

Since plastics contain stabilizers that prevent them from degrading quickly, plastics last a long time. The properties of plastics whether chemical, physical or biological because of which they are considered non-reactive, and are durable are precisely the reason they become a nuisance when they are released into the environment. Although all common plastics including polyethylene (PE), polypropylene (PP), polystyrene (PS), and polyethylene terephthalate (PET) are durable, they eventually undergo fragmentation photo-, physical, and biological degradation over centuries and particles and fibres are released into the environment [10]. Plastic materials which degrade into micron-sized particles (microplastics), persistent in the environment are pollutants and cause harm to all organisms in the environment [11].

Hence plastics, because of their persistence in the environment, traverse all over the globe and can be found on the tops of mountains and the bottom of the oceans and in stomachs of smallest of fishes and copepods to largest animals on land. In fact, their shredding and size reduction makes them even more dangerous since then they can be ingested, inhaled, and even adsorbed as microplastics and remain firmly lodged in niches and locations from which they cannot be extricated [12]. The seas near Mumbai, Kerala and the Andaman and Nicobar Islands are among the worst polluted in the world.

3 Issues Regarding Plastics and Plastic Recycling

3.1 Toxicity

As explained above plastics are synthetic polymers made from petroleum derivatives like naphtha. These synthetic polymers are non-reactive to acid, alkali although they are reactive to alcohols and petroleum solvents [13]. The guidelines for making food grade and medical grade plastics have been regulated by international and national standards. Although the innate nature of these polymers is such that in its virgin and newly made state, it is non-reactive and safe, however even under best conditions of complying with standards, a certain percentage of unpolymerized monomers remain, which over a period get released into the food and body through the medical device causing toxicity.

In addition, there are plastics like Styrofoam or Polystyrene which depolymerize at low temperatures and release the styrene monomer, which is a known carcinogen. Similarly, Poly Vinyl Chloride or PVC by itself is so toxic that it releases hydrogen chloride at low temperatures either in gaseous or liquid form, which is highly corrosive and in the presence of organic matter, leads to the formation of dioxins and furans. The release of its monomer is also known under certain conditions, which is again carcinogenic and mutagenic. "When plastics are broken down, either physically into fragments or chemically due to environmental factors, they eventually end up as microplastic particles. However, long before this process has been completed, chemical additives in the plastics may leach out into the natural environment" [14].

Similarly, there are several plastics, both thermoplastics (those which can be melted and remoulded like HDPE, LDPE, PP, PVC, PET, PS) and thermosets (those which cannot be melted and remoulded like melamine, bakelite, etc.), which give out monomers and toxic compounds over a period and hence their prolonged use is dangerous and harmful for human and animal health and the environment.

Polyurethanes and polyacrylonitriles (PAN) are also made up of monomers known to be carcinogenic and mutagenic [15].

Furthermore, it is known from the manufacture of plastics that different types of plastics are associated with different additives. For instance, chlorinated paraffin, cadmium or zinc phthalates are added to polyvinyl chloride (PVC) for softening it to make flexible products. Bis phenol A is added in the manufacture of polycarbonates and epoxy resins and several other products. Pigments containing heavy metals to colour high-density and low-density polyethylene and polypropylene packaging and hydro fluorocarbons (HCFCs) used for foaming Polystyrene, and polyurethane are toxic as well. Endocrine disrupting chemicals (EDCs) like bisphenol A (BPA) and related substances, flame retardants, phthalates, per and polyfluoroalkyl substances (PFAS), dioxins, UV stabilizers and toxic metals such as lead and cadmium used as additives are harmful to health too [16].

Unfortunately, all these chemicals and heavy metals are highly toxic to humans as well as animals. When these plastics are recycled, these additives are carried in the granules extruded during the thermal recycling process and in their degradation products. Furthermore, when toxic substances like petroleum-based products such as lube oil, pesticides, fertilizers, and several others are stored in thermoplastics like those mentioned above, the toxic molecules enter the plastic and become an integral part of it. When they are recycled, these toxic molecules like poly-aromatic hydrocarbons (PAHs) in the granules become a part of the product like a water container or even a carry bag. These chemicals leach out at room temperature into the new product and affect the consumer [17] "Many of the chemicals we identified may have contaminated the materials during use. For example, a jug used for storing pesticides will absorb some of the pesticides and will contaminate the recycling waste stream. Plastics in the environment are also known to absorb organic pollutants" [18].

3.2 Longevity

Plastics are so persistent in the environment and remain inside life forms and bioaccumulate because life forms and ecosystems can utilize and break down only those molecules which can be utilized as a carbon source for which they have the necessary enzymes to assimilate them to build their own body tissue. However, plastics are made from petroleum and individual molecules of polymers generally referred to as monomers are not recognized by biochemical systems and as of now enzymes do not exist except in a few bacterial species that are capable of breaking them down into simpler substances like they do starches and sugars, which are natural polymers [19]. Some plastics have fibres which shorten when recycled and get released in the environment. Thus, a plastic can be recycled 2–3 times or more depending on different blending techniques before it is no longer recyclable [5].

4 Environment and Health Impacts of Plastics

Health concerns regarding the effects of brominated flame retardants (BFRs), which are added to electronics or bis phenol A as additives in polycarbonate (PC) and different types of plasticizers which are added in PVC have begun to emerge as a matter of concern because of their possible association with endocrine disruption such as between BFRs and childhood autism. The Childhood Autism Risk from Genetics and the Environment (CHARGE), is a UC Davis study addressing a wide spectrum of environmental exposures, endogenous susceptibility factors, and interplay between these two (CHARGE, 2006) [20]. According to a new study led by researchers at the University of California, Berkeley, pregnant women with higher blood levels of a common flame-retardant show altered thyroid hormone levels, a result that could have implications for fetal health. The study found that the odds of subclinical hyperthyroidism were increased 1.9 times for each tenfold increase in PBDE concentrations [21].

5 Plastic Pollution

According to the Hong Kong-based NGO OceansAsia, approximately 300 million tons of plastic is produced worldwide every year, with more than 8 million entering oceans annually-ultimately threatening the ecosystems of marine wildlife. The COVID-19 pandemic further exacerbated the scourge of plastic pollution [12].

"The highest abundance of macro and microplastic contaminants is found on beaches in Maharashtra at the high tide line as compared to Karnataka and Goa," says the study conducted by the Goa-based National Institute of Oceanography (NIO). Such a phenomenon of finding macro and micro-plastics in abundance on Maharashtra beaches suggests that the contaminants are land-based, like (coming from) near-shore plastic industries, port areas, petroleum industries and high tourism activities, it said (Fig. 3).

The study conducted by Dr. Mahua Saha and Dr. Dusmant Maharana of NOI further states that microplastics are those fragments of plastics that are less than 5 mm and larger particles of more than 5 mm are called macroplastics. The researchers conducted the assessment of macro and microplastic contamination for two years on 10 beaches along the western coast of India and their toxic effects on marine organisms. The plastic contaminants on these beaches were found in various colours like white, pale yellow, dark brown, green, blue, and red, said the study "Assessment of macro and microplastics along the west coast of India: abundance, distribution, polymer type and toxicity", published in the Netherlands-based journal "Chemosphere" [23].

In another study "Abundance and quantification of microplastics in water and sediments of Sal River and an approach to study the degradation of microplastics"

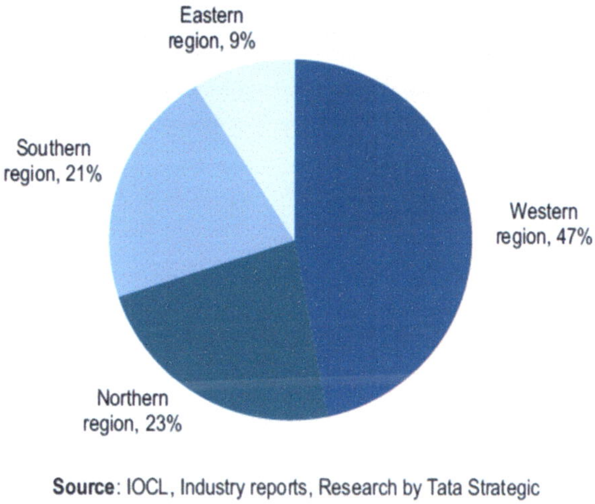

Fig. 3 India—Region wise plastic consumption, 2013 [22]

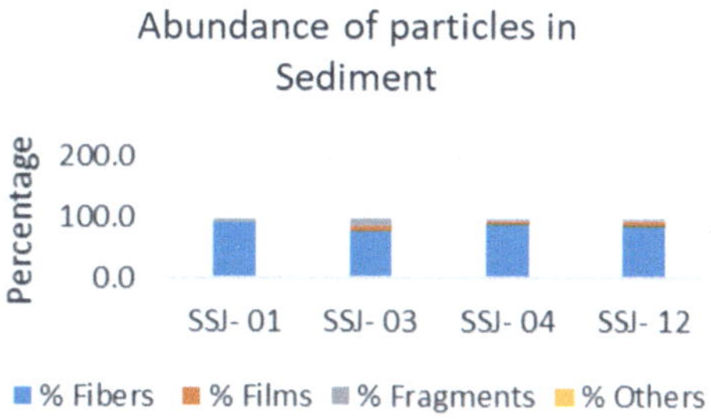

Fig. 4 Abundance of particles in sediments [25] (*Source* file:///D:/User's%20Data/Downloads/375_Extendedabstract.pdf)

[24] conducted by NIO scientists, Dr. Kumar, Dr. Mahua and others, found abundance of particles, both micro and macro in water and sediments (Figs. 4 and 5).

Polyacrylamide and Polyacetylene were the dominant polymers found in both water and sediment. These polymers are majorly used in Water Treatment Plants (WTPs), microbeads, textile industry, domestic plastics, fishing boats and accessories, etc. [25].

6 Plastic Pollution and Impact of COVID 19 on Its Consumption

"Plastic pollution was already one of the greatest threats to our planet before the coronavirus outbreak," said Pamela Coke-Hamilton, UNCTAD's director of international trade. "The sudden boom in the daily use of certain products to keep people safe and stop the disease is making things much worse" [26]. In this report brought out by United Nations Conference on Trade and Development [26], it is stated that global

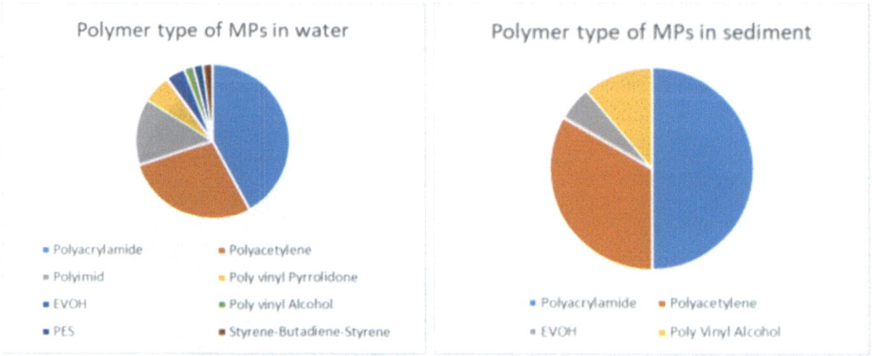

Fig. 5 Different types of polymers found in water and sediment in Sal river, Goa in 2019 (*Source* file:///D:/User's%20Data/Downloads/375_Extendedabstract.pdf) [24]

sales of disposable face masks skyrocketed from an estimated US$800 million in 2019 to US$166 billion in 2020, according to business consulting firm Grand View Research. Regarding the products delivered daily to homes to aid social distancing such as packaging and online shopping besides take out services, the report says that the plastic waste due to COVID-19 would be enormous.

6.1 Plastic Pollution Amidst COVID-19

Naturally, the priority over the past few years has been towards mitigating the contagion of the virus, which includes the implementation of collaborative protective measures. Essential components to controlling the spread include the wearing of surgical face masks and the frequent disinfection of hands. Although these two measures alone cannot stop the pandemic, they play an important role in preventing infection and are therefore vital to controlling and eliminating COVID-19.

6.2 Why is Plastic Relevant to the Pandemic?

Both face masks and hand sanitizer production include the use of plastic. Face masks typically contain polypropylene (PP), which, due to the microfibers' hydrophobic composition, acts as a protective layer against bodily fluid droplets. Other more intricate and expensive face masks include polyurethane (PUR) and/or polyacrylonitrile (PAN) [27].

Plastics and Personal Protective Equipment (PPE) made from woven and non-woven plastic including full suits, caps, shoe covers, masks, eye glasses, face shields, gloves, and splash-proof aprons are necessarily required by healthcare workers especially doctors, nurses, and technicians who come directly in contact with positive patients in any pandemic for treating them as also ambulance drivers, helpers, housekeeping staff and waste workers who deal with transportation and the waste of such patients [28]. However, there are many other hospital workers who do not directly deal with positive patients in pandemics or are involved in treating and serving negative patients for whom the PPE could be rationalized and made appropriate so that their level of risk is addressed but PPEs are not as elaborate as those that are required for those in the positive locations during any pandemic [29].

Directives and suggestions for such rationalization were issued very early during the Pandemic by the World Health Organization (WHO) and subsequently promulgated by the Ministries and Departments of Health in all countries and states with their countries. Similarly, Donning and Doffing (Putting on and taking off) of the PPE was also demonstrated and adhered to, to avoid the spread of the infection from the PPE to non-infected persons. Despite such measures, the use of plastics especially in the form of protective equipment including those used by domestic caregivers and used in the final resting places, transportation hubs and the commercial sector increased 4-fold especially in India.

Although many healthcare establishments saw marked reduction in regular cases and mostly operated only COVID wards, the set offs of waste generation were not nil and there was a 15–20% increase in waste especially in the incinerable category [30]. Although central, state, and local governments emphasized the use of reusable cloth masks and regular hand washing to reduce irrational use of disposable PPEs, this category of waste has increased two to three folds placing a great burden on the municipal system and the environment [30].

The gains that were made before the COVID pandemic, as regards reduction of use of single-use plastics for marketing, packaging and disposal faced a major setback and those who were reluctant to give up plastics before the pandemic because of a certain lackadaisical attitude are now feeling vindicated. Hence reintroduction of the circular economy would have to look for alternatives and a move towards more naturally reusable and biodegradable materials even during pandemics [28].

India must address a demand–supply gap in the commodity sector as seen in Fig. 6. This gives us a good opportunity to introduce more sustainable and reusable materials supporting a circular economy.

If the opportunity to push towards using material which lend themselves to circular economy [32] rather than single-use products largely made of material, which are non-durable, then the danger of not only legal imports but also illegal imports as seen operating now using a loophole in the EXIM policy would proliferate and lead to massive import and dumping of hazardous post-consumer plastics. In a study conducted in New Delhi, India in 1996, the proportion of plastic recycling units in the MSME and informal sector catering to recycling post-consumer plastics was five times higher than India's consumption at that time [33]. As per DPCC in a survey conducted in 2023–24, it is evident that while Delhi requires recycling capacity of 200,000 TPD, it has ten times higher capacity in the city.

Fig. 6 Demand–supply scenario of plastics, Financial Year 2013 [31]

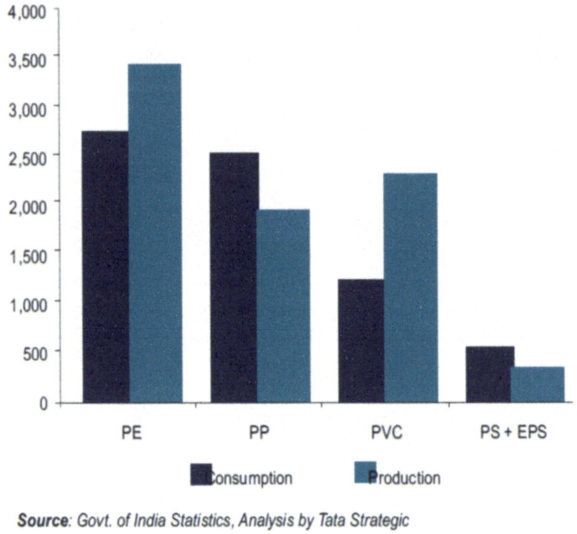

Major application of plastics in food processing is packaging. Since packaging standards have become stringent in India towards food and beverage packaging, recycled plastics are not allowed to be used for these unless they are super-clean meaning they are of food grade [34]. Illegally imported and dumped hazardous plastics would end up as marine litter since most of the illegal post-consumer plastics enter through ports in containers which carried our exports [33]. Such illegally dumped plastics would be a hindrance to circular economy concepts since hazardous plastics cannot be recycled and used for consumer articles.

6.3 EPR Contribution Fees Can Be Used to Incentivize Upstream Innovation

The French and Italian schemes (CITEO and CONAI) have introduced new eco-modulated fee structures to focus on recyclability and eco-design [35]. These fee structures reward more recyclable products, based on features such as: clear labelling, easily separable materials, available recycling technology, and markets for secondary materials. Penalties are applied on features such as complex packaging design or disruptive additives, such as dark colorants or opacifiers, which hinder recycling. These schemes are promising examples of how to encourage industry behaviour change. However, more should be done through EPR schemes to incentivize industries to reduce their plastic consumption overall.

7 How Do We Make Plastic Economy a Circular Economy?

Developing countries have a big stake in the global plastics economy. Their share of global plastics production jumped from 43.5% in 2009 to 58% in 2018 and two out of three plastic manufacturing jobs are in the global south says the UNCTAD analysis report [36]. Developing countries have a big stake in the plastic recycling activities. However, as can be seen from the issues of plastic persistence and recycling, plastics would have to be designed and manufactured differently to include circularity.

8 Plastics and Livelihood

Developing countries are key suppliers of raw materials to the global plastics industry. This is especially true for countries with large crude oil and natural gas sectors, as 99.5% of plastics are derived from petrochemical feedstocks [37]. Notably, 27 developing countries and 5 transition economies derived more than a third of total merchandise export revenues from fossil fuels in 2017 [38, 39]. Now more than ever before, fossil fuel dependent countries face the urgent challenge of diversifying their economies towards more sustainable and inclusive sectors which earlier continued due to historically low oil prices, less travel demand due to the COVID-19 pandemic, and national policies to curb greenhouse gas emissions [36].

9 Plastics Recycling and Livelihood

Plastic production in the developing countries grew at rates significantly above the world average in the first two decades of the twenty-first century [40]. Two in every three jobs in the manufacture of plastic products are now concentrated in the Global South: the industry is estimated to generate 7.7 million direct jobs in developing countries, compared to 2.9 million jobs in developed nations and 500,000 in transition economies [36].

Exports of plastic waste, parings and scrap were estimated at US$4.3 billion in 2017 (excluding re-exports), with 71% originating in developed countries, most notably in the European Union (40%), the United States (15%) and Japan (12%) [38].

As imported waste plastics can cause serious and irreversible environmental pollution, China banned the import of non-industrial plastic waste in 2018. Later other trading partners in developing Asia, including Indonesia, Malaysia, the Republic of Korea, Thailand, Viet Nam, and Taiwan Province of China also imposed their import restrictions due to the limited capacity of local waste management systems and following public outcry, when some incoming shipments were also revealed to contain hazardous plastic waste. In 2019, global trade in plastic waste was 46% lower than before the introduction of those import bans. This has also led to reduction in employment in the plastic recycling sector although the number of jobs in waste collection remains the same in India because of domestic consumption as well as the introduction of EPR and co-processing in cement plants in India [38, 39].

10 Challenges of Recycling in the Informal Sector

One of the biggest challenges for recycling in the informal sector is the compulsion of the informal sector who segregate the waste manually whether in India or in many other developing countries. The issues are:

(a) waste is unsegregated and people put wet waste in plastics, whether bags or other containers and informal sector, even if they are in a material recovery sector and have masks, gloves, etc., must endure the rotting smell and continue to separate the wet waste from the dry waste, which in the process lose their recyclability or resale because of being highly soiled;

(b) furthermore, most of the recycling is happening in the micro or mini sector with investments less than INR 10 million and turn over not exceeding INR Fifty Million [41], which do not come under the regular purview of the State or Central Pollution Control Board, which are continuing to use poor technologies, plasticizers and pigments of poor quality which increases exposure and endangers not only the workers' health but also their families since toxins and particles are carried on clothes. EWS consumers of such products are also exposed to the consequences of such poor quality recycling. MNCs are investing only to meet their EPR requirements and stay in business and are not concerned with the state of poor recycling in the country which is resulting in increased percentage of wastage that end up in WtEs and co-incineration in cement plants at the expense of "super-clean" yield, which they must reuse for meeting their legal mandate as per an UNDP study [42] (Fig. 7).

Fig. 7 Workers' exposure to poly-brominated diphenyl ethers (PBDE) and required OSH compliance. *Source* Department of Health and Human Services, ASTDR, CDC, USA [43]

PBDEs can occur during the production of commercial PBDE mixtures and of PBDE-containing plastic products. Workers involved in recycling plastic products, or who repair computers in confined workplaces can also be exposed to PBDEs. If you are exposed to PBDEs while at work, you may carry them home on your clothes or body. Your occupational health and safety officer at work can tell you whether the products you work with may contain PBDEs and whether those are likely to be carried home. If this is the case, you should shower and change clothing before leaving work. Your work clothes should be kept separate from other clothes and laundered separately.

Workplace exposure

Excerpt from Department of Health and Human Services, Public Health Service, Agency for Toxic Substances and Disease Registry (ATSDR), CDC, USA September 2004

(c) the products made using such unregulated technology are often purchased and used by persons in the lower economic communities, which further endangers the consumers through exposure to monomers (due to inappropriate thermal and mechanical processes), additives and toxins in the plastics due to previous stored material, which enter the plastics and become an integral part of the plastic. This also included additives like plasticizers and pigments that contain heavy metals and persistent poly-aromatic hydrocarbons (PAHs) [33].

11 Challenges to Recycling in the Formal Sector

Here too, the challenges are because of unsegregated waste from wet and other dry waste. In the formal sector, since very often recycling is done by mechanical means using shredders and centrifugal equipment to separate the lightweight plastics from the heavy ones, each category of plastic needs to be separated and hence the formal sector must employ people who can recognize it either by touch or by the markings on it. These are very often erstwhile ragpickers or waste workers and therefore the issue of providing better working conditions and lesser exposure to them as well as to the industrial workers is always in question. Toxic chemicals have been studied in the particulate matter and volatile emissions and have shown to be causing endocrine disrupting diseases in adults and diseases like autism in children due to these chemicals on workers' clothes [44].

Secondary markets for plastic remain broadly unprofitable and unable to scale, hindering the development of a proper circular economy system. Losses result from mixed plastics entering the same recycling process, plastic additives increase contamination making the material unsafe to recycle [45]. For example, in Europe, almost half the material currently collected for recycling is lost during the recycling process [45].

11.1 Ways to Safer Use of Plastics

It has been established that replacing synthetic plastics made from petroleum with either starch filled polymers, which affords biodegradability to the plastics, will only lead to formation of microplastics. These will then contaminate soil and water and enter the food chain through uptake by fresh water and marine animals and will eventually enter the muscular tissue of humans as well [11].

However, although there is a requirement for plastic like material in the medical (heart valves, prosthetics) and aviation and engineering field (light weight but strong, water-resistant equipment) as well as in food packaging sector for safer storage and transport, unregulated use of even compostable plastics including single-use plastics where reusable material are available, can cause similar disasters as now, since compostable plastics if dumped in landfills would generate methane.

Recycling is good—every ton of plastic waste recycled results in a saving of approximately 3.8 barrels of petroleum. Technologies are available in India that can convert 1 kg of plastic to 750 ml of automotive grade gasoline. Shredded plastic waste can be used in laying roads. Jambulingam Street in Chennai was one of India's first plastic roads built in 2002. In 2015–16, the National Rural Road Development Agency laid around 7,500 km of roads using plastic waste. Co-processing of plastic in cement kilns offers a viable mechanism to process non-recyclable, combustible plastic waste. However, none of the above methods have been validated as to whether they are useful and are safe for human health and the environment.

Managing plastics use in the COVID pandemic: Initially, as the COVID-19 caused a lockdown in several countries, China's supply of PPEs to several countries, did not meet hygiene and quality standards and caused a lot of wastage. These were eventually discarded and created a greater panic than the pandemic itself.

In India too, use of seconds, which are granules made from extruding melted post-consumer thermoplastics for using them in the manufacture of downcycled products, do not meet the required standards of packaging or toys. These downcycled products are likely to be contaminated with once or multiple recycled seconds (melting and extrusion many times), however which are difficult to test and establish [46].

12 What Are the Available Alternatives to Plastics and How Can These Help Promote Circular Economy?

The UNCTAD [38, 39] Report says "The list of non-toxic, biodegradable or easily recyclable materials that could replace plastic includes many well-known materials, such as glass, ceramics, natural fibres, paper, cardboard, rice husk, natural rubber and animal proteins. As developing countries are key suppliers of many plastic substitutes, increased global demand could create new, greener trade and investment opportunities for them. Developing nations supply, for example, 92% of the world's jute, with the main suppliers being Bangladesh (74%) and India (9%). They also accounted for 94% of global natural rubber exports in 2019, with Thailand (31.5%), Indonesia (30%) and Côte d'Ivoire (8.5%) leading the pack". They also believe that these can provide many more jobs in these countries [36].

A raft of new bioplastic innovations is starting to catch up, though unlike unsustainable fossil fuels, they are derived from renewable sources. There are many alternatives to plastic packaging such plant based, mushroom root, bagasse, seaweed water bubbles, shower-friendly paper, stone paper and plastic, palm leaves, corn starch and sorghum loose fill, silver board metallized paper, wood pulp cellophane, prawn shell plastic bags, milk plastic and several others. These can be made into products that can fit the circular economy since they can be reused or upcycled and finally composted once their life cycle is completed. However, chemical additives used in their manufacture can make them hazardous.

While bioplastics have the same molecular structure as petroleum-based plastics, which take hundreds of years to decompose, research [47] shows that biomass-based polymers are also more likely to biodegrade and break down, only in industrial compost facilities. Bioplastic proponents believe they are key to making plastic a part of a circular economy [48] while others feel that their biodegradability will add to the difficulties.

Circular economy business models fall into two broad groups: those that propose reusing and extending service life through repair, remanufacture, upgrades and retrofits and those that turn old goods into new resources by recycling materials.

13 International Initiatives in Circular Economy

SDG 12 on sustainable consumption and production is of relevance to the plastics economy. A change of paradigm away from linear production models and towards a circular economy entails both reduced production and increased reuse and recycling of plastics and plastic products. To promote international circular supply chains and prevent developing countries from becoming a dumping ground for undesirable wastes, challenges at hand include greater transparency on the treatment of waste exports, improved measures to prevent illegal or unsustainable waste shipments, and international agreement on definitions of key terms, such as "recyclable" and "biodegradable" products [8].

For companies, the circular economy means a new business model, where you do not earn on the sale of material goods designed for a short life, goods which you use and dispose of, Czyzewski says. The source of cash for companies, he adds, will be services provided for customers, connected with maintenance, and upgrading purchased products. The model is called PaaS (Product as a Service) [48].

At the end of 2018, the Ellen MacArthur Foundation, an advocacy organization for the circular economy, announced the New Plastics Economy Global Commitment, which brought together over 250 businesses, governments, NGOs, and academic organizations pledging to eliminate unnecessary plastic packaging and making the new packaging 100% reusable, recyclable or compostable [48].

According to Bio-garmentry, there is a spectrum of microorganisms that can be cultivated in a lab to help make textiles and fabric. There is a plastic made by de-sequestering carbon from the air and there are algae embedded organic fabric and there are even marine algae which can be made into shiny sequins which are fully compostable [49].

Therefore, even Personal Protective Equipment for protecting oneself from not only COVID pandemic but different types of viruses like SARS, H1N1 and several others can be made and used. These can be reused if made in such a way that it is reusable after sterilizing and composted when it needs to be destroyed and which does not become a burden on the environment nor accumulate and become the reason for spread of infection.

14 Conclusion

Plastics which are made from petroleum products must be minimized since their manufacture impacts the environment right from the extraction of the resources to its manufacture of products from the primary products like pellets. While the use of such products, followed by collection, sorting and recycling and final degradation through incineration or fracking puts tremendous burden on the environment and health of humans and animals, they need alternatives.

On the other hand, replacement of all types of plastics with degradable or even compostable plastics would also cause problems of microplastics and microfibers as well as set off plastics versus food/fodder/fuel debates in several parts of the world besides having impacts on the environment like methane from landfills because of inappropriate waste management and indiscriminate dumping.

Therefore, circular economy should place greater emphasis on reduction, reuse, appropriate manufacture, and use of bioplastics in essential sectors, while using other reusable material for other less critical uses, upcycling or composting as per specific norms [48].

Green bonds, used exclusively for financing projects that contribute towards climate change measures and environmental preservation, have been the fastest-growing new asset class with USD 37 billion and USD 42 billion issued in 2014 and 2015 respectively. In India too, a beginning must be made by the authorities and the industry to increase the issuance of green bonds. With general fiscal reforms gaining momentum in the country, the time is ripe to make the environment a part of the process [50].

Plastic-selective policies made up 86% of plastic-related measures notified to the WTO in the biennium 2017–18 compared to 56% in 2015–16. In recent times following international and national policy developments, the recent Basel Convention's amendments related to plastic waste, especially related to medical materials and devices are noteworthy [51]. However, with only 1% of the plastic chemicals being regulated through international environmental agreements, policies addressing the issue of reporting chemicals are unlikely to be achieved even through the negotiations for the International Treaty on plastic pollution; hence we must reduce plastics in our lives to safeguard health [18].

References

1. Zhu, D., Asnani, P. U., Zurbrügg, C., Anapolsky, S., & Mani, S. (2008). Improving municipal solid waste management in India. WBI.
2. Deere Birkbeck, C. (2020). Strengthening international cooperation to tackle plastic pollution: Options for WTO. Graduate Institute 2020.
3. IBEF, Plastics. (2020). The Indian plastic industry. Retrieved November 5, 2020, from https://www.ibef.org/exports/plastic-industry-india.aspx#:~:text=The%20Indian%20plastics%20industry%20made,hosts%20more%20than%202%2C000%20exporters
4. Directorate General of Commercial Intelligence and Statistics (DGCIS). (2019, September 22). How plastic ban will affect businesses and consumers. The Economic Times, Industry. Retrieved November 4, 2020, from https://economictimes.indiatimes.com/industry/indl-goods/svs/paper-/-wood-/-glass/-plastic/-marbles/how-plastic-ban-will-affect-businesses-and-consumers/articleshow/71236532.cms?utm_source=contentofinterest&utm_medium=text&utm_campaign=cppst
5. TERI. (2018). Factsheet on plastic waste in India. World Environment Day 2018. Retrieved November 2020, from https://www.teriin.org/sites/default/files/files/factsheet.pdf
6. Plast India Foundation. 2022–2023 update. Plastic Industry Status Report—India—2021–2022 & 1H 2022–2023. Retrieved February 8, 2024, from https://plastindia.org/plastic-industry-status-report/report-india/
7. Krishna (nee Mani), S. (1997). Life cycle study of plastic waste, Central Institute of Plastic Engineering Technology, Chennai. Final report (National seminar on recycling and plastics waste management) (p. 130). Retrieved November 6, 2020, from https://open.unido.org/api/documents/4808701/download/FINAL%20REPORT.%20(NATIONAL%20SEMINAR%20ON%20RECYCLING%20AND%20PLASTICS%20WASTE%20MANAGEMENT)%20(21915.en
8. UNCTAD. (2020, July). Growing plastic pollution in wake of COVID-19: How trade policy can help. UNCTAD report. Retrieved October 30, 2020, from https://unctad.org/news/growing-plastic-pollution-wake-covid-19-how-trade-policy-can-help
9. UNEP. (2014). Valuing plastic: The business case for measuring, managing and disclosing plastic use in the consumer goods industry. See: file:///C:/Temp/-Valuing per cent20plastic_ per cent20the per cent20business per cent20case per cent20for per cent20measuring, per cent20managing per cent20and per cent20disclosing per cent20plastic per cent20use per cent20in per cent20the per cent20consumer per cent20goods per cent20industry-2014Valuing per cent20plasticsF.pdf
10. Ojeda, T. (2013). Polymers and the environment. In Yılmaz, F. (Ed.), *Polymer science* (pp. 1–34). In Tech: Rijeka.
11. Kubowicz, S., & Booth, A. M. (2017). Biodegradability of plastics: Challenges and misconceptions. *Environmental Science & Technology, 51*, 12058–12060. Retrieved November 5, 2020, from https://pubs.acs.org/doi/pdf/10.1021/acs.est.7b04051
12. OceanAsia. (2020). Microplastic survey of Soko islands. OceanAsia Foundation Website. Retrieved November 1, 2020, from https://oceansasia.org/plastic-pollution/
13. Kameel, N. I. A., Daud, W. M. A. W., Patah, M. F. A., & Zulkifli, N. W. M. (May 2022). Influence of reaction parameters on thermal liquefaction of plastic waste into oil: A review. *Energy Conversion and Management: X*, 14, 100196. Elsevier. https://doi.org/10.1016/j.ecmx.2022.100196
14. Benjaminsen, C. (2024). Researchers surprised at levels of toxicity in standard plastic products. Phys.Org, Dec 2023. Retrieved February 6, 2024, from https://phys.org/news/2023-12-toxicity-standard-plastic-products.html
15. Blastic, EU, European Regional Development Fund. (2020). Toxicity of plastics, an introduction to plastic and marine litter. Retrieved November 27, 2020, from https://www.blastic.eu/knowledge-bank/impacts/toxicity-plastics/

16. Endocrine Society. (2020, December). Press release. Plastics pose threat to human health. Retrieved February 6, 2024, from https://www.endocrine.org/news-and-advocacy/news-room/2020/plastics-pose-threat-to-human-health
17. Krishna (nee Mani), S. (1997). Life cycle study of plastic waste, Central Institute of Plastic Engineering Technology, Chennai. Final report (National seminar on recycling and plastics waste management) (p. 127). Retrieved November 6, 2020, from https://open.unido.org/api/documents/4808701/download/FINAL%20REPORT.%20(NATIONAL%20SEMINAR%20ON%20RECYCLING%20AND%20PLASTICS%20WASTE%20MANAGEMENT)%20(21915.en
18. Almroth, B. C., & Martinez, E. C. (2024). Dangerous chemicals found in recycled plastics, making them unsafe for use—Experts explain the hazards. *The Conversation*, January 2024. Retrieved February 6, 2024, from https://theconversation.com/dangerous-chemicals-found-in-recycled-plastics-making-them-unsafe-for-use-experts-explain-the-hazards-220498; Accessed on February 15, 2024.
19. Krishna (nee Mani), S. (1997). Life cycle study of plastic waste, Central Institute of Plastic Engineering Technology, Chennai. Final Report (National seminar on recycling and plastics waste management) (p. 120). Retrieved November 6, 2020, from https://open.unido.org/api/documents/4808701/download/FINAL%20REPORT.%20(NATIONAL%20SEMINAR%20ON%20RECYCLING%20AND%20PLASTICS%20WASTE%20MANAGEMENT)%20(21915.en
20. Hertz-Picciotto, et al. (2006). The CHARGE study: An epidemiologic investigation of genetic and environmental factors contributing to autism, July 2006.
21. Chevrier, J., et al. (2010). University of California—Berkeley, Flame retardant linked to altered thyroid hormone levels during pregnancy, June, 2010.
22. FICCI—A report on Plastic Industry. (2014). Potential of plastics industry in northern India with special focus on plasticulture and food processing-2014 a (p. 10).
23. Maharana, D., Saha, M., Yousuf, J., Rathore, C., Rayadurga, S., Xu, X., Koongolla, B., & Li, H.-X. (2019). (1 &2), Assessment of micro and macroplastics along the west coast of India: Abundance, distribution, polymer type and toxicity. Chemosphere, December 2019. https://doi.org/10.1016/j.chemosphere.2019.125708. Retrieved November 13, 2020, from https://www.researchgate.net/publication/338070852_Assessment_of_micro_and_macroplastics_along_the_west_coast_of_India_Abundance_distribution_polymer_type_and_toxicity
24. Kumar, M. M., Saha, M., et al. (2019). Abundance and quantification of microplastics in water and sediments of Sal River and an approach to study the degradation of microplastics. In *Proceedings of the Ocean Society of India Conference (OSICON–19), CMLRE (MoES)*, Kochi, 12–14 December, 2019. Retrieved November 13, 2020, from file:///D:/User's%20Data/Downloads/375_Extendedabstract.pdf
25. Kumar, M., Saha, M., Rathore, C., Gupta, P., Naik, A. P., Desai, A., & Srimuruganandam, B. (2019). Abundance and quantification of microplastics in water and sediments of Sal River and an approach to study the degradation of microplastics, Conference: Sixth Biennial Conference of Ocean Society of India (OSICON-19) At: Centre for Marine Living Resources & Ecology, Kochi, Kerala, India, December 2019. (1&2). Retrieved November 13, 2020, from https://www.researchgate.net/publication/338187441_Abundance_and_quantification_of_microplastics_in_water_and_sediments_of_Sal_River_and_an_approach_to_study_the_degradation_of_microplastics?enrichId=rgreq-0411148bf7229a4e7fb89536bdca22d4-XXX&enrichSource=Y292ZXJQYWdlOzMzODE4NzQ0MTtBUzo5MzEzMzIxMjUzNzY1MTNAMTU5OTA1ODI5MDI0Nw%3D%3D&el=1_x_3&_esc=publicationCoverPdf
26. UNCTAD. (2020). Communication on trade in plastics, sustainability and development by the United nations conference on trade and development (UNCTAD), June 2020. Retrieved October 31, 2020, from https://unctad.org/system/files/information-document/wto_unctad_CTE2020_en.pdf; UNCTAD. (2020). (July 27th). Growing plastic pollution in wake of COVID-19: How trade policy can help, UNCTAD report https://unctad.org/news/growing-plastic-pollution-wake-covid-19-how-trade-policy-can-help. Accessed on Oct 30th, 2020.

27. Express Polymer Letters. (2020, April). Editorial corner—A personal view the coronavirus and plastics. *Express Polymer Letters, 14*(6), 510–511. Retrieved November 1, 2020, from http://real.mtak.hu/108160/1/EPL-0010678_article%20%281%29.pdf
28. Konynasia, C. (2020). Another side effect of COVID-19: The surge in plastic pollution EUROPE July 6th 2020. Retrieved October 31, 2020.
29. WHO. (2020). Rational use of personal protective equipment for coronavirus disease 2019 (COVID-19), February 2020.
30. CPCB. (2020). https://CPCB.nic.in accessed on Oct. 31st, 2020.
31. FICCI—A report on Plastic Industry. (2014). Potential of plastics industry in northern India with special focus on plasticulture and food processing-2014 b (p. 14).
32. Harper, J. (2019, May). Circular economy in the loop, DW Business, DW in Focus. Retrieved November 1, 2020, from https://www.dw.com/en/circular-economy-what-goes-around-comes-around/a-48854512
33. Krishna (nee Mani), S. (1997). Life cycle study of plastic waste, Central Institute of Plastic Engineering Technology, Chennai. Final Report (National seminar on recycling and plastics waste management) (p. 135). Retrieved November 6, 2020, from https://open.unido.org/api/documents/4808701/download/FINAL%20REPORT.%20(NATIONAL%20SEMINAR%20ON%20RECYCLING%20AND%20PLASTICS%20WASTE%20MANAGEMENT)%20(21915.en
34. Ministry of Environment, Forests and Climate Change, Government of India, Plastic Waste Management Rules 2016, March 2016, amended February 2022. Retrieved February 7, 2024.
35. Watkins, E., Gionfra, S., Schweitzer, J. P., Pantzar, M., Janssens, C., & Brink, P. (2017). EPR in the EU plastics strategy and the circular economy: A focus on plastic packaging Institute for European Environmental Policy, 2017. In *EPR in the EU Plastics Strategy and the Circular Economy*, December 2017.
36. UNCTAD. (2020). The development dimension of trade in plastics.
37. European Bioplastics. (2020, June). Bioplastics market data. Retrieved October 31, 2020, from https://unctad.org/system/files/information-document/wto_unctad_CTE2020_en.pdf. WTO, Committee on Trade and Environment.
38. UNCTAD. (2019). Advancing sustainable development goal 14: Sustainable fish, seafood value chains, trade and climate. UNCTAD Report. Retrieved November 1, 2020, from https://unctad.org/system/files/official-document/ditcted2019d3_en.pdf
39. UNCTAD. (2019). Commodities and Development Report 2019: Commodity Dependence, Climate Change and the Paris Agreement.
40. Plastics Europe. (2019). Plastics—The facts 2019. Figures include thermoplastics, polyurethanes, thermosets, elastomers, adhesives, coatings and sealants and PP-fibres, but exclude PET-fibres, PA-fibres and polyacrylic-fibres. Retrieved November 2020. https://plasticseurope.org/knowledge-hub/plastics-the-facts-2019/
41. RBI. (2020, July). Credit flow to micro, small and medium enterprises sector. Notifications. Retrieved November 16, 2020, from https://www.rbi.org.in/Scripts/NotificationUser.aspx?Id=11934&Mode=0
42. UNDP. (2023). A decentralized community led approach for plastic waste recycling in India, Mid-term Evaluation. https://erc.undp.org/evaluation/documents/detail/22627, August 2023.
43. Department of Health and Human Services, Public Health Service, Agency for Toxic Substances and Disease Registry (ATSDR). https://www.atsdr.cdc.gov/ToxProfiles/tp2007-c1.pdf, CDC, USA, September 2007.
44. Mani, S. (2012). WEEE management in the USA and India: Research and education for a responsible approach to managing WEEE. Waste Electrical and Electronic Handbook, Woodhead Publishing Ltd. (pp. 550–575).
45. MESAB. (2020). The circular economy—A powerful force for climate mitigation. Retrieved November 1, 2020.
46. Blood, et al. (2020). Why the world's recycling system stopped working. Accessed on November 1, 2020.

47. Clift, et al. (2019). Article—Special issue on designing law and policy towards managing plastics in a circular economy. *Managing Plastics: Uses, Losses and Disposal. Chalmers, 15*(0), 93–107. Law Environment and Development Journal (LEAD). Retrieved November 1, 2020, from https://research.chalmers.se/publication/512835/file/512835_Fulltext.pdf
48. Harper, J. (2019, May). Circular economy in the loop, DW Business, DW in Focus. Retrieved November 27, 2020, from https://www.dw.com/en/circular-economy-what-goes-around-comes-around/a-48854512
49. Manchanda, S. (2020). Are you ready to be in a symbiotic relationship. *Times Life*, Times of India, October 11, 2020.
50. Deshpande, A. (2018). Financing the circular economy: Green bonds, circular economy for SQ—Raising the Sustainability Quotient, Version 4, Published by E-Konnect Knowledge Foundation and Bombay Chamber of Commerce.
51. WTO. (2020). Notifications on COVID-19 at https://www.wto.org/english/tratop_e/Covid19_e/Covid19_e.htm
52. Rafey, A., & Siddiqui, F. Z. (2023). A review of plastic waste management in India—challenges and opportunities. *International Journal of Environmental Analytical Chemistry, 103*(16), 3971–3987. https://doi.org/10.1080/03067319.2021.1917560
53. CSE. (2020). Plastic waste is India's and the world's most formidable environmental challenge today, and the COVID-19 pandemic has made matters worse: CS. Waste—Press Releases, September 24, 2020.
54. Endocrine Society. (2020, December). Press release, plastics pose threat to human health. Washington DC. Retrieved February 15, 2024, from https://www.endocrine.org/news-and-advocacy/news-room/2020/plastics-pose-threat-to-human-health
55. UN Comtrade Database. Retrieved November 2020. https://comtradeplus.un.org/

Plastic Debris, Microplastics, and Marine Pollution

Satish Sinha

1 Introduction

Marine pollution, for long has been a matter of concern and debate. While the sources of pollution were largely identified as ocean-related commercial activities such as shipping or fishing, the sources of marine pollution are better understood today and include waste generated from humans' activities, chemicals, plastics, wastewater, oil, etc. Marine pollution has been defined by article 1(1)(4) of the 1982 UN Convention on the Law of the Sea as: "the introduction by man, directly or indirectly, of substances or energy into the marine environment which results or is likely to result in such deleterious effects as harm to living resources and marine life". Sources of marine pollution are generally categorized into sea and land-based origins. International shipping and other ocean-related commercial activities have long been recognized as a source of marine pollution while other land-based anthropogenic activities also contribute to marine pollution. There have been efforts by the International Maritime Organization to address some of the sea-based pollution sources through the adoption of the MARPOL in 1973 with successive amendments and subsequently other conventions too. The ground-breaking 2019 UN IPBES (Intergovernmental Science-Policy Platform on Biodiversity and Ecosystem Services) assessment, states that more than 80% of global wastewater is being discharged back into the environment without adequate treatment, while 300–400 million tonnes of heavy metals, solvents, toxic sludge and other industrial waste are dumped into the world's waters every year [1]. Goal 14 of the UN Sustainable Development Goals, named Life below Water, focuses on "conserving and sustainably using the oceans, seas, and marine resources," [2]. It might be extremely difficult to achieve the SDG goals by the target timelines. The oceans are under increasing stress from multiple urban and rural sources of pollution, notably

S. Sinha (✉)
Toxics Link, E-224, 1st Floor, East of Kailash, New Delhi 110065, India
e-mail: satish@toxicslink.org

M. Goel and N. G. Tripathi (eds.), *Plastic Pollution*,
https://doi.org/10.1007/978-981-97-5528-8_5

among those are chemicals, POPs, and heavy metals apart from these wasteplastics and microplastics emerging as major pollutants to marine ecology. Plastics and microplastics are both highly complex due to their abundance, invisibility and non-degradable properties that make plastic in the ocean a material of concern for marine ecology and the planet. Microplastics are plastics of smaller size (up to 5 mm) that are intentionally designed small or get fragmented into smaller pieces. Plastics and microplastics constitute a very large percentage of marine litter that are transported over long distances by ocean currents causing pollution to areas that are relatively pristine and not the point sources of such pollution necessitating global action to address these concerns.

2 Plastic and Microplastic

Plastic is one of the most versatile materials known to humankind owing to its amazing properties such as light weight, long life, weight-to-strength ratio, waterproof, plasticity and inexpensive making it the most widely used material across diverse sectors pushing its demand and production higher year on year.

The word plastic is derived from the Greek word 'plastikos', meaning the substance that can be molded into any shape [3]. Plasticity during manufacture, allows it to be cast, pressed, or extruded into a range of shapes such as films, fibres, plates, tubes, bottles, boxes, and much more. Plastics are mainly derived from petrochemicals and are composed of long-chain hydrocarbon polymers with high molecular weights [4]. Polymers are often mixed with additives during the manufacturing process to bring about certain changes in the characteristics of the polymer [5]. Over 10,000 chemicals are used in plastics as additives. Of those, 2,486 substances have been identified as hazardous with toxic effects [6]. These additives are chemicals used in plastic production to achieve requisite properties and end-use material specifications. Many of these chemicals (additives) have been categorized as endocrine disruptors and carcinogens that leach or are released into the environment over their complete lifecycle.

Based on thermal properties, plastics are categorized as thermoplastics and thermosetting polymers. Thermoplastics are polymers that can be molded into any shape upon heating, e.g., Polyethylene (PE), Polystyrene (PS), Polypropylene (PP), Polyvinyl chloride (PVC), etc. Polyethylene (PE) is a very commonly used plastic with applications ranging from film packaging and electrical insulation to containers and piping. In contrast, Thermosetting polymers once set into an exact shape cannot be molded again. These polymers cannot be recycled as their change is irreversible, e.g., Polyurethanes, Phenol–formaldehyde, etc. [3]. Bakelite, a phenol–formaldehyde resin, is one of the most commonly used thermosets to make electrical switches, handles of utensils, etc.

Within five decades, plastic production has progressed into a global industry of 600 billion USD [7] and only about 9% of the plastic waste is being recycled globally while the remaining continues to exist in the environment in various forms [8].

Around 60% of the globally manufactured plastics are likely to have already accumulated in the environment (UNEP Beat Plastic Pollution, n.d.). These accumulated plastics can be lying in landfills or be transported away from land to oceans through streams and rivers. Much of these accumulated plastics do break down, fragment into smaller pieces and are termed microplastics. Both plastics and microplastics continue to persist in the environment for long periods (Fig. 1).

Plastics are known to cause lethal and sub-lethal effects in marine organisms. Their effects include entanglement, starvation, drowning, laceration of internal tissues, smothering and deprivation of oxygen and light, physiological stress, and toxicological harm [11]. As plastic is non-biodegradable, it lasts in the oceans for hundreds of years, breaking down into smaller particles or microplastics. The small size of microplastics makes them more dispersible, reaching even farther and deeper into the ocean, where they invade more habitats.

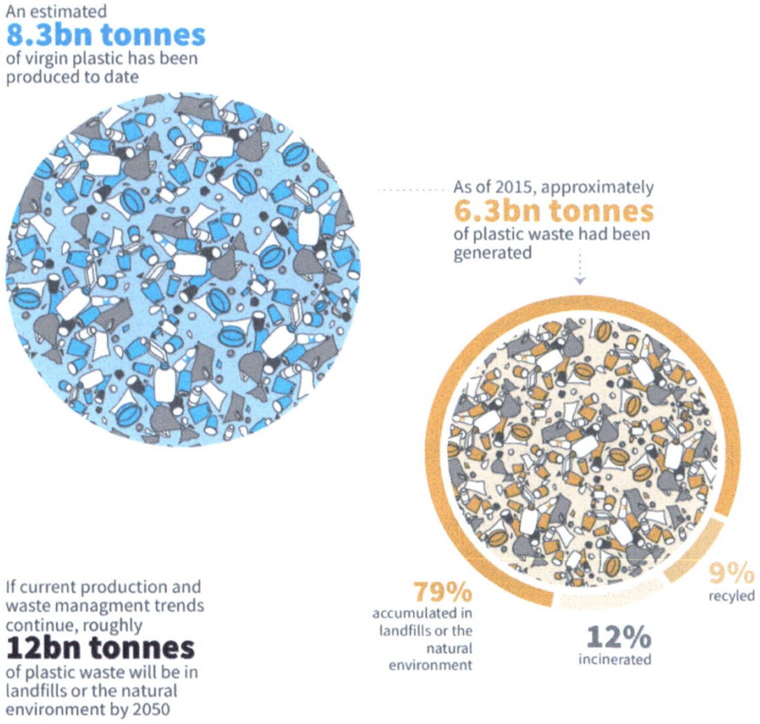

Fig. 1 Global plastic production and its management [9, 10]

Table 1 Categories of plastics based on dimensions (*Source* Our World in Data)

Particle category	Diameter range (mm = millimeters)
Nanoplastics	<0.0001 mm (0.1 μm)
Small microplastics	0.00001–1 mm
Large microplastics	1–4.75 mm
Mesoplastics	4.76–200 mm
Macroplastics	>200 mm

2.1 Microplastic

The presence of small plastic fragments in the open ocean was reported for the first time in the 1970s [12]. However, it was not until 2004 that the term "microplastics" was coined in a paper describing the long-term accumulation of fragments just a few microns in diameter [13]. Plastic debris or tiny fragments or beads of plastic, typically less than 5 mm in size, are called microplastics. The term "microplastics" was coined in 2004 to describe plastic particles that are not easily visible to the naked eye. Microplastics can be either intentionally produced or fragmented into smaller sizes due to chemical, biological or physical reactions and are categorized by size only. The smallest portions of plastic fragments of less than 0.0001 mm are known as "Nanoplastics" and plastic fragments of less than 5 mm are known as "microplastics".

The category based on size is listed in Table 1.

For some scholars, plastics can be grouped into mesoplastics (500 μm–5 mm), microplastics (50–500 μm), and nanoplastics (<50 μm), each with their distinct physical characteristics and biological impacts [14]. For others, the groupings can be Mega >1 m; Macro 25 mm–1 m; Meso 5–25 mm, Micro <5, and Nano 1 μm or less [15].

Microplastics are also classified into primary and secondary based on their origin and source. Primary microplastics are deliberately manufactured as tiny particles in varying shapes, such as microbeads, nurdles, granules, powders, and synthetic microfibers. They are manufactured to serve specific function. About 98% of the primary microplastics in the environment originate from land-based activities [16]. Microbeads and other forms of microplastics are widely used in personal care products.

Secondary microplastics result from the fragmentation or breakdown of larger plastics into smaller ones "(e.g., textile and rope fibres, weathering, and fragmentation of larger litter items, vehicle tyre wear, paint flakes)" [17]. The fragmentation occurs in oceans and beaches when acted upon by UV light, waves-led physical abrasion, availability, and the turbulence of oxygen [18].

Microplastics serve as potential carriers of additives incorporated to attain specific properties, functioning as vectors that convey chemicals from the surrounding environment. The toxicity associated with microplastics is primarily influenced by these additives. The capacity of microplastics to absorb, accumulate both organic and

Table 2 Toxic effects of plastic additives

Chemicals	Toxic effects
Bisphenol A (EDC)	Effects genomic, non-genomic, oestrogen, androgen and thyroid receptor-mediated functions
Phthalates (plasticizers) [di-2-ethylhexyl phthalate (DEHP) and butyl benzyl phthalate (BBzP)]	Damage the liver, kidneys, lungs, and reproductive system
Alkylphenols (e.g. nonylphenols-EDC)	Toxic, estrogenic and carcinogenic effects
Flame retardants (brominated and chlorinated flame retardants, some are POPs) (example: decaBDE)	Neurological damage, hormone disruption, and cancer
UV stabilizer (present in plastic building materials and food packaging)	Metabolic imbalance and developmental toxicity

inorganic contaminants, and subsequently transport them over long distances poses significant threats to marine ecosystems worldwide [19, 20]. If we take into account the presence of at least 0.01–0.27 million metric tons of plastics floating in the oceans at any given time, with an estimated 3–7% of the upper limit (0.27 Mt) containing additives, this could result in the transportation of approximately 8,100–18,900 metric tons of organic additives over vast distances within buoyant plastic matrices [21, 22]. Considering a portion of these chemicals degrades over time due to environmental factors, some may remain unaffected due to low temperatures and the absence of light at ocean depths, potentially leading to benthic organisms redistributing the chemicals and plastic materials in the future [18].

Some of the known toxic effects of plastic additives are placed in the Table 2:

3 Sources and Pathways of Microplastics Pollution in the Oceans

Plastic waste accounts for around 80–85% of the ocean litter [23, 24]. In contemporary times, irrespective of the source, "the marine environment hosts approximately 92% of global marine plastic debris [25], and microplastics were reported with a concentration of 102,000 particles per cubic meter in seawater" [26].

Emerging from a variety of sources, microplastics can generally be grouped into in-land, sea-based, and air-based sources [27]. Activities like shipbuilding, shipbreaking, aquaculture, mariculture, fishing, and deliberate dumping of plastic waste directly into the oceans are sea-based sources or direct sources of microplastic pollution. International shipping, travel, tourism, fishing, ghost fishing and exploration of seas for commercial interests such as oil wells are potential sources of direct dumping of plastic waste. Whereas industrial effluents, toxic sewage discharges, landfill leachate, personal care products, wear and tear of plastic products, abrasion

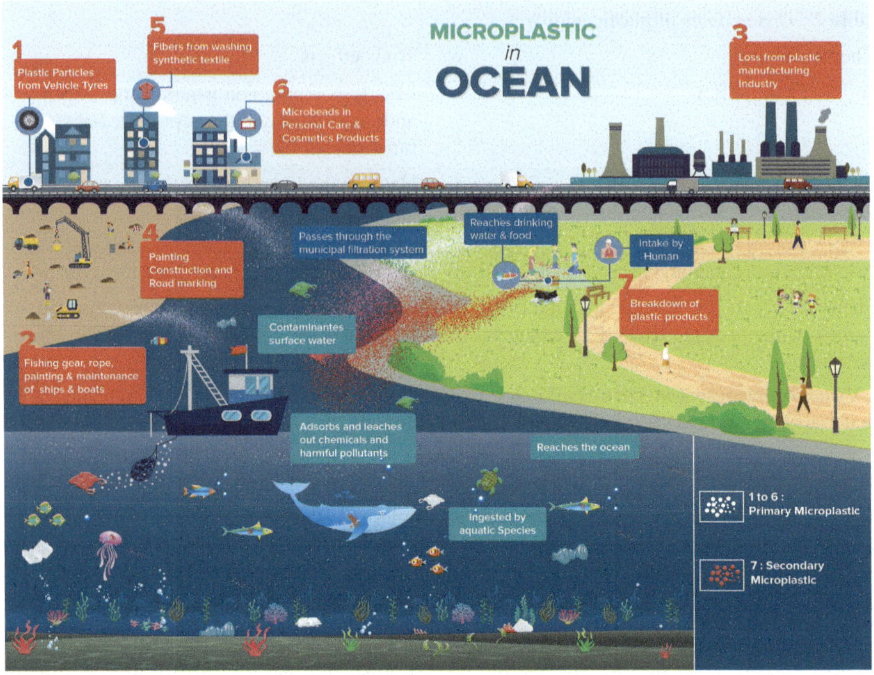

Fig. 2 Exposure and fate of microplastic in ocean

of tyres, beach tourism, and laundering of synthetic textiles, are land-based sources or indirect sources of microplastic pollution [28] (Fig. 2).

In the category of direct dumping in oceans, fishing accounts for a significant proportion of plastic in the oceans. About 1.15 million tons per year of plastic debris that are released into seas are lost or deliberately abandoned in fishing nets [29]. These nets are abandoned in the ocean by the fishing community either because of the high cost of recovery, emergencies or intentionally. These materials are of environmental concern as they breakdown into microplastics that cause damage to the coral reefs and enter the food web when ingested by microorganisms. The term "ghost fishing" is used when abandoned nets and fishing gear in general continue to catch marine vertebrates and invertebrates that may become tangled in them [30].

Travel and tourism are another important contributor to plastic pollution. Eight out of ten tourists travel to coastal cities and beaches adding to the burden of local municipalities in managing additional waste at the same time tourism provides additional revenue to the local population and is viewed as an opportunity. Providing additional comfort, safety and hygiene are key selling points of this industry. There is always an additional demand for personal care products, packaged food, and disposable items for safety reasons even bottled water is all single-use items and disposed of post-use. Cruise ships also account for significantly large volumes of plastic waste since it has the capacity to carry a large number of tourists 6000–7000 for a long duration.

The industry witnessed a setback during the pandemic and is on its way to reviving itself and is predicted to grow at about 12% by 2026. The cruise industry apart from transportation and accommodation provides many onboard services including spas, pools, entertainment, sports, etc., thus generating plastic and wastewater and adding to the burden of microplastics in the ocean.

Nearly 35% of primary microplastics in the oceans are textile microfibers (MFs). These microfibers could be comprised of cotton, cellulose, or synthetic material with polyester being the predominant fibre [31]. The major synthetic fibres in shorelines include polyester (56%), acrylic (23%), polypropylene (PP) (7%), polyethylene (PE) (6%), and polyamide (PA) (3%) [31]. The textile industry during the process of manufacturing, washing and dying fabric uses various chemicals and surfactants that induce the release of microfibers in the wastewater systems [32]. Waste water and sewage treatment facilities both in domestic and industrial settings lack the capability to effectively segregate and remove microfibers. Instead, these microfibers end up being transferred to either sludge or in the treated wastewater. Through various pathways, these microfibers ultimately make their way into the ocean.

Car tyres account for a significant 28% release of microplastics. This usually happens due to friction between the tyres and the surface on which it is plying. These materials are also termed as TWP (Tire Wear Particle) consisting of rubber polymers (about 50%) and other constituents like fillers and softeners. These usually get deposited on road surfaces or open surfaces along the road which then get dispersed by air and finally carried to the ocean either by rainwater, storm drains or through surface water.

City dust is another major contributor to the microplastic in the ocean and accounts for 24% of the total release. It comprises particles or fragments generated from plastic recycling activities, landfill weathering, plastic production facilities, microfibers from textiles, marine building coatings, artificial turfs, harbours, household dust and construction activities. All these fibers and fragments also contribute to the poor air quality in cities apart from their deposition in the final sink "the Ocean" through the medium of air and indirectly through rivers or flood water (Fig. 3).

Alternative routes for microplastics to enter the environment include the direct release of small pieces of plastics that are used as abrasives in industrial and domestic cleaning applications (e.g. shot blasting or scrubbers used in proprietary hand cleansers) and spillage of plastic pellets and powders that are used as a feedstock for the manufacture of most plastic products. Personal care products are also documented to contain primary microplastics to provide abrasive and exfoliating properties. Most consumer items such as toothpaste, creams and facials especially rinse-off products, shampoos and soaps contain microplastics that are released into domestic waste. Before the invention of plastic microbeads natural abrasive products such as cocoa beans, ground almonds, ground apricots, sea salt and ground pumice were used. Post the invention of plastic microbeads the cosmetic industry preferred to use plastic microbeads as these provided several advantages over the natural material. Microbeads are easier to rinse off, less rough, offer smoother exfoliating properties, lower density and are more compatible with other materials in the product. As per

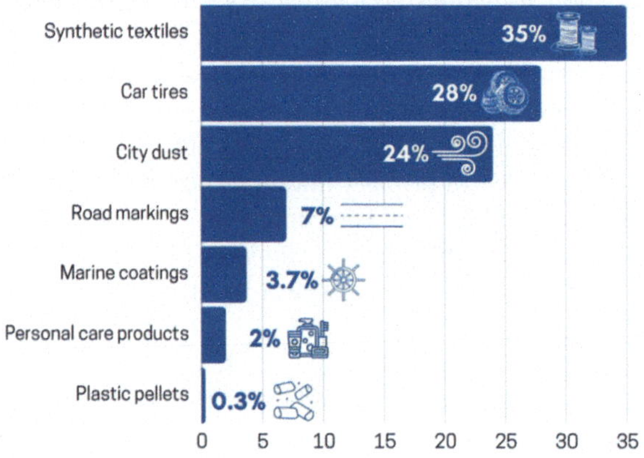

Fig. 3 Microplastic mass across the world's ocean [33, 34]

an estimate, approximately 1500 tons/year of microplastics enter wastewater treatment globally and are released into aquatic environments. Microplastics contained in personal care products invariably get washed down into the drains and transported to wastewater treatment plants. These treatment plants are not effective in filtering out the microplastics, which are later carried to the ocean through rivers and streams. It is estimated that about 12 × 100,000 tons/year of microplastics are released into the environment from personal care products [35].

There has been increased usage of plastic in agricultural practices too for multiple purposes. Plastic sheets are used as mulch in agriculture fields to preserve moisture, conserve water and improve crop fertility. These mulch sheets degrade due to wind, sunlight and other natural elements while in use in agricultural fields and are fragmented into microplastics. At times, the sheets are left behind since their retrieval adds to the financial burden of the farmers and over time these degrade into smaller fragments and microparticles. These fibres are also deposited in the soil and are transported to the ocean through rainwater and floods [36]. Data on microplastic generation from agriculture mulch is limited and hard to quantify.

An important source of microplastics also happens to be plastic litter and waste generated from domestic households and commercial establishments both in urban and rural settings. A significantly large part of this plastic is used for packaging goods and commodities and is essentially used once and discarded into the waste stream. Such packaging is also termed single-use plastics or SUP. Almost 40% of the total plastic produced is used for packaging that is mostly single-use items and discarded post-one-use only. This waste comprising of single-use items is essentially lightweight and of low value hard to collect if not segregated at the point of generation. A large volume of this single-use plastic is discarded in municipal waste streams and dumped in landfills in countries that do not have a good waste management system. Plastic waste dumped in landfills is hard to retrieve and process and continues to be part of the waste volume that is washed down by rainwater and later breaks down into microplastics. Landfills account for a significantly large percentage of plastic and microplastics generation and the quantities can be at variance between countries largely depending on the state of waste management practices adopted and its efficacy.

Total secondary microplastic emission to the marine environment is estimated to be 68,500–275,000 tonnes per year. According to the OECD, 6.1 million tonnes (Mt) of plastic waste leaked into aquatic environments and 1.7 Mt flowed into oceans in 2019 [37]. A study estimated that 16,888–167,431 tonnes of microplastics are generated by pre-production of plastic pellets per year [38].

Figure 4 illustrates the pathway for the plastic waste being dumped in the ocean by the population of about 2 billion living in coastal areas.

The pathway for microplastic pollution in marine ecosystems has been identified as direct dumping into the ocean or human activities directly executed within the ocean. Rivers, lakes and streams accumulate plastic waste generated from industrial and domestic sources and finally discharged into the oceans. Rivers are one of the major pathways for transporting microplastics from land-based sources to the oceans. The top twenty polluting rivers of the world with the annual quantities of plastic loading into the oceans are placed below. A large number of these rivers are located in the Asian region and account for the most pollution to the marine system (Fig. 5).

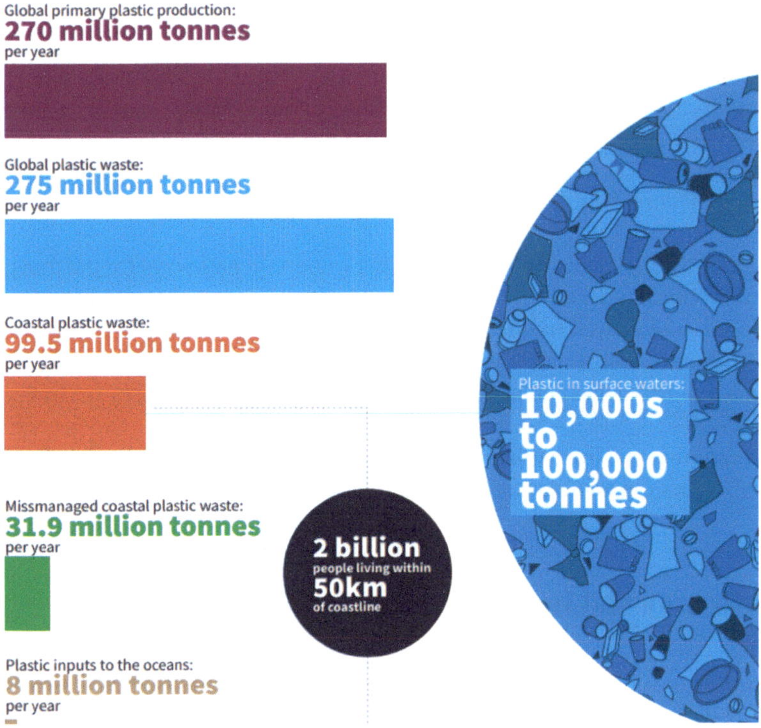

Fig. 4 Pathways of plastic entry into the world's ocean [23, 39]

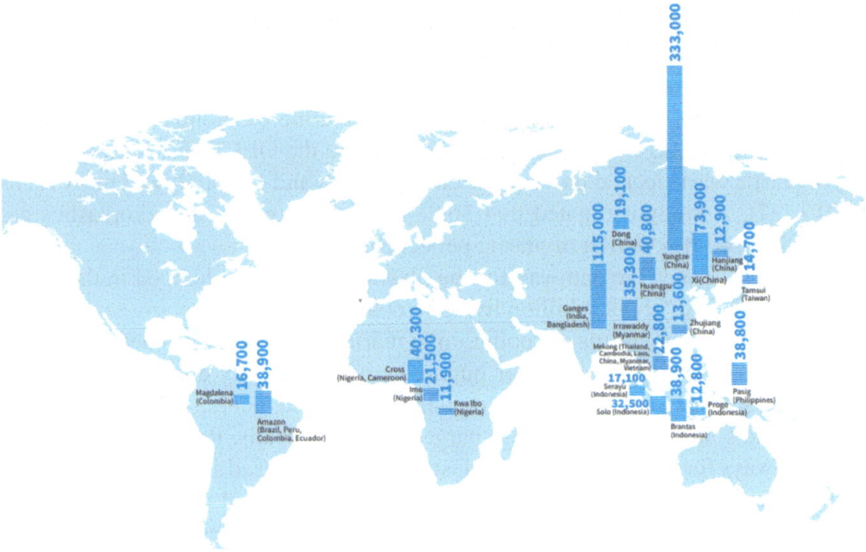

Fig. 5 Rivers with the highest annual plastic loading into the ocean (in tonnes) [40]

The top 20 polluting rivers in the world represented 67% of the global annual river input. In total, 103 out of 122 top polluting rivers are located in Asia [41]. Asia alone represents 86% of the annual global plastic inputs from rivers into the ocean; therefore, Asia is a hotspot and a major emitter of global plastic pollution [42]. The predominant factor influencing the total annual waste generation was the size of the population. That is why the countries with the largest populations are at the top of the list.

Floods and aerial deposition are also other important pathways for the transportation of microplastics. Floods are a regular phenomenon across the world and are being experienced at increased frequency. Apart from other adverse impacts it also carries all the trash both solids and liquids with it to the rivers and onwards to the ocean. Microplastics also deposited along the fold plains and other surfaces are transported to the oceans.

Air is another medium though less discussed. It can pick up smaller fragments and deposit them in another geography and into the oceans.

4 Environmental Impacts

Plastics and microplastics are recognized as major sources of pollution to the marine environment since these materials are finally deposited into the oceans and transported over long distances by ocean currents thus contaminating the ecosystem. It poses a serious threat to the entire ecosystem, affecting all living beings. Many studies have attested that more than 80% of the adverse impacts caused by marine debris are associated with plastic waste [43].

Unfortunately, pollution is now ubiquitous in the world's oceans, whether in the depths, the surface, or within marine life that inhabits the water. Marine pollution significantly impacts marine life. The majority of waste disposed of in oceans takes many years to decompose, leading to a reduction in oxygen levels. The low level of oxygen affects the health of the marine plants and animals. Pollutants like oil spills act as a barrier between the sun and the seafloor. These pollutants restrict the sunlight from reaching the seafloor affecting the photosynthesis process of plants. These also result in disrupting the coral reef cycle. Rising pollution can also lead to an increase in the concentration of nitrogen and phosphorous through various sources like fertilizer runoff or other forms of nutrient pollution. Although plants require Nitrogen and phosphorous to grow, too high a concentration can cause algae blooms, in which algae overrun the ecosystem. Once these organisms start to sink and decompose, oxygen is depleted and the area becomes a dead zone because marine life cannot survive in that environment. Fish and other forms of life that can swim away leave; other species that cannot move die off.

Microplastics have penetrated the marine ecosystem so intensely that they can be detected throughout the food chain from the lowest trophic level to the highest. Studies have reported the presence of MPs in organisms like Zooplankton, Chaetognatha, ichtyo-plankton, copepods and salps belonging to the lower trophic levels, and in crustaceans, echinoderms, bivalves and vertebrates such as fish, seabirds and mammals belonging to the higher trophic levels [44, 47].

Microplastic exhibits deleterious effects towards the marine ecosystem and possesses the ability to disrupt ecosystem services. The toxicity stems primarily from 'chemicals of concern' associated with the plastic debris. Manifestation of toxicity can occur by—(a) Plasticizers with ecotoxicological effects. (b) A composite of Persistent organic matter, other chemicals of concern and MPs that enhance toxicological effects [45]. (c) Heavy metal toxicity due to the interaction of charged microplastic surfaces and heavy metals in the marine environment.

The extent of the adverse effects of microplastic contamination on the marine ecosystem defines the ecological risks it poses [46, 47]. In the marine ecosystem, living organisms can be exposed to microplastics directly or indirectly, for example, through the respiration or egestion of prey contaminated with microplastics [48, 49].

Microplastics have adverse physical and chemical impacts on marine organisms. Here, parameters such as bioaccumulation and biomagnification play critical roles in determining the extent of pollutants within food webs [50]. Bioaccumulation happens when an organism absorbs a contaminant faster than it can eliminate it [51]. When this contaminant is absorbed and passed up in the food chain, it is termed biomagnification [52]. This is where the concentration of the contaminant progresses as the trophic level increases [53, 54]. It is important to note that the foundation of the process is the trophic transfer i.e. contamination at higher trophic levels due to the consumption of contaminated prey at the lower trophic levels.

The micron-sized plastic particles are often mistakenly consumed as food material by marine organisms including coral, phytoplankton, zooplankton, sea urchins, lobsters, fish and many more [55]. Consequently, they end up in organisms at higher trophic levels, transferring and magnifying toxic chemicals across the food web [56]. In the marine environment, organic contaminants such as polychlorinated biphenyls (PCBs), polycyclic aromatic hydrocarbons, petroleum hydrocarbons, organochlorine pesticides (2,20-bis(p-chlorophenyl)-1,1,1 trichloroethane (DDT) and its metabolites; together with hexachlorinated hexane (HCH)), polybrominated diphenyl ethers (PBDEs), alkylphenols may be found in plastic debris like pellets, fragments and microplastics. Most of these contaminants are either added to plastics during manufacture or are adsorbed to plastic debris from the environment [57].

Microplastic fragments, originating from terrestrial sources, often get deposited in the marine ecosystem where they are available to pose significant threats to the ecosystem, including organisms like the coral reefs [58]. In response, corals employ a 'microplastic feeding' mechanism involving ingestion, retention and digestion of plastic fragments [59]. The adverse effects of microplastic on corals include the retention of plastic fragments in mesenteric tissue, leading to a decrease in feeding capability and energy reserves. Additionally, microplastics may exacerbate the negative impact on coral reefs by facilitating pathogen transmission through associated microbial biofilm formation [60]. Microplastics also have detrimental effects on plankton, a crucial food source in the marine ecosystem. The intrusion of microplastics into the cell walls of phytoplankton results in reduced chlorophyll absorption [61], impairing their photosynthetic efficiency and ability to utilize solar energy for essential metabolic processes. When exposed to microplastics, heterotrophic plankton consume and retain plastic fragments in their tissues through the process of phagocytosis [62].

Zooplankton, a valuable marine invertebrate and primary consumers of phytoplanktons have a range of feeding mechanisms and utilize the mechanism of chemo-mechano receptors for prey selection [63]. The widespread presence of microplastics in marine habitats leads to interactions with zooplankton, as both share similar dimensions (>333 μm), making interactions highly likely [64]. Experimental studies have revealed that zooplankton ingest latex beads when exposed to microplastics [65]. Additionally, research has shown that zooplankton tend to ingest polystyrene beads ranging from 1.7 to 30.6 μm in size.

5 Health Impacts

Microplastics have been found in all kinds of habitats and marine life and recent studies have found these tiny particles in human bodies as well. Over 260 species, including invertebrates, turtles, fish, seabirds and mammals, have been reported to ingest or become entangled in plastic debris, resulting in impaired movement and feeding, reduced reproductive output, lacerations, ulcers and death [66, 67]. Therefore, the effects on organisms may vary depending on the exposure to plastic litter in coastal and marine regions (land-based litter or ship-based litter) [68].

The toxicological risk assessment of ingested MPs explores three main aspects: particles, polymer additives and adsorbed pollutants. Many species with different feeding habits, such as filter feeders, deposit feeders, and detritivores, consume plastic. However, plastic ingestion can be particularly problematic for species that mistakenly select plastic items as their food. As a result, some populations have an extremely high incidence of ingestion. For instance, 95% of dead fulmars found in the North Sea have plastic in their digestive system. Other birds, including albatross and prions, have also been reported to have significant amounts of plastic in their guts [69].

Certain features of microplastics such as the microscopic size, attractive colors and high buoyancy make these tiny fragments easily available for fish. Fish ingest microplastics by mistaking these fragments as plankton or other natural prey. A study on microplastic ingestion in planktivorous fish like Acanthochromispolyacanthus was able to detect microplastics in the gut cavity of individual fish [70].

Other experiments have demonstrated that ingestion of microplastics by fishes can cause histopathological modification. This includes the musca epithelial lining becoming detached from the laminopropia, which causes the villi to widen, reduce, and puff up. Additionally, the number of globet cells in the intestine increases, and there are certain changes to the normal structure of the fish's serosa [71].

When different concentrations of MPs (10,000 and 80,000 particles per m^3) were exposed to study the effect of polystyrene European fish (*Percafluviatilis*) eggs and larvae on the Swedish coast, the study found that eggs exposed to high concentrations of microplastics had a slower hatching rate compared to the control group. Larvae that were exposed to microplastics incurred developmental issues as they grew slower and smaller than normal larvae and had a lower response to chemical alarms indicating the presence of predators. This consequently had a negative effect on the survival rate of fish [54]. Similar effect studies have indicated that microplastic also severely alters the metabolic functions in living organisms, specifically, the up-regulation of fatty acids and down-regulation of amino acids [72]. Disbalance in the ratio of triglycerides and Cholesterol in the blood-serum level of fish, as well as the significant alternation in the delivery of cholesterol between the muscle and liver of fish [73], are additional observable effects of the plastic contaminant. This, for instance, has been well illustrated in *Centropagestypicus*, a copepod, which upon ingesting microplastics of size 7.3 μm, eventually lost its feeding ability [64]. Similarly, the benthic organism *Hyalellaazteca* and marine lugworm *Arenicola marina* showed a decrease in growth, loss of appetite and reproductive process when exposed to MPs [74, 75].

Large marine creatures like sharks, whales, seals, sea turtles, and polar bears, are similarly at high risk of ingesting microplastic as seen in the stomachs and intestines of harbour seals, such as *Phocavitulina* [76]. More often these marine mammals are filter feederswhich consume significant amounts of microplastics directly from the ocean water or indirectly by eating prey that contains microplastics. The regular detection of microplastics in the stomachs of sharks and whales in the Mediterranean Sea are further indicators of the burden carried by the marine environment and the significant threat marine life faces [77].

Microplastics have been found in all kinds of marine species and habitats, recent studies have found these tiny particles in human bodies as well. Microplastic enters in human body via different pathways like ingestion, dermal exposure and inhalation. A study published in 2019 [78] involving eight healthy volunteers residing in different geographical regions with varied dietary trends, detected microplastics (not larger than 500 μm) in all eight fecal samples.

Microplastics, which are tiny particles of plastic, have been found in various marine species and habitats. Recent studies have also discovered the presence of these particles in the human body. Microplastic can enter the human body through different pathways such as ingestion, dermal exposure, and inhalation. In 2019, a study was conducted involving eight healthy volunteers from different geographical regions with varied dietary habits. The study found microplastics (not larger than 500 μm) in all eight faecal samples. Another study conducted in Italy examined the presence of microplastics in human urine samples collected from six volunteers (three men and three women) from different cities in the south of Italy [79]. In another, six human placentas were collected from consenting women with physiological pregnancies for the investigation. In total, 12 microplastic fragments (ranging from 5 to 10 μm in size) with spherical or irregular shapes were found in 4 placentas. The investigation indicated that the Microplastics could reach placenta tissues at all levels once these enter the human body [80]. Another study has found microplastics in 26 out of the 34 human breast milk samples that were tested [81]. A recent study has reported microplastics in human lung tissue samples and indicated inhalation as a route for exposure [82].

Cellular and animal experiments have successfully demonstrated that microplastics do adversely affect various systems in the human body, including the digestive, respiratory, endocrine, reproductive, and immune systems.

In the case of the digestive system, microplastics can cause physical irritation to the gastrointestinal tract, leading to inflammation and various gastrointestinal symptoms [83]. They may also disrupt the intestinal microbiome, causing an imbalance between beneficial and harmful bacteria, resulting in symptoms such as abdominal pain, bloating, changes in bowel habits and poor immunity [84]. Additionally, MPs can induce chemical toxicity by absorbing and accumulating environmental toxins like heavy metals and polycyclic aromatic hydrocarbons. When ingested orally, these toxic substances can enter the body through the gastrointestinal tract, causing symptoms such as nausea, vomiting, and abdominal pain [85].

Regarding the effects on the respiratory system, microplastics may cause oxidative stress in the airways and lungs when inhaled, leading to respiratory symptoms such as coughing, sneezing, and shortness of breath due to inflammation and damage, as well as fatigue and dizziness due to a low blood oxygen concentration. A recent study showed that nano-sized plastics were associated with mitochondrial damage in human respiratory cells.

Microplastics can also impact the respiratory system by causing oxidative stress in the airways and lungs when inhaled. This oxidative stress can lead to respiratory symptoms such as coughing, sneezing, and shortness of breath due to inflammation and damage, as well as fatigue and dizziness due to low blood oxygen concentration

[86]. Studies have shown that exposure to even a nano-sized plastic can lead to mitochondrial damage in human respiratory cells [87].

Furthermore, microplastics can disrupt the functioning of the endocrine system by altering the production, release, transport, metabolism, and elimination of hormones, which further leads to disorders that may include metabolic aberrations, developmental disorders and reproductive issues such as infertility, miscarriage and congenital malformations [88]. Microplastics can serve as a conduit for environmental toxicants like bisphenol A and phthalates which will be absorbed into the body and contribute to diseases of the endocrine and reproductive systems [89].

6 Conclusions

Plastic as a material has been in conversation for its benefits and its being a wonder material and for its adverse impacts on the environment too. Scientific communities, citizens and institutions have long discussed and debated plastics leading to growing demand for some mechanism to regulate plastics. The downside of plastics is better understood today evidenced with more data and research. Individual countries, states and municipalities have initiated several measures to contain and regulate this material. Citizens across the world are much better informed and aware of the adverse impacts of plastics and would be at ease to reduce their consumption if only they had an alternative. The Indian government in 2021 announced a ban on the manufacture, sale and use of some of the single-use items. There have been several attempts by regional governments also to ban the use of some of these products in their jurisdictions and have been welcomed by the local population.

The issue of microplastic is not well understood perhaps because of its invisibility and not been discussed well. Microplastics is more complex due to its small size, invisibility and its abundance. Its presence in all medium is of serious concern, though the ocean emerges as the final depository seriously affecting the marine ecosystem. Microplastics are today being documented for contaminating the food chain both in sea and over land. There is growing evidence of the presence of microplastic in water, fish, salt, sugar and in human body too. Microplastic has been increasingly detected in various organs of human body and even in blood. There has been growing evidence of microplastics impacting the reproduction rate in fish directly attributed to some of chemicals adsorbed in microplastics and ingested by the fish. The data on impacts of microplastics in human health is yet to be comprehensively understood but it is certainly worrying to know that much of the food chain is showing evidence of the presence of microplastics in them.

The threat to health and environment being real requires immediate attention else the cost of inaction could be extremely high. Nations are initiating some measures towards reducing waste generation and banning of products made of plastics while some of these measures might produce limited results, the issue of plastics and microplastics requires very comprehensive action covering the complete lifecycle of plastics. The use of synthetic chemicals during plastic production needs very careful evaluation for their toxicity. The other suggested actions can be.

- Information and awareness among citizens and communities.
- Substitution of toxic chemicals used in plastic production.
- Reduce plastic production and consumption.
- Improve waste management practices, regulate plastic waste trade.

Improved recycling technologies. Transition to circular economy model use of alternate material, information and education on the adverse impacts of plastics and microplastics can aid in the process of behavior change of communities leading to reduction in the use of plastic. The other challenge with plastic as a material is to find an alternative or substitute to plastic with almost similar properties, which would be a big driver in changing practices and behavior. There have been several initiatives across many nations to address the concerns of plastics and microplastics but with limited success. There is an ongoing global initiative also to address the concerns of plastics over its complete lifecycle and these measures are expected to be legally binding to nations.

Ecosystems like coral reef and microorganisms play an important role in maintaining the balance in marine ecology and the planet. Oceans are also extremely important ecosystem supporting very diverse organisms and plant life. It is a major resource of protein in the form of fish and other animals consumed by humans and any deterioration in quality and quantity of food from the ocean will have a direct impact on humans and on the planet.

References

1. IPBES. (2019). Summary for policymakers of the global assessment report on biodiversity and ecosystem services of the intergovernmental science-policy platform on biodiversity and ecosystem services. In: Díaz, S., Settele, J., Brondízio, E. S., Ngo, H. T., Guèze, M., Agard, J., Arneth, A., Balvanera, P., Brauman, K. A., Butchart, S. H. M., Chan, K. M. A., Garibaldi, L. A., Ichii, K., Liu, J., Subramanian, S. M., Midgley, G. F., Miloslavich, P., Molnár, Z., Obura, D., Pfaff, A., Polasky, S., Purvis, A., Razzaque, J., Reyers, B., Roy Chowdhury, R., Shin, Y. J., Visseren-Hamakers, I. J., Willis, K. J., Zayas, C. N. (Eds.), IPBES secretariat, Bonn, Germany (56 pp.).
2. Goal 14: Conserve and sustainably use the oceans, seas and marine resources. UN. Retrieved February 1, 2023, from https://www.un.org/sustainabledevelopment/oceans/
3. Ghosh, S. K., Pal, S., & Ray, S. (2013). Study of microbes having potentiality for biodegradation of plastics. *Environmental Science and Pollution Research, 20*, 4339–4355.
4. Shimao, M. (2001). Biodegradation of plastics. *Current Opinion in Biotechnology, 12*(3), 242–247.
5. Hahladakis, J. N., et al. (2023). An overview of chemical additives present in plastics: migration, release, fate and environmental impact during their use, disposal and recycling. *Journal of Hazardous Materials, 344*(2018), 179–199. Retrieved February 20, 2023, from https://doi.org/10.1016/j.jhazmat.2017.10.014
6. Wiesinger, H., Wang, Z., Hellweg, S. (2021). Deep dive into plastic monomers, addi-tives, and processing aids. *Environmental Science & Technology*.

7. Uheida, A., Mejía, H. G., Abdel-Rehim, M., Hamd, W., & Dutta, J. (2021). Visible light photocatalytic degradation of polypropylene microplastics in a continuous water flow system. *Journal of Hazardous Materials, 406*, 124299.
8. Brooks, A. L., Wang, S., & Jambeck, J. R. (2018). The Chinese import ban and its impact on global plastic waste trade. *Science Advances, 4*(6), eaat0131.
9. Geyer, R., Jambeck, J. R., & Law, K. L. (2017). Production, use, and fate of all plastics ever made. *Science Advances*, 1700782.
10. https://www.bbc.com/news/science-environment-42264788
11. United Nations Environment Programme. (2021). From pollution to solution: A global assessment of marine litter and plastic pollution. *Nairobi.*
12. Carpenter, E. J., & Smith, K. L., Jr. (1972). Plastics on the Sargasso Sea surface. *Science, 175*(4027), 1240–1241.
13. Thompson, R. C., Olsen, Y., Mitchell, R. P., Davis, A., Rowland, S. J., John, A. W., Russell, A. E., et al. (2004). Lost at sea: Where is all the plastic?. *Science, 304*(5672), 838–838.
14. Andrady, A. L. (2011). Microplastics in the marine environment. *Marine Pollution Bulletin, 62*, 1596–1605. Retrieved February 7, 2023, from https://doi.org/10.1016/j.marpolbul.2011.05.030
15. Lusher, A., Hollman, P., Mendoza-Hill, J., & Lusher. (2017). *Microplastics in fisheries and aquaculture: Status of knowledge on their occurrence and implications for aquatic organisms and food safety* (p. 27). Food and Agriculture Organization. Retrieved February 1, 2023, from https://www.fao.org/3/I7677E/I7677E.pdf
16. Toxics Link. (2018). Personal eco care product. In *Toxics Link* (p. 7). Retrieved February 9, 2023, from https://toxicslink.org/docs/Microplastic.pdf
17. UNEP. *Pollution to solution* (p. 12).
18. Note 40. Microplastics as contaminants in the marine environment: A review.
19. Scopetani, C., Cincinelli, A., Martellini, T., Lombardini, E., Ciofini, A., Fortunati, A., Ugolini, A., et al. (2018). Ingested microplastic as a two-way transporter for PBDEs in Talitrussaltator. *Environmental Research, 167*, 411–417.
20. Andrade, H., Glüge, J., Herzke, D., et al. (2021). Oceanic long-range transport of organic additives present in plastic products: An overview. *Environmental Sciences Europe, 33*, 85. https://doi.org/10.1186/s12302-021-00522-x
21. Andrady, A. L., & Neal, M. A. (2009). Applications and societal benefits of plastics. *Philosophical Transactions of the Royal Society B: Biological Sciences, 364*, 1977–1984. https://doi.org/10.1098/rstb.2008.0304
22. Obbard, R. W. (2018). Microplastics in polar regions: The role of long-range transport. *Current Opinion in Environmental Science & Health, 1*, 24–29. https://doi.org/10.1016/j.coesh.2017.10.004
23. Barnes, S. J. (2019). Understanding plastics pollution: The role of economic development and technological research. *Environmental Pollution, 249*, 812–821.
24. Ostle, C., Thompson, R. C., Broughton, D., Gregory, L., Wootton, M., & Johns, D. G. (2019). The rise in ocean plastics evidenced from a 60-year time series. *Nature Communications, 10*(1), 1622.
25. Eriksen, M., et al. (2014). Plastic pollution in the world's oceans: More than 5 trillion plastic pieces weighing over 250,000 tons afloat at sea. *PLoSONE*, 9, e111913. Retrieved February 8, 2023, from https://pubmed.ncbi.nlm.nih.gov/25494041/
26. Prata, J. C., et al. (2020). Environmental exposure to microplastics: An overview on possible human health effects. *Science of the Total Environment, 702*, 134455. Retrieved February 8, 2023, from https://pubmed.ncbi.nlm.nih.gov/31733547/
27. Yang, H., Chen, G., & Wang, J. (2021). Microplastics in the marine environment: Sources, fates, impacts and microbial degradation. *Toxics, 9*(2), 41. Retrieved February 8, 2023, from https://www.mdpi.com/2305-6304/9/2/41
28. Veiga, J. M., et al. (2016). Identifying sources of marine litter. In *MSFD GES TG marine litter thematic report*; JRC Technical Report; EUR 28309 (p. 11). https://doi.org/10.2788/018068

29. Lebreton, L. C., Van Der Zwet, J., Damsteeg, J. W., Slat, B., Andrady, A., & Reisser, J. (2017). River plastic emissions to the world's oceans. *Nature Communications, 8*(1), 15611.
30. https://www.ghostfishing.co.uk/how-much-ghost-gear-is-in-the-ocean/
31. Henry, B., Laitala, K., & Klepp, I. G. (2019). Microfibres from apparel and home textiles: Prospects for including microplastics in environmental sustainability assessment. *Science of the Total Environment, 652*, 483–494.
32. Revel, M., Châtel, A., & Mouneyrac, C. (2018). Micro (nano) plastics: A threat to human health? *Current Opinion in Environmental Science & Health, 1*, 17–23.
33. https://www.statista.com/chart/17957/where-the-oceans-microplastics-come-from/
34. Boucher, J., & Friot, D. (2017). *Primary microplastics in the oceans: A global evaluation of sources* (p. 43). IUCN.
35. https://toxicslink.org/publications/reports/microplastics-in-cosmetic
36. https://toxicslink.org/publications/reports/plastic-mulching-microplastics-in-agricultural-soils
37. Retrieved February 20, 2023, from https://www.oecd.org/environment/plastic-pollution-is-growing-relentlessly-as-waste-management-and-recycling-fall-short.htm
38. Hann, S., et al. (2018). *Investigating options for reducing releases in the aquatic environment of microplastics emitted by (but not intentionally added in) products* (Eunomia, 2018). Retrieved February 18, 2023, from https://www.eunomia.co.uk/reports-tools/investigating-options-for-reducing-releases-in-the-aquatic-environment-of-microplastics-emitted-by-products/#:~:text=Reports-,Investigating%20Options%20for%20Reducing%20Releases%20in%20the,of%20Microplastics%20Emitted%20by%20Products&text=This%20report%20has%20been%20prepared,product%20from%20wear%20and%20tear
39. Jambeck, J. R., Geyer, R., Wilcox, C., Siegler, T. R., Perryman, M., Andrady, A., Narayan, R., & Law, K. L. (2015). Marine pollution. Plastic waste inputs from land into the ocean. *Science, 347*(6223), 768–771. https://doi.org/10.1126/science.1260352. PMID: 25678662.
40. Lebreton, L. C., Damsteeg, J., Slat, B., Andrady, A., & Reisser, J. (2017). River plastic emissions to the world's oceans. *Nature Communications, 8*(1), 1–10. https://doi.org/10.1038/ncomms15611
41. van Calcar, C. V., & van Emmerik, T. V. (2019). Abundance of plastic debris across European and Asian rivers. *Environmental Research Letters, 14*(12), 124051.
42. Kim, J. S., Lee, H. J., Kim, S. K., & Kim, H. J. (2018). Global pattern of microplastics (MPs) in commercial food-grade salts: Sea salt as an indicator of seawater MP pollution. *Environmental Science & Technology, 52*(21), 12819–12828.
43. Hong, S. H., Shim, W. J., & Jang, M. (2018). Chemicals associated with marine plastic debris and microplastics: Analyses and contaminant levels (Chapter 9). In *Microplastic contamination in aquatic environments: An emerging matter of environmental urgency* (pp. 271–315). Elsevier Inc.
44. Von Moos, N., Burjhardt, P., & Kohler, A. (2012). Uptake and effects of microplastics on cells and tissue of the blue mussel Mytilus edulis L. after an experiment exposure. *Environmental Science & Technology, 46*, 113227–11335.
45. Ahmad, M., Li, J. L., Wang, P. D., Hozzein, W. N., & Li, W. J. (2020). Environmental perspectives of microplastic pollution in the aquatic environment: A review. *Marine Life Science & Technology, 2*, 414–430.
46. Kershaw, P., Carney Almroth, B., Villarrubia-Gómez, P., Koelmans, A. A., & Gouin, T. (2020). *Proceedings of the GESAMP International Workshop on Assessing the Risks Associated with Plastics and Microplastics in the Marine Environment.*
47. Assessment, E. (1992). Guidelines for exposure assessment. *Federal Register, 57*(104), 22888–22938.
48. Miller, M. E., Hamann, M., & Kroon, F. J. (2020). Bioaccumulation and biomagnification of microplastics in marine organisms: A review and meta-analysis of current data. *PLoS ONE, 15*(10). https://doi.org/10.1371/journal.pone.0240792
49. Hollman, P. C., Bouwmeester, H., & Peters, R. J. B. (2013). Microplastics in aquatic food chain: Sources, measurement, occurrence and potential health risks.

50. Mackay, D., & Boethling, R. S. (Eds.). (2000). *Handbook of property estimation methods for chemicals: Environmental health sciences*. CRC Press.
51. Wang, J., Tan, Z., Peng, J., Qiu, Q., & Li, M. (2016). The behaviors of microplastics in the marine environment. *Marine Environmental Research, 113*, 7–17.
52. Kelly, B. C., Ikonomou, M. G., Blair, J. D., Morin, A. E., & Gobas, F. A. (2007). Food web specific biomagnification of persistent organic pollutants. *Science, 317*(5835), 236–239.
53. Gerber, K. (2009). User's guide and technical documentation KABAM version 1.0 (K OW (based) aquatic bioaccumulation model). In *Report of environmental fate and effects division office of pesticide programs*. US Environmental Protection Agency, Washington, DC.
54. Simpson, S., & Batley, G. (Eds.). (2016). *Sediment quality assessment: A practical guide*. CSIRO Publishing.
55. Institut Veolia. (n.d.). *Special Issue 19 | 2019 Reinventing Plastics*. https://journals.openedition.org/factsreports/4993
56. Teuten, E. L., Saquing, J. M., Knappe, D. R., Barlaz, M. A., Jonsson, S., Björn, A., Takada, H., et al. (2009). Transport and release of chemicals from plastics to the environment and to wildlife. *Philosophical Transactions of the Royal Society B: Biological Sciences, 364*(1526), 2027–2045.
57. Thompson, R. C., Moore, C. J., & Swan, S. H. (2009). Plastics, the environment and human health: Current consensus and future trends. *Philosophical Transactions of the Royal Society B: Biological Sciences, 364*(1526), 2153–2166. https://doi.org/10.1098/rstb.2009.0053
58. Sharma, S., & Chatterjee, S. (2017). Microplastic pollution, a threat to marine ecosystem and human health: A short review. *Environmental Science and Pollution Research, 24*, 21530–21547.
59. Lusher, A. L., Hernandez-Milian, G., O'Brien, J., Berrow, S., O'Connor, I., & Officer, R. (2015). Microplastic and macroplastic ingestion by a deep diving, oceanic cetacean: The True's beaked whale Mesoplodonmirus. *Environmental Pollution, 199*, 185–191.
60. Reichert, J., Schellenberg, J., Schubert, P., & Wilke, T. (2018). Responses of reef building corals to microplastic exposure. *Environmental Pollution, 237*, 955–960.
61. Nerland, I. L., Halsband, C., Allan, I., & Thomas, K. V. (2014). Microplastics in marine environments: Occurrence, distribution and effects.
62. Laist, D. W. (1987). Overview of the biological effects of lost and discarded plastic debris in the marine environment. *Marine Pollution Bulletin, 18*(6), 319–326.
63. DeMott, W. R. (1988). Discrimination between algae and detritus by freshwater and marine zooplankton. *Bulletin of Marine Science, 43*(3), 486–499.
64. Cole, M., Lindeque, P., Fileman, E., Halsband, C., Goodhead, R., Moger, J., & Galloway, T. S. (2013). Microplastic ingestion by zooplankton. *Environmental Science & Technology, 47*(12), 6646–6655.
65. Lin, V. S. (2016). Research highlights: Impacts of microplastics on plankton. *Environmental Science: Processes & Impacts, 18*(2), 160–163.
66. Laist, D. W. (1997). Impacts of marine debris: entanglement of marine life in marine debris including a comprehensive list of species with entanglement and ingestion records. In *Marine debris: sources, impacts, and solutions* (pp. 99–139). Springer.
67. Derraik, J. G. (2002). The pollution of the marine environment by plastic debris: A review. *Marine Pollution Bulletin, 44*(9), 842–852.
68. Ryan, P. G., Moore, C. J., & Moloney, C. L. (2009). Monitoring the abundance of plastic debris in the marine environment. *Philosophical Transactions of the Royal Society B: Biological Sciences, 364*(1526), 1999–2012. https://doi.org/10.1098/rstb.2008.0207
69. Gregory, M. R. (2009). Environmental implications of plastic debris in marine settings: Entanglement, ingestion, smothering, hangers-on, hitch-hiking and alien invasions. *Philosophical Transactions of the Royal Society B: Biological Sciences, 364*, 2013–2025.
70. Critchell, K., & Hoogenboom, M. O. (2018). Effects of microplastic exposure on the body condition and behaviour of planktivorous reef fish (Acanthochromispolyacanthus). *PLoS ONE, 13*(3), e0193308.

71. Peda, C., Caccamo, L., Fossi, M. C., Gai, F., Andaloro, F., Genovese, L., Maricchiolo, G., et al. (2016). Intestinal alterations in European sea bass Dicentrarchuslabrax (Linnaeus, 1758) exposed to microplastics: Preliminary results. *Environmental Pollution, 212*, 251–256.
72. Lu, Y., Zhang, Y., Deng, Y., Jiang, W., Zhao, Y., Geng, J., Ren, H., et al. (2016). Uptake and accumulation of polystyrene microplastics in zebrafish (Danio rerio) and toxic effects in liver. *Environmental Science & Technology, 50*(7), 4054–4060.
73. Cedervall, T., Hansson, L. A., Lard, M., Frohm, B., & Linse, S. (2012). Food chain transport of nanoparticles affects behaviour and fat metabolism in fish. *PLoS ONE, 7*(2), e32254.
74. Au, S. Y., Bruce, T. F., Bridges, W. C., & Klaine, S. J. (2015). Responses of Hyalellaazteca to acute and chronic microplastic exposures. *Environmental Toxicology and Chemistry, 34*(11), 2564–2572.
75. Besseling, E., Wegner, A., Foekema, E. M., Van Den Heuvel-Greve, M. J., & Koelmans, A. A. (2013). Effects of microplastic on fitness and PCB bioaccumulation by the lugworm Arenicola marina (L.). *Environmental Science & Technology, 47*(1), 593–600.
76. Rebolledo, E. L. B., Van Franeker, J. A., Jansen, O. E., & Brasseur, S. M. (2013). Plastic ingestion by harbour seals (Phocavitulina) in the Netherlands. *Marine Pollution Bulletin, 67*(1–2), 200–202.
77. Gabbatiss, J. (2018). Microplastics 'pose major threat' to whales and sharks, scientists warn. *INDEPENDENT, Monday, 6*, 15.
78. Schwabl, P., Köppel, S., Königshofer, P., Bucsics, T., Trauner, M., et al. (2019). Detection of various microplastics in human stool: A prospective case series. *Annals of Internal Medicine, 171*(7), 453–457. https://doi.org/10.7326/M19-0618
79. Pironti, C., Notarstefano, V., Ricciardi, M., Motta, O., Giorgini, E., et al. (2022). First evidence of microplastics in human urine, a preliminary study of intake in the human body. *Toxics, 11*(1), 40. https://doi.org/10.3390/toxics11010040
80. Ragusa, A., et al. (2021). Plasticenta: First evidence of microplastics in human placenta. *Environment International, 146*, 106274. Retrieved February 7, 2023, from https://doi.org/10.1016/j.envint.2020.106274
81. Ragusa, A., et al. (2022). Raman microspectroscopy detection and characterisation of microplastics in human breastmilk. *Polymers, 14*(13), 2700. Retrieved February 7, 2023, from https://doi.org/10.3390/polym14132700
82. Jenner, L. C., et al. (2022). Detection of microplastics in human lung tissue using MFTIR spectroscopy. *The Science of the Total Environment, 831*, 154907. Retrieved February 8, 2023, from https://doi.org/10.1016/j.scitotenv.2022.154907
83. Bouwmeester, H., Hollman, P. C., & Peters, R. J. (2015). Potential health impact of environmentally released micro-and nanoplastics in the human food production chain: Experiences from nanotoxicology. *Environmental Science & Technology, 49*(15), 8932–8947.
84. Jin, Y., Lu, L., Tu, W., Luo, T., & Fu, Z. (2019). Impacts of polystyrene microplastic on the gut barrier, microbiota and metabolism of mice. *Science of the Total Environment, 649*, 308–317.
85. Abbasi, S., Moore, F., & Keshavarzi, B. (2021). PET-microplastics as a vector for polycyclic aromatic hydrocarbons in a simulated plant rhizosphere zone. *Environmental Technology & Innovation, 21*, 101370.
86. Wright, S. L., & Kelly, F. J. (2017). Plastic and human health: A micro issue? *Environmental Science & Technology, 51*(12), 6634–6647.
87. Lin, S., Zhang, H., Wang, C., Su, X. L., Song, Y., Wu, P., Zheng, C., et al. (2022). Metabolomics reveal nanoplastic-induced mitochondrial damage in human liver and lung cells. *Environmental Science & Technology, 56*(17), 12483–12493.
88. Vandenberg, L. N., Luthi, D., & Quinerly, D. A. (2017). Plastic bodies in a plastic world: Multi-disciplinary approaches to study endocrine disrupting chemicals. *Journal of Cleaner Production, 140*, 373–385.
89. Campanale, C., Massarelli, C., Savino, I., Locaputo, V., & Uricchio, V. F. (2020). A detailed review study on potential effects of microplastics and additives of concern on human health. *International Journal of Environmental Research and Public Health, 17*(4), 1212.

Current status of marine litter and microplastic contamination across India's coastal boundaries: Challenges and Management

Pravakar Mishra, K. Dhineka, T. Kaviarasan, M. Sambandam, and Sunanda Sahu

Abbreviations

AFM	Atomic Force Microscope
ALDFG	Abandoned Lost or Discarded Fishing Gear
ATR-FTIR	Attenuated Total Reflectance–Fourier Transform Infrared Spectroscopy
EVOH	Ethylene Vinyl Alcohol
ETP	Effluent Treatment Plant
FESEM-EDX	Field Emission Scanning Electron Microscopy with Energy-Dispersive X-ray Analysis
FTIR	Fourier-Transform Infrared Spectroscopy
μ-FT-IR	Micro-Fourier Transform Infrared Spectroscopy
HDPE	High-Density Polyethylene
ICP-OES	Inductively Coupled Plasma–Optical Emission Spectrometry
ML	Marine Litter
MLMP	Marine Litter and Microplastics
MPA	Marine Protected Area
MPs	Microplastics
MPW	Mismanaged Plastic Waste
Mt	Metric Tons
NOR	Norfloxacin
NEM	Northeast Monsoon
pNEM	Pre-northeast Monsoon
PA	Polyamide

P. Mishra (✉) · K. Dhineka · T. Kaviarasan · M. Sambandam · S. Sahu
National Centre for Coastal Research, Ministry of Earth Sciences (MoES), Government of India, NIOT Campus, Pallikarani, Chennai 600100, India
e-mail: mishra@nccr.gov.in

M. Goel and N. G. Tripathi (eds.), *Plastic Pollution*,
https://doi.org/10.1007/978-981-97-5528-8_6

PD	Palladium
PAHs	Polycyclic Aromatic Hydrocarbons
PAM	Polyacrylamide
PP	Polypropylene
PS	Polystyrene
PU	Polyurethane
PVC	Polyvinyl Chloride
SEM	Scanning Electron Microscopy
STP	Sewage Treatment Plant
SUPs	Single-Use Plastics
UNEP	United Nations Environment Programme

1 Introduction

1.1 Global Scenario About Plastic Pollution in the Ocean

Plastic has grown to become a more dominating commercial material in the consumer market, between the 1930s and 1940s [1]. Worldwide 8.3 billion metric tons of plastic have been produced since 1950; half of that amount has been manufactured in the last 13 years, and future production is expected to rise [2]. Globally, plastic waste is recycled by about 9%, and 12% is incinerated and the remains end up in the natural environment [2]. Between 60 and 99 million metric tons (Mt) of mismanaged plastic waste (MPW) were created globally in 2015, under typical conditions, this amount might triple up to 155–265 Mt y^{-1} by 2060 [3]. These MPW are transported from land through rivers and enter into the ocean [1]. An estimated 0.8 million to 2.7 million metric tons per year, or more than 1000 rivers, are responsible for 80% of MPW emissions. In particular, small rivers are major contributors to the transportation of MPW to the ocean [4]. According to a recent report, plastic input to the ocean environment has decreased [5, 6], but lower input and longer residence time in the marine environment will continue to rise, potentially harming the ecosystem if countermeasures and prevention strategies are not implemented [5].

1.2 Sources of Marine Litter

The United Nations Environment Programme [7] defines Marine litter (ML) as "any persistent, manufactured, or processed solid material discarded, disposed of, or abandoned in the marine and coastal environment." Those materials include plastics, cloth, glass, metal, paper, rubber, wood, and ceramics [8]. Numerous studies conducted worldwide [9–13] have shown that marine litter originates primarily from

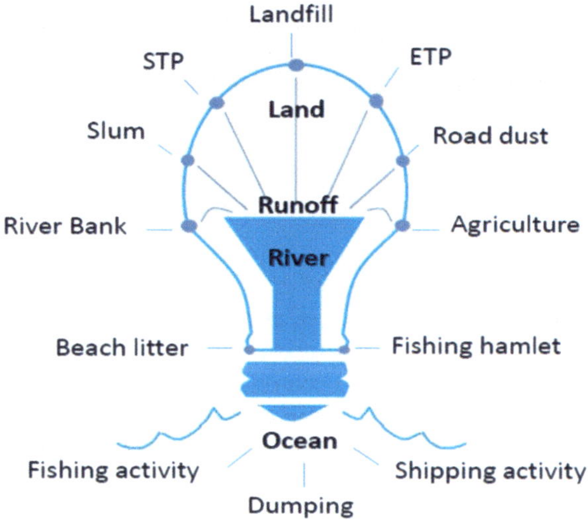

Fig. 1 Source to sink pathway of microplastics

land (80%) and maritime (20%) sources [14, 15]. The source-to-sink pathway of the microplastics is depicted in Fig. 1.

Land-based sources include MPW transported from hinterland, landfill, river banks through water bodies, untreated sewage outfalls, Sewage Treatment Plant (STP), Effluent Treatment Plant (ETP), and beach litter, while the majority of ocean-based sources arise from fishing and shipping, including recreational and commercial vessels [16].

1.3 Marine Litter and Its Impact on Marine Organisms

The biological impacts of marine litter pertain to entanglement, ingestion and serving as a vector; social effects include the loss of beach aesthetics and indigenous values; economic impacts involve clean-up costs, decreased tourism, and losses to fisheries and aquaculture; and public safety effects include leaching of harmful chemicals and cuts, abrasions, and punctures. Over time, a range of degradation mechanisms break down the bigger plastics in the marine environment into smaller particles known as microplastics (MPs), which are less than 5 mm [17]. MPs are categorized into two types based on their origin: (i) primary MPs, which are synthetic clothes made in factories, beads sewn or attached to personal care products, etc., and (ii) secondary MPs, which are formed when larger plastic particles break down as a result of a variety of physical and chemical factors [18]. Since these particles resemble food for fish and other species, they are consumed by marine creatures. Microplastics have been discovered in several marine organisms that are intended for human consumption, such as fish and shellfish [19], and other processed seafood like canned [20] and dry fish [21]. Consumption of whole seafood, including shellfish, is particularly hazardous to humans, it is most likely determined by hydrophobicity, surface chemistry, dose, size, and polymer type of the ingested MPs [22].

1.4 Risks Associated with Occurrence and Human Exposure

Microplastics can be found everywhere: in the atmosphere, ocean, remote islands, urban areas, Mount Everest, and the Mariana Trench. As a result of their pervasive nature, humans are easily exposed to microplastic contamination. Microplastic exposure is a common occurrence for humans due to its extensive dissemination. The various pathways of microplastics entry into the ocean, and transfer to humans through MP ingested seafood consumption are shown in Fig. 2. A recent study discovered that microplastic in humans, i.e., blood [23], lung tissues [24], and placenta [25]. However, the hazardous consequences of microplastics, which can differ based on their kind, size, shape, and concentration, are little understood about human health [26].

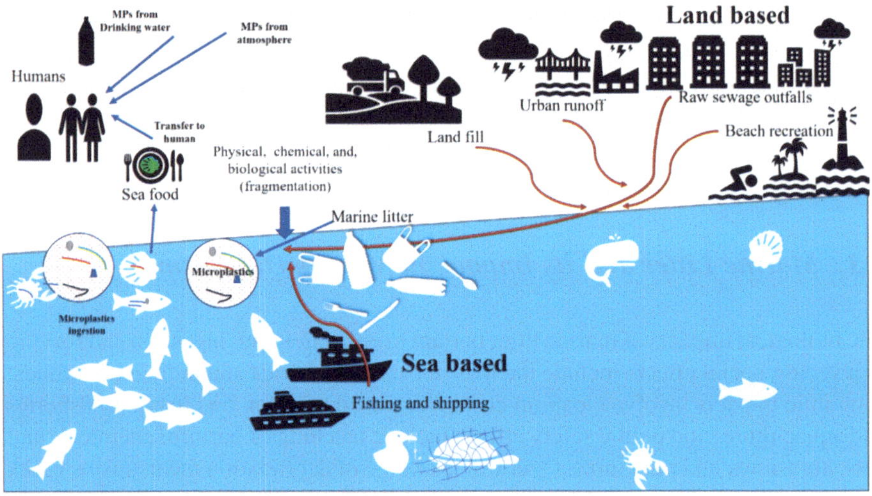

Fig. 2 Trophic transfer of microplastics in the food chain

2 Bibliometric Analysis on Worldwide Microplastic Research in the Marine Environment

The Scopus database (www.scopus.com) was searched using the following keywords: microplastics, OR marine, OR biota, OR seafood, OR sediment, OR seawater till November 23, 2023. The terms were found in the title, abstract, and keywords. Contamination of fish, seafood, sediment (beach and seafloor), and water from the coast was the main selection criteria employed for the literature. Studies conducted in controlled contexts, reviews, and conference proceedings were also included. The data used for this analysis are articles published from 2020 to 2023. The obtained results were examined using software tools like VOS Viewer to perform statistical and mathematical analysis in the subject area quantitatively. The process for analysing literature data, including titles, authors, institutions, journals, keywords, and references, provides citation networks, co-occurrence networks, and coupling networks. This is made possible by the objective and quantitative nature of the scientific research done through examination of publication trends, analysis of source journals, contributions and collaborations from authors, co-occurrence analysis of keywords, and co-citation analysis.

2.1 *Annual Quantitative Distribution of the Literature*

Globally, there were 820 papers on microplastics found in total. A comprehensive picture of scientific production patterns and the continued growth and development of the marine microplastics research field may be seen from a statistical study of the number of papers published between 2020 and 2023 (Fig. 3).

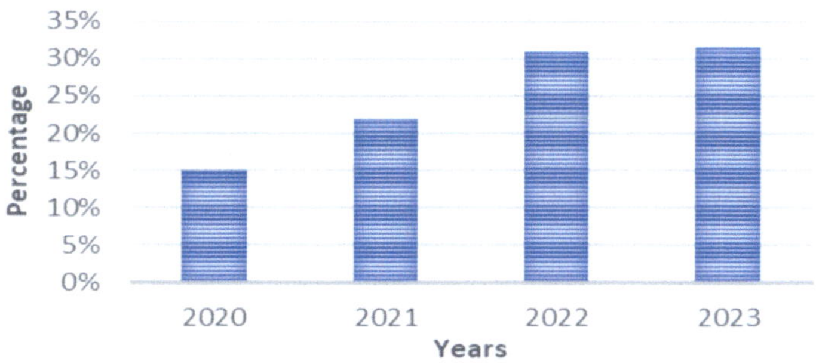

Fig. 3 Percentage of quantitative distribution of the literature

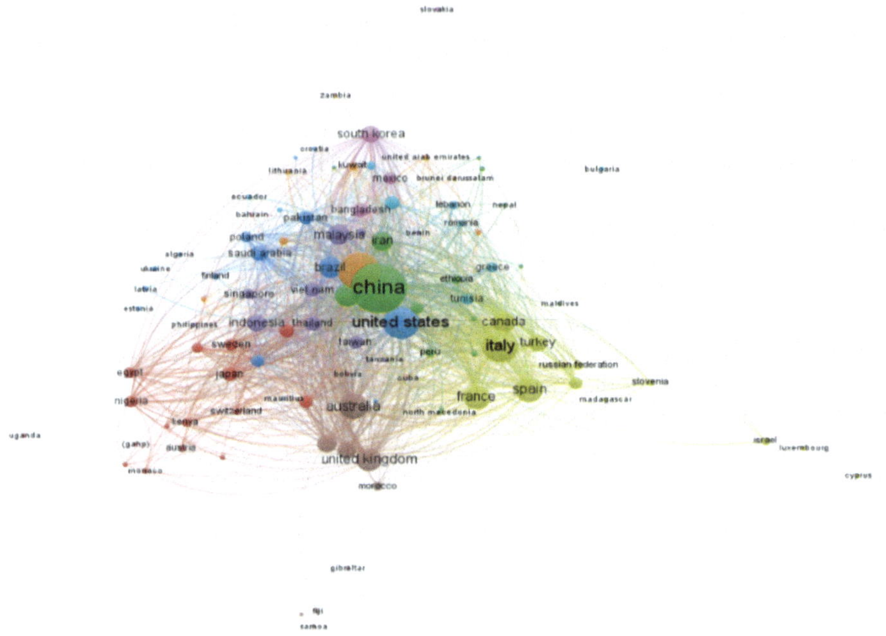

Fig. 4 The nation's flagship on investigating marine microplastics

2.2 Top Nations in Terms of Papers Published

A distribution analysis was conducted to examine the cooperation between the countries/regions conducting microplastic research in marine environments. 108 countries participated in the marine microplastic study, 51 of which contributed to more than five publications. The top ten nations/regions with the most articles published were China (188), India (111), Italy (82), the United States (81), Australia (53), Spain (50), Germany (40), the United Kingdom (36), and Canada (35).

Figure 4 displays the knowledge area map for the co-authored nations and regions. The map's nodes, which stand for various nations and regions, are scaled according to the total number of articles. A working partnership between two nodes is implied by their link.

2.3 Top Journals that Published More Research Articles on Microplastics

Of the 820 publications, 101 articles were published in the Science of the Total Environment, followed by Marine Pollution Bulletin (98), Environmental Pollution (48), Chemosphere (39), and Journal of Hazardous Materials (30) (Fig. 5).

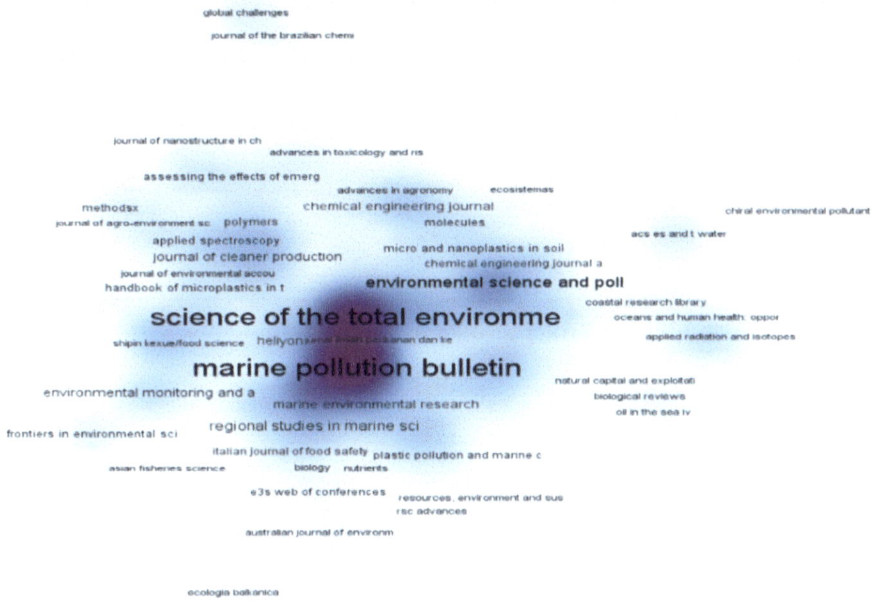

Fig. 5 Journals that published more research articles on microplastics

2.4 *Analysis of Keywords Co-occurrence*

Keyword co-occurrence analysis aims to identify the limits of research in an area by examining its fundamental ideas and organizational framework. The analysis software produced clusters, as shown by the outcomes in Fig. 6. A node reflects a keyword, its size indicates how often it occurs, and the density of connecting lines between nodes indicates the degree of relationship among keywords. During the analysis a minimum number of occurrence of keywords was set as 5, and the results show a total of 131 keywords, among the top keywords are as in the following orders microplastics (340) > microplastic (101) > pollution (44) > nanoplastics, Plastic pollution, fish (43) > human health (33) > sediment (30) > toxicity (29) > water (21).

The node microplastic found co-occurred along with almost all keywords. The node fish co-occurred with shellfish, biota, trophic transfer, human health, ecotoxicology, ecotoxicology risk assessment, ingestions, bioaccumulation, and health risk assessment, which indicates the relationships of each study and their importance.

2.5 *Development of Microplastic Research in Marine Biota*

A total of 475 papers were found after limiting the keywords such as bioaccumulation, fish, toxicity, health risks, health hazard, mollusks, bivalve, human health, animal

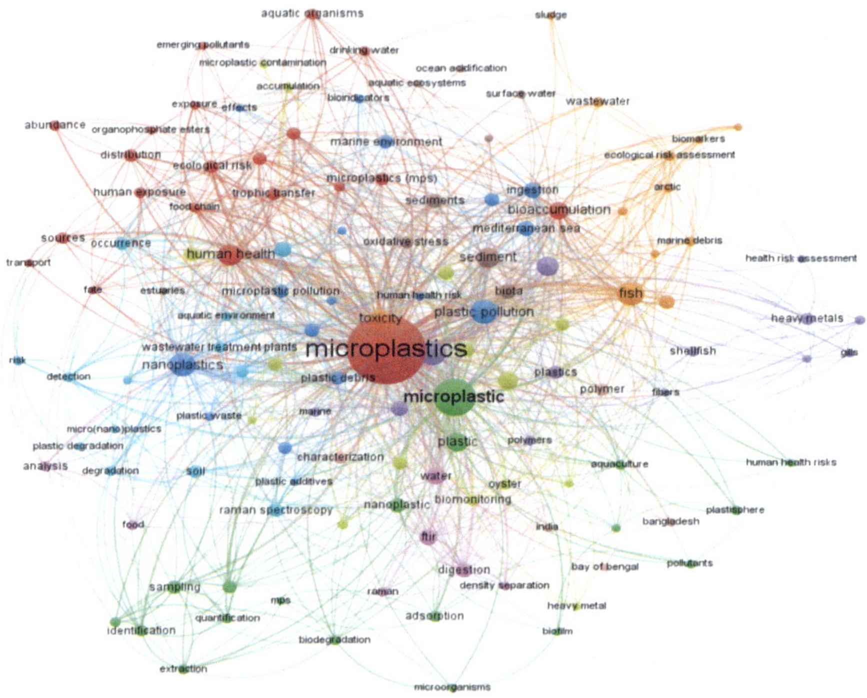

Fig. 6 Keyword co-occurrence cluster revealing the top keywords

tissue, ingestion, seafood, environmental exposure, biota, oxidative stress from the published article. Furthermore, an ecotoxicology and environmental exposure study [27] revealed that 3 μm Polystyrene (PS) MP contamination poses a potential risk to the environment and human health by revealing impaired gill function in mussels (*Mytilus galloprovincialis*) after a brief exposure in laboratory conditions. Similarly, the toxicity of Palladium (Pd) in dissolved water has revealed that marine mussels (*M. galloprovincialis*) had 2–3 times higher accumulation when exposed to the MPs adsorbed Pd, and are susceptible to oxidative stress and feeding inhibition [28]. Another study on the impact of microplastics on bivalves concluded that exposure to MPs can potentially cause variations in the development of larvae and lower their survival chances. Beyond physiological alterations, MPs may have an impact on the structure and functioning of aquatic habitats, which may affect the provision of ecosystem services for bivalves. The problem is made worse by the possibility of negative consequences from other pollutants building up on the surface of microplastics [29]. Biochemical and microbiological responses were studied in mussels and exposed to environmentally suitable doses of nanoparticles (NPs, which are 80 nm PS beads) and the antibiotic norfloxacin (NOR), both individually and in combination. After 15 days of treatment, the combined toxic effects of NOR and NPs had been found to change enzyme activity, disturb gut homeostasis, and cause physiological damage in mussels [30]. The above investigations demonstrate an increasing number of research being conducted in experimental settings at the global level.

3 Marine Litter and Microplastic (MLMP) Research in India

Until November 3, 2023, a systematic search was conducted in the Scopus database and Web of Science utilizing the keywords “microplastics,” “marine,” “debris,” “litter,” and “India.” The search scope encompassed the title, abstract, and keywords fields. The key selection criteria for the literature were marine litter, and microplastics in marine environments in India. In addition, controlled studies, reviews, and conference proceedings were included. This analysis relied on articles published till 2023.

The first publication on marine plastic pellets in the intertidal region of Caranzalem, Goa in the year 1982. The outline of the marine litter and microplastics research in India is shown in Fig. 7. The West Coast (48%) had more number of publications than the East Coast (42%) and islands (8%). There is an uneven distribution among state-wide studies since Tamil Nadu (n = 57) carried out the maximum research on this subject followed by Kerala (n = 29). Totally 157 research articles and 17 review articles related to MLMP were published in India. In terms of habitat, research was mainly carried out on the beaches due to ease of access for sampling. Other studies included the coast (22%), estuaries, lagoons, and backwater (12%), biota (12%), mangroves (2%), and saltpan (2%).

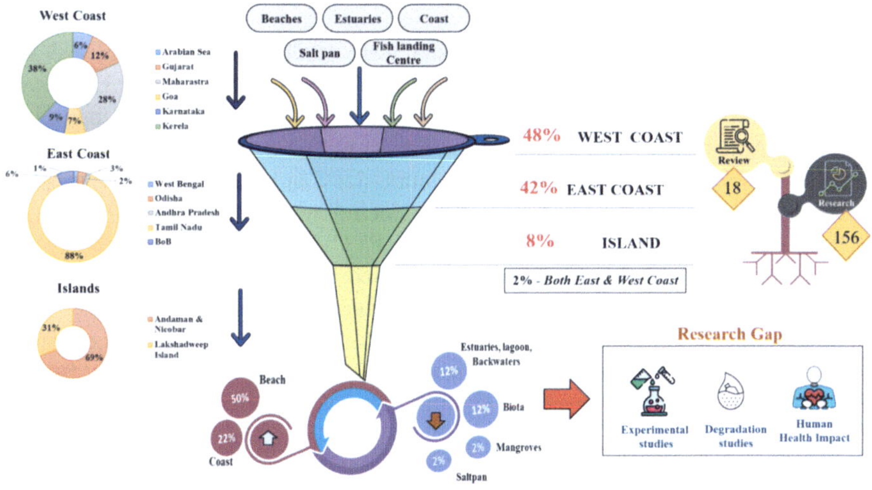

Fig. 7 An overview of marine litter and microplastic research publication in India

Experimental and degradation studies of plastics in different marine environments, and MPs impact on human health are the few research areas that are not covered in India. The MLMP distribution in the Indian marine environment of each coastal state and island is discussed in this study. The state-wise distribution of MLMP density range in different sample types is depicted in Table 1.

3.1 East Coast

3.1.1 West Bengal

The marine litter pollution along the Hooghly Estuary, within the Gangetic delta, was investigated focusing on 17 beaches [31]. Litter samples were collected along 100-m transects during monsoon and post-monsoon seasons, (OSPAR monitoring standard), and categorized 16,597 items into 6 types and 44 categories. Monsoon season exhibited higher litter abundance (1.10 ± 0.39 items/m^2) compared to post-monsoon (0.86 ± 0.32 items/m^2). Most beaches were classified as having low cleanliness. Hazardous litter constituted 6.5% of the total items. Modelling studies indicated that high discharge from rivers significantly contributed to anthropogenic litter on north-east beaches. Litter flux decreased with distance from the shore, acting as a sink to the sea floor. Results highlighted the representation of household, tourism, and fishing-related litter, emphasizing the necessity for improved regional litter management measures.

3.1.2 Odisha

The coastal beaches (n = 9) from Odisha were studied and found an average microplastic abundance of 258.7 ± 90.0 particles/kg of beach sediment [32]. Swargadwara station had the highest microplastic abundance (378.3 ± 39.7 particles/kg), while Abhayachandpur exhibited the lowest (153.3 ± 27.3 particles/kg). The study identified microplastics of diverse colours, sizes, and shapes, and Raman spectroscopic analysis revealed 12 polymer types, including High-Density Polyethylene (HDPE), Polystyrene (PS), Polyvinyl chloride (PVC), and Acrylonitrile copolymer.

The microplastic distribution study was carried out in Chilika Lake, one of the world's largest brackish water lagoons [33]. Microplastic abundance in Chilika Lake sediment ranged from 25.2 ± 9.8 to 7 ± 2.2 items/kg, and in water, it ranged from 110.7 ± 35.6 to 26 ± 8.1 items/100 L. Anthropogenic activities, especially fishing and tourism, were identified as major sources, alongside upstream input from rivers. Fragments and foams were the prevalent shapes in sediment and water, respectively. The sediment contained smaller microplastics (<1 mm), while water had white-coloured foams in the 2–5 mm size range. Ten microplastic polymers were identified, with Polypropylene (PP) and PS being the most dominant in both sediment and water samples. The study also noted weathering-related surface features on microplastics.

Table 1 State-wise distribution of MLMP density range

State	Number of studies	Type of study	Sample type	Density range
Gujarat	10	ML-3, MP-6, MLMP-1	ML	1–128 g/m^2
			Sediment	1.4–26 MPs/kg
			Biota	0.73 ± 0.41 MPs/g and 44.54 ± 9.7 MPs items/g
Maharashtra	21	ML-12, MP-7, MLMP-2	ML	1.46 ± 0.14 to 11.67 ± 8.83 items/m^2; 157–820 items/m^2
			Sediment	108.24 ± 28.19 to 10,568.3 ± 3053.3 items/kg
			Water	372 ± 143 items/L to 537.5 ± 95 items/L
			Biota	5.62 ± 2.27 to 6.6 ± 2.98 items/individual and 198.82 items/individual
Goa	5	ML-1, MP-4	ML	0.27–10.08 g/m^2
			Sediment	320–3950 particles/kg
			Water	48 ± 19 MP particles/L
			Biota	0.7 ± 0.3 to 4 ± 2 MPs/g
Karnataka	8	ML-3, MP-5	ML	7.3–2985 g/m^{-2}
			Sediment	264 ± 62 to 1275 ± 532 particles/kg
			Water	0.11 ± 0.07 MPs/l to 755.03 ± 400.96 particles/L
			Biota	44.54 ± 9.7 MPs items/g
Kerala	29	ML-7, MP-20, MLMP-2	ML	33.9 ± 28.93 to 120.05 ± 106.28 items/100 m^2; 53–159 particles m^{-2}
			Sediment	27 ± 13 to 1275 ± 532 particles/kg
			Water	8.49 ± 4.26 to 101.6 ± 24 particles/L
			Biota	0.01 ± 0.014 to 0.4 ± 0.6/MPs/individual; 15.9 items/g; 44.54 ± 9.7 MP items/kg
Tamil Nadu	58	ML-14, MP-42, MLMP-2	ML	1.15 ± 0.01 to 5.14 ± 0.04 items/m^3 and 60–820 items/m^3; 8.2–11,439 marine litter items/m^2
			Sediment	6 ± 1.73 to 505 ± 121 items/kg
			Water	5.67 ± 3.1 to 154.7 ± 4.16 items/L
			Biota	0.64 ± 0.31 to 1.8 ± 0.54 MPs/g; 0.00–22.19 items/individual
Andhra Pradesh	1	MP-1	Sediment	1834–2116 pellets

(continued)

Table 1 (continued)

State	Number of studies	Type of study	Sample type	Density range
Odisha	2	MP-2	Sediment	14.9 ± 6.5 to 258.7 ± 90.0 particles/kg
			Water	56.7 ± 28.2 items/100 L
West Bengal	1	ML-1	ML	0.86 ± 0.32 items/m^2 to 1.10 ± 0.39 items/m^2; ML: 0.09–695 items/m^2;
Andaman and Nicobar Island	9	ML-6, MP-2, MLMP-1	Water	0.45 ± 0.32 to 0.93 ± 0.59 particles/m^3
Lakshadweep Islands	4	ML-4 (Pellets-2)	ML	4.53–31.73 g/m^2
Review	17	ML-4, MP-6, MLMP-7		–

ML—Marine Litter; MP—Microplastics; MLMP—Marine litter and Microplastics

3.1.3 Andhra Pradesh

The occurrence, spatial and seasonal distributions, adsorbed contaminants, polymer composition, and ecological risks associated with MPs at eight renowned beaches on the central east coast of Andhra Pradesh were investigated [34]. The research, conducted during the pre-northeast monsoon (pNEM) in October 2020 and northeast monsoon (NEM) in January 2021, collected 3950 MPs. Results revealed a higher abundance of MPs during the NEM compared to pNEM. Favorable wind and currents during the NEM contributed to increased MP abundance. Spatial variations among beaches were significant. The study emphasizes a considerable polymer hazard risk to the ecosystem due to MP presence, with surface weathering features indicating potential future toxicity. Commonly identified polymers include PE, PP, PE and PP co-polymer, and PS.

3.1.4 Tamil Nadu

In recent years, numerous studies (n = 58) have been conducted in various regions of Tamil Nadu, to assess the presence and impact of marine litter, particularly MPs, in different environments. These investigations span various marine habitats like beaches (n = 24), coastal areas (n = 18), estuaries (n = 4), salt pans (n = 3), and biota (n = 9). The studies employ diverse methodologies for sampling and analysis, including sediment and water sampling, handpicking, and the use of techniques like Fourier-transform infrared spectroscopy (FT-IR) and scanning electron microscopy (SEM). The findings provide valuable insights into the density, shapes, colours, and types of plastics present, shedding light on potential sources and their environmental implications.

In the beach habitat, studies related to litter (7), MLMP (1), MPs (11), meso-plastic (1), macroplastic (1), and biofouling (1) were carried out. The majority of the studies were carried out in Chennai and Thoothukudi beaches, and sediment was the common sample type. The invertebrate species Bivalvia-*Donax cuneatus*, was studied in the Gulf of Mannar site and it was found that 0.29–2.7 items/individual (MP) were recorded, with fibre and transparent being the most dominant shape and colour respectively [35]. PE and PP were the major polymers. In the estuarine environment, 3.28 ± 0.87 MPs/individual of Bivalvia: *Perna viridis* was reported [36].

Another study conducted in the Gulf of Mannar beach from March 2006 to February 2008, focused on beach litter and identified plastic pollution resulting from fishing activities, tourism, and sewage-related debris. The study reported a density of 2088–5409 g of plastic per 100 m^2 [37]. An investigation in Chennai beach, spanning March 2015, revealed MPs of 3–5 mm in sediment, indicating potential sources from rivers during floods. The plastic types identified included PE and PP, sourced mainly from the Cooum and Adyar rivers [38]. Similarly, the spatio-temporal studies (2019–2020) carried out in the coastal sediments from Chennai to Puducherry also reported Adyar station to be the most polluted site indicating riverine input as the major source [39].

In the coastal matrices, fibre contributed the major shape, and in terms of colour, transparent and blue were the highest. PE and PP were the predominant polymers. Multiple sources including riverine runoff, fishing activities, and effluent discharge are mainly responsible for the MP entry into the oceans.

Notably, studies on fish species and invertebrates demonstrated the ingestion of plastics by marine organisms, with detection methods ranging from stereoscopic microscopes to Raman spectroscopy. The presence of plastic particles in fish, mussels, and oysters highlights the potential transfer of MPs through the food chain. The first report of MPs in biota was reported in 2018 [40]. Out of the 20 individuals collected from the fish landing site, 30% of the fish analysed showed the presence of plastics. Microfibres were more than irregularly shaped plastic fragments. Black, red, and translucent were reported to be the common colours with PE and PP as the dominant polymer type. The industrial and commercial waste washed through the Buckle Canal, Thoothukudi was found to be the major source of MP contamination.

Another study focused on the Gulf of Mannar coast, examining microplastics and their trace metal composition in the muscle and intestine of five commercially important fish species. The study utilized micro-Fourier Transform Infrared Spectroscopy (μ-FT-IR) and atomic force microscope (AFM) to investigate MP abundance and morphology, while Inductively coupled plasma optical emission spectrometry (ICP-OES) was employed to assess the adsorption/leaching of trace metals. The study revealed varying abundances of total MPs in different fish species, with PE, PP, Polyamide (PA), and fibrous shape as predominant. The estimated possible consumption of MPs by adults and children did not suggest immediate health problems. The presence of trace metals reflected the influence of sewage waste, although non-carcinogenic risk evaluations did not indicate significant health risks for consumers [41].

Recent studies in 2021 and 2022 continued to investigate marine litter, exploring its presence in demersal sharks, pelagic and benthic fishes, and estuarine waters. These studies further identified specific polymers such as PP, PE, and Polyacrylamide (PAM), indicating the diversity of plastic types in the marine environment.

Overall, in Tamil Nadu, the maximum litter recorded was 1361–2295 nos/100 m and 26.64–41.66 kg/100 m; and Microplastics in beaches were 304–896 items (post-flood). Out of four estuarine studies, higher MP values accounted for water is 31.7 ± 3.78 to 154.7 ± 4.16 items/L and 51.7 ± 3.51 to 171.3 ± 9.02 items/kg for sediment. In the coastal matrix, water recorded a maximum of 60–820 items/m^3 and 60–1620 items/kg in sediment. In saltpan studies, 35 ± 15 to 72 ± 40 items/kg were recorded. In biota, 3.2–7.6 MPs/fish was the maximum reported value.

3.2 West Coast

3.2.1 Gujarat

Gujarat analysed ten studies concentrating on microplastics and marine litter between 2004 and 2023. Out of these, six papers delved into microplastics, three into marine litter, and one into both. Seasonal sampling was consistently applied, revealing diverse sources of microplastics, including sea-based, land-based, and ship-breaking activities. Polymer characterization using FT-IR and Raman spectrophotometer identified major components such as PP, ABS, PS, PET, PE, PVC, and Polyurethane (PU) in microplastics commonly found in bottles, bags, toys, and food packaging.

In 2004, the sediment samples were collected during three different periods at a depth of 5cm from the Alang-Sosiya beach, revealed the presence of microplastics (MPs) with a density ranging from 66 to 89 mg kg^{-1}, primarily attributed to the influence of ship-breaking activities [42]. Along multiple beaches, including Porbandar, Madhavpur, Chorwad, and Veraval, sediment samples were collected through hand-picking in a 100 m area, resulting in litter density ranging from 19 to 158 items m^{-2} and 4.17 to 40.18 g m^{-2}, predominantly consisting of plastic [43]. The Narmada Estuary study focused on MPs in sediments (0.1–5 mm), investigating adsorption capacity and leaching of polycyclic aromatic hydrocarbons (PAHs) [44]. Another study in 2021, examined marine litter on beaches, reporting a density of 1 to 128 g/m^2 from January to March 2020 [45]. A study conducted in the Gulf of Kachchh in 2022 examined marine litter on Marine Protected Area (MPA) Island beaches from January to March 2020 [46]. Additionally, a 2022 beach study in Gujarat reported MP densities ranging from 1.4 to 26 MPs/kg sediment, with transparent fibrous shapes, and identified major polymers such as PP, PE, and PS [47]. A Fishing Harbour study in 2023 investigated MPs, revealing densities of 0.82 ± 0.58 MPs/g and 2.02 ± 1.48 MPs/individual in gut and gills, with prevalent polymers including PE, PA, PU, and PS [48]. Two studies in 2023 focused on the coast and muddy shore, respectively, with the former exploring the interaction between MPs and bacteria [49] and the latter reporting MP densities ranging from 0.32 ± 0.18 particles/g to 2.81 ± 0.50 particles/g, with predominant polymers such as PP, PS, PE, PVC, PU, ABS, and PET [50].

3.2.2 Maharashtra

Between 2012 and 2023, a total of 21 papers were published, with a focus on MLMP in various habitats. The studies covered beaches, creeks, mangrove stands, and the coast. Noteworthy findings include the highest MP abundance in sediments at 220 ± 50 MPs kg^{-1} sediment and an annual average of 49.41 ± 12.77 particles/100 L in water [51]. Biota studies revealed the presence of 17–96 items/haul and 48–4352 g/haul [52], while mangrove litter density was 10.62 ± 0.48 items/tree [53]. Major sources of MPs were identified as domestic sewage, recreational activities, religious practices, fishing, and water channels. The MP characterization indicated the presence of PE, PP, and PA.

The floating plastic debris in a water channel flowing from the creek to the sea was assessed during the inter-monsoon period, and over 106 microplastic particles weighing 9.9 g were obtained during low tide conditions and 95 particles weighing 8.7 g during high tide conditions [54]. The microplastics were categorized into plastic fragments, microfibres, microbeads, and foamed plastics. The study concludes that the insignificant quantity of microplastics entering the sea from the landward side indicates the primary origin is the fragmentation and degradation of macro- and mega-plastics transported through inland water channels connected to the sea. It emphasizes the importance of preventing macro- and megaplastic entry from the landward side to reduce microplastic pollution in oceans.

3.2.3 Goa

From Goa, a total number of 5 papers based on microplastics and marine litter in beaches and estuaries were published. Out of these, four articles were on microplastics and one on marine litter. Seasonal sampling was carried out, revealing the sources of microplastics as fishing activities, fishing vessels, and tourism. The type of sampling ranged from the traditional handpicking method to the usage of Van Veen grab and Niskin sampler for sediment and water sampling respectively. Polymer characterization using Attenuated Total Reflectance–Fourier transform infrared spectroscopy (ATR-FTIR) identified major components such as PAM, PVC, PA, polyacetylene, and ethylene vinyl alcohol (EVOH). Significant findings of the studies include the highest ML abundance in beach sediments at 0.27–3.73 g/m^2 (May 2020); 0.97–10.08 g/m^2 [55] and the highest MP abundance in beach sediments was 8–17/25 g [56]. The highest MP abundance in the estuarine water column and sediment were 48 ± 19 MP particles/L and 3950 ± 930 MP particles/kg respectively. In shellfish (whole soft tissue), the average concentrations of MPs were 4 ± 2 (*Crassostrea sp.*), 3.2 ± 1.8 (*Perna viridis*) and 0.7 ± 0.3 (*Paphia malbarica*) MPs/g body weight [57].

3.2.4 Karnataka

A total number of 8 studies were carried out in this state within the year 2007 to 2023 out of which 5 were on microplastic and 3 on marine litter. Coasts, beaches, and estuarine waters were sampled and analysed. Some of the techniques involved in the characterization of isolated microplastics include optical fluorescence microscopy, Raman spectroscopy, and Field emission scanning electron microscopy with Energy-dispersive X-ray analysis (FESEM-EDX). Polymers like PAM, PA, PVC, EVOH, and polyacetylene are commonly reported.

The beach litter study on Tannirbhavi Beach in Mangaluru revealed the highest ML abundance ranging from 29.7 to 392 items m^{-2} and 108.4 to 2985 g m^{-2} [58]. The collected litter included items such as bottle caps, lids, cutlery, syringes, and plastic nuggets, with fishing activities, food packaging, domestic plastic debris, recreation, religious activities, and ports identified as contributors to marine litter pollution. The MP study on the sediment from five beach locations along the Arabian Sea coast in Karnataka revealed the highest MP abundance ranging from 264 ± 62 n/kg to 1002 ± 174 n/kg of dry sand among the beach studies [59]. The study associated MP levels with factors like population, tourism, and industries, and the microscopic analysis indicated weathering effects on the particles. The coastal waters from St. Mary's Island had an MP abundance of 0.218 ± 0.33 particles/L [60]. Blue fibres and white foams were prominent, with fishing activities and packaging identified as major pollution sources. The Udyavara River Estuary, exhibited an MP abundance of 320.83 (±98.30), 514.55 (±352.16), and 755.03 (±400.96) particles/m^3 in the surface, middle, and bottom layers, respectively [61]. The study linked MP distribution to salinity, indicating the influence of port, fishing, industrial activities, monsoon rain, and tidal effects. In the case of marine dried fish products, the highest MP abundance in biota was 44.54 ± 9.7 MPs items/g [62]. Eviscerated fish exhibited higher MP levels than whole fish, posing potential risks to consumer safety based on the microplastics polymer-induced risk index. Overall, these studies highlight the pervasive nature of microplastic pollution in diverse coastal ecosystems and emphasize the need for comprehensive management strategies.

3.2.5 Kerala

The first study from Kerala on Mesoplastics contaminated in *Thunnus albacares* (Yellow Tuna) was published in the year 2013 [63], followed by more studies focusing on marine litter, microplastics and Abandoned, Lost, or discarded fishing gear (ALDFG) with a total 29 publications. Out of these, 6 belong to Marine litter (2 ALDFG), 20 microplastics, 2 Marine litter and microplastics, and 1 mesoplastic study. The study areas include locations such as Kochi, Allapuzha, Vembanad Lake, Nattika Kerala Coast, Cochin, and Mangalavanam mangrove. The habitats studied range from deep-sea environments (fish) off Kochi to mud banks, brackish water lakes, coastal areas, beaches, estuaries, and fish landing centres. The sampled groups consist of diverse marine life, including fish species like *Sardinella longiceps* (Oil sardine), and *Stolephorus commersonnii* (Anchovies). Sediment, stake nets, invertebrates, and fish are among the sample types collected for analysis. MPs were found in 6 out of 16 anchovies [64] and brackish water sediments (March–April 2016) across 10 sites using a Van Veen grab revealed particle densities ranging from 96 to 496 particles m^{-2}, shapes covering fragments, fibres, foams, and films [65]. Debris collected from stake nets over 15 days in August 2014 amounted to 476–1503 g day^{-1} net^{-1}, sourced from plastics [66].

The maximum MP in beach sediment was reported as 120.5 ± 36.8 particles per 100 g (surface sediments); and 109.9 ± 35.6 particles per 100 g (bottom sediments) [67]. In Mangalavanam mangrove, an average MP abundance of 1275 ± 532 particles/kg d.w., (sediment), and 101.6 ± 24 particles/litre,(water) was reported [68]. A higher count of 15,360 number (497.6 kg) was recorded from the six beaches [69] and in another study, the total average debris concentration was 120.05 ± 106.28 items/100 m^2 with a plastic debris concentration of 86.53 ± 79.38 items/100 m^2 [70]. In terms of Biota, a total of 128 particles with an average of 0.4 ± 0.6 MPs/shrimp was found to be the highest reported value [71]. PP and PE were the most commonly reported polymer types. Stream drainage, net mending yard, fishing activities, and runoff were found to be the sources of plastic pollution in the marine environment.

3.3 Islands

3.3.1 Andaman and Nicobar Islands

A total of nine marine litter and microplastic studies were reported in the beaches and bay of Andaman and Nicobar Islands. The first litter observational study reported in 2003 [72] highlighted the improper handling of solid waste from fishing/mariculture activity and ship traffic resulting in the litter from Plastic and fishing buoys. The MPs in water was 0.93 ± 0.59 per m^3 (Avg.), with a highest value of 1.88 per m^3 [73]. The maximum litter was recorded as 695 items/m^2 and PVC, PS and white-coloured litter were dominant.

3.3.2 Lakshadweep Islands

Out of the four studies, two were on Marine litter and two were on plastic pellets. A maximum of 2702 plastic resin pellets was reported on the Agatti island beaches [74]. An average of 10.57 to 193 ± 97 items/400 m^2 was reported in the case of marine litter collected from Kadmat, Suheli Paar, and Minicoy [75]. Plastic carry bags and containers of household products, fishing, and local littering were considered as the major sources of pollution.

3.4 Other Studies

Seven studies were conducted in different study areas combining more than two states. Three studies involved beach litter collected from multiple coastal states and islands. Out of the other four, one study was on MP contamination in zooplankton, one on dry fishes, and two were on microplastic studies in coastal water, sediment, beach, and fish. A collaborative study across Gujarat, Karnataka, and Kerala in 2023 analysed MPs in dry fish, identifying PP as a major component [62]. The maximum reported value was 5.3×10^4 particles km^{-2}, 209 ± 99 particles kg^{-1} of dry weight [76]. Microplastics in the deep sea sediment (225, 230, and 1070 m) off Krishna Godavari basin were reported with an average abundance of 12 MPs/50 g DW (230 m) followed by 10 MPs/50 g DW (1070 m) and 2 MPs/50 g DW (225 m) respectively [77].

3.5 Recent Experimental Studies of Microplastic Research in Marine Biota in India

Studies on the experimental approach of microplastics in marine biota were scarce as of 2020. But recently, Athulya et al. [78] investigated the effects of 0.1 mg/mL PE, PS, and a combination of both microplastics on marine clams (*Donax faba*) at two distinct pHs (8 and 7.8). Observed that exposure to microplastics at low pH levels may cause physiological stress and have an adverse effect on marine species' ability to survive. Similarly, a study conducted on edible scallops (*Amusium pleuronectes*) using virgin LDPE microplastics to transfer heavy metals in a known concentration (30 μg/l) of chromium (IV) into a culture tank containing edible scallops revealed that LDPE microplastics can adhere to and transfer biotic tissues when exposed to contaminants like heavy metals [79].

4 Coastal Clean-Up and Awareness Drives

The government of India has initiated a number of programs for PAN India coastal clean-up and awareness events, since 2017. A mobile app and dashboard have been created by NCCR for the beach litter-field data collection through the citizen science approach (Fig. 8). About 150 tons of litter was removed from ~200 beaches through

Fig. 8 PAN India coastal clean-up and awareness campaign

various programs such as Swachh Sagar Surakshit Sagar, International Coastal Clean-up Day, World Oceans Day, and Swachhata Pakhwada. The field litter data reveals that the majority of litter was single-use plastics (SUPs) implying that recreational/ tourist activities were the major sources.

Drastically reducing consumption of the low utility and high littering potential SUPs with the legislative leverage of its ban (notified in PWM rules, 2021) and the dilation of the ban to other frequently littered SUPs would effectively eliminate a significant source of marine pollution. In recent years, a reduction in the average percentage of plastic litter was observed compared to previous years due to a ban on the production and usage of identified SUPs, improved cleaning frequency of beaches, and increasing awareness among the public due to various campaigns and the Clean India mission.

5 Research Gap

Despite significant progress in understanding marine microplastics, there exist notable research gaps that warrant attention for a more comprehensive comprehension of their ecological and human health implications. Experimental studies exploring the biological and physiological responses of marine organisms to microplastic exposure are still limited, hindering a nuanced understanding of the direct impacts on various species. Additionally, there is a scarcity of research investigating the degradation patterns of plastic under diverse environmental conditions, such as varying temperatures, salinities, and microbial compositions. Such studies are crucial for predicting the long-term fate and persistence of microplastics in different ecosystems. Furthermore, while considerable attention has been given to microplastic contamination in beach habitats, sediment, and water samples, there is a relative dearth of research concerning other marine environments like open water and deep-sea ecosystems. Expanding the focus beyond coastal areas is essential for a more holistic understanding of the global distribution and dynamics of microplastics. Moreover, the human health impacts of microplastics, particularly through seafood consumption and other exposure pathways, remain an area requiring comprehensive investigation to assess potential risks to human populations. Bridging these research gaps is imperative for developing effective mitigation strategies and safeguarding both marine ecosystems and human health from the pervasive influence of microplastics.

6 A Clarion Call for the National Marine Litter Policy and Tweak in the Management Aspects

Even though several waste management legislations are enforced, a huge quantum of mismanaged plastics is still entering the marine environment. Due to this, a geometrical increase of microplastics occurs in the marine environment. Strengthening the waste management infrastructure complemented with proper source segregation and collection will help to abate the mismanaged plastics. Measures such as taxes on unsustainable products, and market incentives for sustainable alternatives will discourage the production and consumption of a good that damages the environment. Strengthening the circular economy, sustainable production, mapping the sources of generation complementing with the sector-specific guidelines, and enlightening the ill effects of dumping are the areas to look out for pollution reduction. It is reasonable to employ preventive measures such as installing floating trash barriers in the rivers, creeks, and waterways to break off the transport of plastics into the marine environment. Improving the frequency of beach clean-up activities, and adoption of beaches by the regional educational institutes, NGOs will not only restrain the litter entering into coastal waters but also enhance the aesthetics of beach and tweaks tourism revenue. Of late, it is realized that a National Marine Litter Policy based on a holistic approach is the day's call.

Acknowledgements The authors are thankful to the Secretary, Ministry of Earth Sciences, Government of India and Director, National Centre for Coastal Research for encouraging and implementing the Marine Litter and Microplastic Programme at NCCR, Chennai.

References

1. Jambeck, J. R., Geyer, R., Wilcox, C., Siegler, T. R., Perryman, M., Andrady, A., Narayan, R., & Law, K. L. (2015). Plastic waste inputs from land into the ocean. *Science, 347*(6223), 768–771. https://www.science.org/doi/10.1126/science.1260352
2. Geyer, R., Jambeck, J. R., & Law, K. L. (2017). Production, use, and fate of all plastics ever made. *Science Advances, 3*(7), 25–29. https://doi.org/10.1126/sciadv.1700782
3. Lebreton, L., & Andrady, A. (2019). Future scenarios of global plastic waste generation and disposal. *Palgrave Communications, 5*(1), 1–11. https://doi.org/10.1057/s41599-018-0212-7
4. Meijer, L. J. J., van Emmerik, T., van der Ent, R., Schmidt, C., & Lebreton, L. (2021). More than 1000 rivers account for 80% of global riverine plastic emissions into the ocean. *Science Advances, 7*(18), 1–14. https://doi.org/10.1126/sciadv.aaz5803
5. Kaandorp, M. L. A., Lobelle, D., Kehl, C., Dijkstra, H. A., & van Sebille, E. (2023). Global mass of buoyant marine plastics dominated by large long-lived debris. *Nature Geoscience, 16*(8), 689–694. https://doi.org/10.1038/s41561-023-01216-0
6. Zhang, Y., Wu, P., Xu, R., Wang, X., Lei, L., Schartup, A. T., Peng, Y., Pang, Q., Wang, X., Mai, L., Wang, R., Liu, H., Wang, X., Luijendijk, A., Chassignet, E., Xu, X., Shen, H., Zheng, S., & Zeng, E. Y. (2023). Plastic waste discharge to the global ocean constrained by seawater observations. *Nature Communications, 14*(1), 1–12. https://doi.org/10.1038/s41467-023-37108-5
7. UNEP. (2009). Marine litter: A global challenge. United Nations Environment Programme, Nairobi
8. Cheshire, A. C., Adler, E., Barbière, J., Cohen, Y., Evans, S., Jarayabhand, S., Jeftic, L., Jung, R. T., Kinsey, S., Kusui, E. T., Lavine, I., Manyara, P., Oosterbaan, L., Pereira, M. A., Sheavly, S., Tkalin, A., Varadarajan, S., Wenneker, B., & Westphalen, G. (2009). UNEP/IOC guidelines on survey and monitoring of marine litter. UNEP Regional Seas Reports and Studies No. 186, IOC Technical Series No. 83, 120 pp.
9. Browne, M. A., Galloway, T. S., & Thompson, R. C. (2010). Spatial patterns of plastic debris along estuarine shorelines. *Environmental Science and Technology, 44*(9), 3404–3409. https://doi.org/10.1021/es903784e
10. Galgani, F., Burgeot, T., Bocquéné, G., Vincent, F., Leauté, J. P., Labastie, J., Forest, A., & Guichet, R. (1995). Distribution and abundance of debris on the continental shelf of the Bay of Biscay and in Seine Bay. *Marine Pollution Bulletin, 30*(1), 58–62. https://doi.org/10.1016/0025-326X(94)00101-E
11. Kaviarasan, T., Dhineka, K., Sambandam, M., Sivadas, S. K., Sivyer, D., Hoehn, D., Pradhan, U., Mishra, P., & Ramana Murthy, M. V. (2022). Impact of multiple beach activities on litter and microplastic composition, distribution, and characterization along the southeast coast of India. *Ocean and Coastal Management, 223*. https://doi.org/10.1016/j.ocecoaman.2022.106177
12. Mishra, P., Kaviarasan, T., Sambandam, M., Dhineka, K., Murthy, M. V. R., Iyengar, G., Singh, J., & Ravichandran, M. (2023). Assessment of national beach litter composition, sources, and management along the Indian coast—A citizen science approach. *Marine Pollution Bulletin, 186*. https://doi.org/10.1016/j.marpolbul.2022.114405
13. Pruter, A. T. (1987). Sources, quantities and distribution of persistent plastics in the marine environment. *Marine Pollution Bulletin, 18*(6 Suppl. B), 305–310. https://doi.org/10.1016/S0025-326X(87)80016-4
14. Allsopp, M., Walters, A., Santillo, D., & Johnston, P. (2006). Plastic debris in the world's oceans. *Greenpeace International*, 44 pp.
15. GESAMP. (2021). Sea-based sources of marine litter. In Gilardi, K. (Ed.), (IMO/FAO/UNESCO-IOC/UNIDO/WMO/IAEA/UN/UNEP/UNDP/ISA Joint Group of Experts on the Scientific Aspects of Marine Environmental Protection). Rep. Stud. GESAMP No. 108, 109 pp.
16. Galgani, F., Hanke, G., & Maes, T. (2015). Global distribution, composition and abundance of marine litter. *Marine Anthropogenic Litter*, 29–56.

17. Arthur, C., Baker, J., & Bamford, H. (2009). Proceedings of the international research workshop on the occurrence, effects, and fate of microplastic marine debris. *Group, January*, 530.
18. Cole, M., Lindeque, P., Halsband, C., & Galloway, T. S. (2011). Microplastics as contaminants in the marine environment: A review. *Marine Pollution Bulletin, 62*(12), 2588–2597. https://doi.org/10.1016/j.marpolbul.2011.09.025
19. Kaviarasan, T., Mishra, P., Dhineka, K., Sambandam, M., Begum, M., & Murthy, M. V. R. (2022). Microplastic contamination in commercially important fish and shellfish from coastal waters of the Bay of Bengal off Chennai, southeast coast of India. *Oceans Conference Record (IEEE)*, 1–5. https://doi.org/10.1109/OCEANSChennai45887.2022.9775536
20. Diaz-Basantes, M. F., Nacimba-Aguirre, D., Conesa, J. A., & Fullana, A. (2022). Presence of microplastics in commercial canned tuna. *Food Chemistry, 385*(November 2021), 132721. https://doi.org/10.1016/j.foodchem.2022.132721
21. Karami, A., Golieskardi, A., Ho, Y. B., Larat, V., & Salamatinia, B. (2017). Microplastics in eviscerated flesh and excised organs of dried fish. *Scientific Reports, 7*(1), 1–9. https://doi.org/10.1038/s41598-017-05828-6
22. Smith, M., Love, D. C., Rochman, C. M., & Neff, R. A. (2018). Microplastics in seafood and the implications for human health. *Current Environmental Health Reports, 5*(3), 375–386. https://doi.org/10.1007/s40572-018-0206-z
23. Leslie, H. A., Van Velzen, M. J., Brandsma, S. H., Vethaak, A. D., Garcia-Vallejo, J. J., & Lamoree, M. H. (2022). Discovery and quantification of plastic particle pollution in human blood. *Environment International, 163*, 107199.
24. Jenner, L. C., Rotchell, J. M., Bennett, R. T., Cowen, M., Tentzeris, V., & Sadofsky, L. R. (2022). Detection of microplastics in human lung tissue using μFTIR spectroscopy. *Science of the Total Environment, 831*, 154907.
25. Ragusa, A., Svelato, A., Santacroce, C., Catalano, P., Notarstefano, V., Carnevali, O., Giorgini, E., et al. (2021). Plasticenta: First evidence of microplastics in human placenta. *Environment International, 146*, 106274.
26. Lee, Y., Cho, J., Sohn, J., & Kim, C. (2023). Health effects of microplastic exposures: Current issues and perspectives in South Korea. *Yonsei Medical Journal, 64*(5), 301.
27. De Marco, G., Eliso, M. C., Oliveri Conti, G., Galati, M., Billè, B., Maisano, M., Ferrante, M., & Cappello, T. (2023). Short-term exposure to PS microplastics hampers the cellular function of gills in the Mediterranean mussel Mytilus galloprovincialis. *Aquatic Toxicology, 264*(October). https://doi.org/10.1016/j.aquatox.2023.106736
28. Rodrigues, J. P., Santos-Echeandía, J., Chaves-Pozo, E., Campillo, J. A., Rocha-Santos, T., Duarte, A. C., Rivera-Hernández, J. R., Valdés, J., & Albentosa, M. (2023). Interactive effects of palladium (Pd) and microplastics (MPs) on metal bioaccumulation and biological responses in the Mediterranean mussel, Mytilus galloprovincialis. *Marine Pollution Bulletin, 194*(April). https://doi.org/10.1016/j.marpolbul.2023.115284
29. Khanjani, M. H., Sharifinia, M., & Mohammadi, A. R. (2023). The impact of microplastics on bivalve mollusks: A bibliometric and scientific review. *Marine Pollution Bulletin, 194*(PA), 115271. https://doi.org/10.1016/j.marpolbul.2023.115271
30. Xu, R., Li, L., Zheng, J., Ji, C., Wu, H., Chen, X., Chen, Y., Hu, M., Xu, E. G., & Wang, Y. (2023). Combined toxic effects of nanoplastics and norfloxacin on mussel: Leveraging biochemical parameters and gut microbiota. *Science of the Total Environment, 880*(April), 163304. https://doi.org/10.1016/j.scitotenv.2023.163304
31. Mugilarasan, M., Karthik, R., Purvaja, R., Robin, R. S., Subbareddy, B., Hariharan, G., Rohan, S., Jinoj, T. P. S., Anandavelu, I., Pugalenthi, P., & Ramesh, R. (2021). Spatiotemporal variations in anthropogenic marine litter pollution along the northeast beaches of India. *Environmental Pollution, 280*. https://doi.org/10.1016/j.envpol.2021.116954
32. Patchaiyappan, A., ZakiAhmed, S., Dowarah, K., Khadanga, S. S., Singh, T., Jayakumar, S., Thirunavukkarasu, C., & Devipriya, S. P. (2021). Prevalence of microplastics in the sediments of Odisha beaches, southeastern coast of India. *Marine Pollution Bulletin, 167*. https://doi.org/10.1016/j.marpolbul.2021.112265

33. Singh, N., Mondal, A., Abhinav, G., Nagamani, P., & Darbha, G. K. (2023). Microplastics in Indian Brackish Water Lagoon: Occurrence and distribution in the Chilika Lake. *Water, Air, & Soil Pollution, 234*(10). https://doi.org/10.1007/s11270-023-06627-8
34. Raju, M. P., Veerasingam, S., Suneel, V., Saha, M., Rathore, C., Naik, A., Suneetha, P., & Ramakrishna, S. S. V. S. (2023). Seasonal variation and spatial distribution of microplastic pellets and their associated contaminants along the central east coast of India. *Environmental Science and Pollution Research, 30*(26), 68489–68503. https://doi.org/10.1007/s11356-023-27100-3
35. Narmatha Sathish, M., Immaculate Jeyasanta, K., & Patterson, J. (2020). Monitoring of microplastics in the clam Donax cuneatus and its habitat in Tuticorin coast of Gulf of Mannar (GoM), India. *Environmental Pollution, 266*, 115219. https://doi.org/10.1016/j.envpol.2020.115219
36. Dowarah, K., Patchaiyappan, A., Thirunavukkarasu, C., Jayakumar, S., & Devipriya, S. P. (2020). Quantification of microplastics using Nile Red in two bivalve species Perna viridis and Meretrix meretrix from three estuaries in Pondicherry, India and microplastic uptake by local communities through bivalve diet. *Marine Pollution Bulletin, 153*(February), 110982. https://doi.org/10.1016/j.marpolbul.2020.110982
37. Ganesapandian, S., Manikandan, S., & Kumaraguru, A. K. (2011). Marine litter in the northern part of Gulf of Mannar, southeast coast of India. *Research Journal of Environmental Sciences, 5*(5), 471–478.
38. Veerasingam, S., Mugilarasan, M., Venkatachalapathy, R., & Vethamony, P. (2016). Influence of 2015 flood on the distribution and occurrence of microplastic pellets along the Chennai coast, India. *Marine Pollution Bulletin, 109*(1), 196–204. https://doi.org/10.1016/j.marpolbul.2016.05.082
39. Dhineka, K., Kaviarasan, T., Sambandam, M., Mishra, P., & Ramana Murthy, M. V. (2022). Comparison of Microplastic abundance in varying depths of deep-sea sediments, Bay of Bengal. *Oceans Conference Record (IEEE)*, 31–34. https://doi.org/10.1109/OCEANSChennai45887.2022.9775261
40. Kumar, V. E., Ravikumar, G., & Jeyasanta, K. I. (2018). Occurrence of microplastics in fishes from two landing sites in Tuticorin, South east coast of India. *Marine Pollution Bulletin, 135*(June), 889–894. https://doi.org/10.1016/j.marpolbul.2018.08.023
41. Selvam, K., Xavier, K. A. M., Shivakrishna, A., Bhutia, T. P., Kamat, S., & Shenoy, L. (2021). Abundance, composition and sources of marine debris trawled-up in the fishing grounds along the north-east Arabian coast. *Science of the Total Environment, 751*, 141771. https://doi.org/10.1016/j.scitotenv.2020.141771
42. Reddy, M. S., Basha, S., Adimurthy, S., & Ramachandraiah, G. (2006). Description of the small plastics fragments in marine sediments along the Alang-Sosiya ship-breaking yard, India. *Estuarine, Coastal and Shelf Science, 68*(3–4), 656–660. https://doi.org/10.1016/j.ecss.2006.03.018
43. Sukhdhane, K. S., Kripa, V., Shukla, S. P., Sreenath, K. R., Damodaran, D., & Vase, V. K. (2019). Assessment of Marine Litter along Four Sandy Beaches of Saurashtra Coast, Gujarat, India. *International Journal of Current Microbiology and Applied Sciences, 8*(06), 2623–2632. https://doi.org/10.20546/ijcmas.2019.806.315
44. Sharma, M. D., & Krupadam, R. J. (2022). Adsorption-desorption dynamics of synthetic and naturally weathered microfibers with toxic heavy metals and their ecological risk in an estuarine ecosystem. *Environmental Research, 207*. https://doi.org/10.1016/j.envres.2021.112198
45. Behera, D. P., Kolandhasamy, P., Sigamani, S., Devi, L. P., & Ibrahim, Y. S. (2021). A preliminary investigation of marine litter pollution along Mandvi beach, Kachchh, Gujarat. *Marine Pollution Bulletin, 165*. https://doi.org/10.1016/j.marpolbul.2021.112100
46. Baroth, A., Mamgain, S., Sivakumar, K., Hatkar, P. S., & Pathan, S. (2022). Role of protected area in reducing marine and plastic litter: A case study from India's first marine protected area and comparison with non-protected areas. *Journal of Industrial Ecology, 26*(6), 2080–2091. https://doi.org/10.1111/jiec.13248

47. Rabari, V., Patel, K., Patel, H., & Trivedi, J. (2022). Quantitative assessment of microplastic in Sandy beaches of Gujarat state, India. *Marine Pollution Bulletin, 181*. https://doi.org/10.1016/j.marpolbul.2022.113925
48. Rabari, V., Patel, H., Patel, K., Patel, A., Bagtharia, S., & Trivedi, J. (2023). Quantitative assessment of microplastic contamination in muddy shores of Gulf of Khambhat, India. *Marine Pollution Bulletin, 192*. https://doi.org/10.1016/j.marpolbul.2023.115131
49. Kumar, M., Chaudhary, D. R., & Jha, B. (2023). Surface-associated bacterial assemblages on marine anthropogenic litter in the intertidal zone of the Arabian Sea, India. *Marine Pollution Bulletin, 193*. https://doi.org/10.1016/j.marpolbul.2023.115211
50. Rabari, V., Rakib, M. R. J., Trivedi, J., Idris, A. M., & Malafaia, G. (2023). Microplastics occurrence in commercial crab (Portunus segnis) from the western coast of India and pollution indices: First investigation and evidence. *Science of the Total Environment, 905*. https://doi.org/10.1016/j.scitotenv.2023.167500
51. Tiwari, M., Rathod, T. D., Ajmal, P. Y., Bhangare, R. C., & Sahu, S. K. (2019). Distribution and characterization of microplastics in beach sand from three different Indian coastal environments. *Marine Pollution Bulletin, 140*(October 2018), 262–273. https://doi.org/10.1016/j.marpolbul.2019.01.055
52. Selvam, S., Manisha, A., Roy, P. D., Venkatramanan, S., Chung, S. Y., Muthukumar, P., Jesuraja, K., Elgorban, A. M., Ahmed, B., & Elzain, H. E. (2021). Microplastics and trace metals in fish species of the Gulf of Mannar (Indian Ocean) and evaluation of human health. *Environmental Pollution, 291*. https://doi.org/10.1016/j.envpol.2021.118089
53. De, K., Sautya, S., Dora, G. U., Gaikwad, S., Katke, D., & Salvi, A. (2023). Mangroves in the "Plasticene": High exposure of coastal mangroves to anthropogenic litter pollution along the Central-West coast of India. *Science of the Total Environment, 858*. https://doi.org/10.1016/j.scitotenv.2022.160071
54. Manickavasagam, S., Shukla, S. P., Kumar, S., Kumar, K., & Bhuvaneswari, R. (2021). Assessment of marine microplastics in floating plastic debris using a fixed sampling device: The example of South Juhu creek, Mumbai coast, India. *Journal of Coastal Conservation, 25*(1). https://doi.org/10.1007/s11852-021-00810-x
55. Nigam, R., Luis, A. J., Prasad, P., Kuttikar, S., Yadav, R., Vaz, E., & Kotha, M. (2022). Spatio-temporal assessment of COVID-19 lockdown impact on beach litter status and composition in Goa, India. *Marine Pollution Bulletin, 174*. https://doi.org/10.1016/j.marpolbul.2021.113293
56. Balasubramaniam, M., & Phillott, A. D. (2016). Preliminary observations of microplastics from beaches in the Indian Ocean. *Indian Ocean Turtle Newsletter (IOTN)*, January, 13–16. https://www.iotn.org/wp-content/uploads/2017/05/23-05-PRELIMINARY-OBSERVATIONS-OF-MICROPLASTICS-FROM-BEACHES-IN-THE-INDIAN-OCEAN.pdf
57. Saha, M., Naik, A., Desai, A., Nanajkar, M., Rathore, C., Kumar, M., & Gupta, P. (2021). Microplastics in seafood as an emerging threat to marine environment: A case study in Goa, west coast of India. *Chemosphere, 270*. https://doi.org/10.1016/j.chemosphere.2020.129359
58. Sulochanan, B., Veena, S., Ratheesh, L., Padua, S., Rohit, P., Kaladharan, P., & Kripa, V. (2019). Temporal and spatial variability of beach litter in Mangaluru, India. *Marine Pollution Bulletin, 149*(June), 110541. https://doi.org/10.1016/j.marpolbul.2019.110541
59. Yaranal, N. A., Subbiah, S., & Mohanty, K. (2021). Distribution and characterization of microplastics in beach sediments from Karnataka (India) coastal environments. *Marine Pollution Bulletin, 169*. https://doi.org/10.1016/j.marpolbul.2021.112550
60. Khaleel, R., Valsan, G., Rangel-Buitrago, N., & Warrier, A. K. (2023). Microplastics in the marine environment of St. Mary's Island: Implications for human health and conservation. *Environmental Monitoring and Assessment, 195*(9). https://doi.org/10.1007/s10661-023-11651-6
61. Unnikrishnan, V., Valsan, G., Amrutha, K., Sebastian, J. G., Rangel-Buitrago, N., Khaleel, R., Chandran, T., Reshma, S. R., & Warrier, A. K. (2023). A baseline study of microplastic pollution in a Southern Indian Estuary. *Marine Pollution Bulletin, 186*. https://doi.org/10.1016/j.marpolbul.2022.114468

62. Rukmangada, R., Naidu, B. C., Nayak, B. B., Balange, A., Chouksey, M. K., & Xavier, K. A. M. (2023). Microplastic contamination in salted and sun dried fish and implications for food security—A study on the effect of location, style and constituents of dried fish on microplastics load. *Marine Pollution Bulletin, 191*. https://doi.org/10.1016/j.marpolbul.2023.114909
63. Sajhikumar, K. K., Ragesh, N., Remya, R., & Mohamed, K. S. (2013). Occurrence of plastic debris in the stomach of yellowfin tuna (Thunnus albacares) from the Arabian Sea: A cause for concern. *Marine Fisheries Information Service—Technical and Extension Series, 217*(January), 13. https://doi.org/10.13140/RG.2.1.5181.2000
64. Kripa, V. (2014). Microplastics in the gut of anchovies caught from the mud bank area of Alappuzha, Kerala. *Marine Fisheries Information Service—Technical and Extension Series, 219*, 27–28.
65. Sruthy, S., & Ramasamy, E. V. (2017). Microplastic pollution in Vembanad Lake, Kerala, India: The first report of microplastics in lake and estuarine sediments in India*. *Environmental Pollution, 222*, 315–322. https://doi.org/10.1016/j.envpol.2016.12.038
66. Shylaja, G., Kripa, V., Prema, D., & Abhilash, K. S. (2018). Assessment of marine debris in the stake net fisheries of Vembanad Lake, Kerala, India. *Journal of the Marine Biological Association of India, 60*(1), 91–96. https://doi.org/10.6024/jmbai.2018.60.1.2044-14
67. Purushothaman, A., Vishnudattan, N. K., Nehala, S. P., Meghamol, M. D., Neethu, K. V., Joseph, J., Nandan, S. B., Padmakumar, K. B., & Thomas, L. C. (2023). Patterns and variability in the microplastic contamination along the southwest coast of India with emphasis on submarine groundwater discharge sites. *Marine Pollution Bulletin, 194*. https://doi.org/10.1016/j.marpolbul.2023.115432
68. Kannankai, M. P., Alex, R. K., Muralidharan, V. V., Nazeerkhan, N. P., Radhakrishnan, A., & Devipriya, S. P. (2022). Urban mangrove ecosystems are under severe threat from microplastic pollution: A case study from Mangalavanam, Kerala, India. *Environmental Science and Pollution Research, 29*(53), 80568–80580. https://doi.org/10.1007/s11356-022-21530-1
69. Daniel, D. B., Ashraf, P. M., & Thomas, S. N. (2020). Microplastics in the edible and inedible tissues of pelagic fishes sold for human consumption in Kerala, India. *Environmental Pollution, 266*, 115365. https://doi.org/10.1016/j.envpol.2020.115365
70. Daniel, D. B., & Thomas, S. N. (2023). Derelict fishing gear abundance, its causes and debris management practices – Insights from the fishing sector of Kerala, India. *Marine Policy, 148*. https://doi.org/10.1016/j.marpol.2022.105429
71. Daniel, D. B., Thomas, S. N., & Thomson, K. T. (2020). Assessment of fishing-related plastic debris along the beaches in Kerala Coast, India. *Marine Pollution Bulletin, 150*(November 2019), 110696. https://doi.org/10.1016/j.marpolbul.2019.110696
72. Dharani, G., Abdul Nazar, A. K., Venkatesan, R., & Ravindran, M. (2003). Marine debris in Great Nicobar. *Current Science, 85*(5), 574–575.
73. Goswami, P., Vinithkumar, N. V., & Dharani, G. (2020). First evidence of microplastics bioaccumulation by marine organisms in the Port Blair Bay, Andaman Islands. *Marine Pollution Bulletin, 155*. https://doi.org/10.1016/j.marpolbul.2020.111163
74. Mugilarasan, M., Venkatachalapathy, R., Sharmila, N., & Gurumoorthi, K. (2017). Occurrence of microplastic resin pellets from Chennai and Tinnakkara Island: Towards the establishment of background level for plastic pollution. *Indian Journal of Geo-Marine Sciences, 46*(6), 1210–1212.
75. Kaviarasan, T., Naik, S., Sivadas, S. K., Dhineka, K., Sambandam, M., Sivyer, D., Mishra, P., & Ramana Murthy, M. V. (2020). Assessment of litter in the remote beaches of Lakshadweep Islands, Arabian Sea. *Marine Pollution Bulletin, 161*. https://doi.org/10.1016/j.marpolbul.2020.111760
76. Sambandam, M., Dhineka, K., Sivadas, S. K., Kaviarasan, T., Begum, M., Hoehn, D., Sivyer, D., Mishra, P., & Murthy, M. V. R. (2022). Occurrence, characterization, and source delineation of microplastics in the coastal waters and shelf sediments of the central east coast of India, Bay of Bengal. *Chemosphere, 303*. https://doi.org/10.1016/j.chemosphere.2022.135135
77. Dhineka, K., Sambandam, M., Sivadas, S. K., Kaviarasan, T., Pradhan, U., Begum, M., Mishra, P., & Murthy, M. V. R. (2022). Characterization and seasonal distribution of microplastics in the

nearshore sediments of the south-east coast of India, Bay of Bengal. *Frontiers of Environmental Science and Engineering, 16*(1). https://doi.org/10.1007/s11783-021-1444-0

78. Athulya, P. A., & Chandrasekaran, N. (2023). Exposure of true to life microplastics to *Donax faba* under two different pH conditions: A microcosm approach. *Regional Studies in Marine Science, 67*, 103197.
79. Nithin, A., Sundaramanickam, A., Surya, P., & Kumar, T. A. (2023). Chromium (IV) Transfer to Amusium pleuronectes by LDPE Microplastics: An Experimental Study. *Journal of Hazardous Materials,* 131869.

Abundance of microplastics in the mangrove dominated Indian Sundarbans: A mini-review

Tapti Sengupta and Abhijit Mitra

Abbreviations

CFS	Containers Freight Station
ISBR	Indian Sundarban Biosphere Reserve
MPs	Microplastics
POPs	Persistent Organic Pollutants
RF	Reserve Forest

1 Introduction

Plastic is considered as a major waste in the oceans, seas, bays, and estuaries. The adverse impact of plastic on the biotic communities in the marine and estuarine ecosystem is a matter of great concern and is keenly related to widespread use of plastics that started around six decades back [1, 2], and is continuing in every sector of the human life. Plastics that are discharged in the marine and estuarine ecosystem originates from land-based usages like household or personal care products, hospital wastes, industrial effluents, etc., or are aquatic-based like nylon nets used for fishing activities, floats used in nets to maintain the buoyancy, etc. [3]. Because of the current pattern and tidal actions in the marine and estuarine systems, plastics have reached almost every corner of the world from poles to equator [4–7].

T. Sengupta (✉)
Department of Microbiology, West Bengal State University, West Bengal, Barasat, Kolkata 700126, India
e-mail: tapti.sg@wbsu.ac.in

A. Mitra
Department of Marine Science, Calcutta University, Kolkata, India
e-mail: abhijit_mitra@hotmail.com

M. Goel and N. G. Tripathi (eds.), *Plastic Pollution*,
https://doi.org/10.1007/978-981-97-5528-8_7

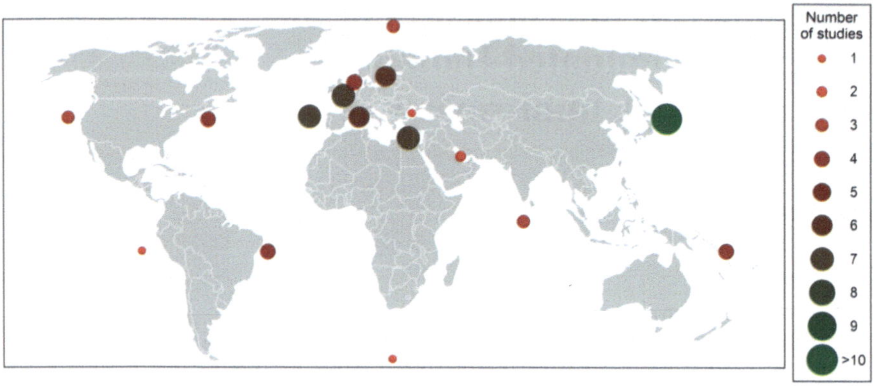

Fig. 1 Distribution of areas where the studies analyzed in the present review are conducted and number of studies (from 1 to >10) for each area is showed by the colored circles

According to Gago et al. [8] "Plastics with tiny dimensions ranging from 1 μm to 5 mm are referred to as microplastics (MPs)". MPs may float in the aquatic phase or are deposited on the sediment substratum or may accumulate within the animal body. MPs usually concentrate several contaminants on their surface such as persistent organic pollutants (POPs) and tend to be ingested by fauna prevailing in various niches. They are also transported through food webs and may adversely impact the human health preferably through consumption of sea foods [9, 10]. A total of 74 reference items have been found dealing in detail with MPs pollution in the last 5 years and most of these studies were carried out in the northern hemisphere. Figure 1 clearly depicts that no studies have been carried out on MPs in the Indian Sundarban Biosphere Reserve (ISBR) although this deltaic complex has a wide spectrum of flora and fauna for which it has been declared as the World Heritage Site.

The Indian Sundarban Biosphere Reserve (ISBR) with an area of 9630 km^2 is exposed to microplastic pollution from several sources like anthropogenic sources. The estuaries are the lifelines of ISBR which exhibit significant spatial variations in terms of MPs pollution.

The Western sector of ISBR is exposed to severe anthropogenic impacts compared to the central sector, which is adjacent to the reserve mangrove forest [11–15]. On this background the present study aims to understand the detailed scenario of microplastic pollution in the Western and Central sectors of ISBR along with its sources.

2 Methodology

Kakdwip and Sagar South stations in the western Indian Sundarbans are noted for maximum anthropogenic activities like fish landing related activities, shrimp culture-oriented activities, tourism, presence of busy market place, pilgrimage, and industrial activities from the adjacent Haldia port cum industrial belt Table 1.

Table 1 List of industries situated at Haldia port-cum-industrial complex

Sl. no	Name of industry	Product
1	Indian Oil Corporation Ltd., Haldia	L.P.G., motor gasoline, Naptha, ATF, MTO, HSD, JBO, kerosene, furnace oil, lubes, bitumen
2	KoPT/Haldia Dock Complex	Port services
3	Tata Chemicals Ltd., Haldia	Industrial phosphate and acids
4	Exide Industries Ltd., Haldia	Automotive batteries, heavy duty batteries, containers, special types of separators, etc.
5	Shaw Wallace, Haldia	Pesticides
6	MCC PTA India Corpn. Pvt. Ltd., Haldia	P.T.A.
7	Haldia Petrochemicals Ltd., Haldia	LLDPE, HDPE, Naptha Cracker, etc.
8	IOCL, Paradip-Haldia Oil Pipeline	Petroleum storage and transportation
9	IOC Petronas Ltd., Haldia	L.P.G.
10	Shamon Ispat Ltd.	Steel rolling
11	Dhunseri Petrochem and Tea Ltd.	Petroleum residues
12	Greenways Shipping Agencies Pvt. Ltd.	Containers freight station (CFS)
13	IOC Ltd., Haldia	Petroleum storage
14	Hindustan Petroleum Corporation Ltd.	Petroleum and allied products
15	Bharat Petroleum Corporation Ltd., Haldia	Petroleum and allied products
16	Hindustan Unilever Limited	Detergents
17	Marcus Oils and Chemical Pvt. Ltd.	Polyethylene waxes
18	IOC Ltd., Haldia	Petroleum storage
19	Ruchi Soya Industries Ltd.	Edible oil
20	Manaksia Ltd.	Aluminum and steel

(continued)

Table 1 (continued)

Sl. no	Name of industry	Product
21	Sanjana Cryogenic Storages Ltd.	Ammonia storage and handling terminal
22	R. D. B. Rasayans Ltd.	PP jumbo bag and small bag
23	Reliance Industries Limited	Storage and handling petroleum product
24	Adani Wilmar Ltd.	Edible oil refinery
25	Electrosteel Castings Ltd.	Coke oven plant, sponge iron plant, power plant
26	URAL India Ltd.	Automobile
27	K.S. Oils Ltd.	Edible oil refinery
28	DPM Net Pvt. Ltd.	Fishing net
29	Hooghly Met Coke and Power Co. Ltd.	Coke oven plant
30	Ruchi Infrastructure Pvt. Ltd.	3rd party liquid storage tank terminal
31	Shree Renuka Sugars Ltd.	Sugar refinery and food complex
32	Gokul Refoils and Solvent Ltd.	Edible oil refinery
33	Emami Biotech Ltd.	Bio-diesel plant
34	Ennore Coke Private Ltd.	Coke oven plant
35	West Bengal Waste Management Ltd.	Industrial waste/municipal waste management complex
36	Lalbaba Seamless Tubes Pvt. Ltd.	Seamless tube
37	Modern India Con-cast Ltd.	Ferro alloy plant
38	Rohit Ferro Tech Ltd.	Ferro alloy

Source Haldia Development Authority—An Autonomous Body under Government of WestBengal, India (www.nltr.org) [16]

The central Indian Sundarbans with the stations Bali and Chotomollakhali have large-scale shrimp culture farms and tourism of medium scale magnitude. Fishing vessels, trawlers and prawn seed catching activities using nets are pivotal source of microplastics in the estuaries of Indian Sundarbans. Based on the vulnerability of the scenario, four stations were selected in the Western and Central sectors of Indian Sundarbans, whose locations are stated here.

Coordinates:

1. Sagar island: 21°38″51.6′N; 88°02″20.9′E (Western Sector).
2. Kakdwip: 21°52″26.5′N; 88°08″04.5′E (Western Sector).
3. Bali: 22°04′35.17″N; 88°44′55.70″E (Central Sector).
4. Chotomollakhali: 22°10′40.00″N; 88°54′26.71″E (Central Sector).

The protocols for determining microplastics from surface water and sediments are given below in the flowchart.

Method of Estimation for MPs in Water Samples:

- Surface water sample were taken by using net with a 300 µm mesh and a 60 cm opening width

⇩

- Trawling was done with a ~2 knots speed for 15 minutes

⇩

- Samples were collected by clean glass jar

⇩

- 70% ethanol was added for preservation

⇩

- Sediment samples were taken by Ekman dredge from the top 5 cm of bottom sediments

⇩

- Samples were kept in clean glass journal Pre-proof 4 jars

⇩

- Water and sediment samples were further processed at the CPCB lab

⇩

- Water samples were sieved using a 300 µm custom sieve

⇩

- Sieved samples were collected in a 1.0 L glass beaker

⇩

- Samples were oven dried for 24 hrs. at 90°C

⇩

- $FeSO_4$& 30% H_2O_2 were poured into each beaker and then waited for 5 mins

⇩

- The mixtures were placed on a hotplate at 75°C until gas bubbles appeared

⇩

- 6 g of NaCl was mixed for 20 ml of sample to increase the concentration of the WPO solution

⇩

- 700 ml of filtered $ZnCl_2$ solution was added to the DS with the WPO solution and left for 24 hrs. with cover

⇩

- Cellulose nitrate filter paper was used to filter the DS solution

↓

- The filtrate was observed under an Olympus CX41 microscope with a camera from 4X to 100X magnification for the detection of MPs items

↓

- The quantity of MPs items in each sample was calculated for the final result

↓

- Normalized MPs particles in each sample was divided by the sampling area

↓

- The sampling area was calculated by multiplying the sampling distance (1 km) by the width of developed net opening (60 cm)

↓4

- The calculated results were normalized as per Km^2

Method for MPs Estimation from Sediment Samples

- 400 g of wet sediments were placed in a 1.0 L glass beaker for oven dried at 90°C for 24 hrs.

↓

- 300 ml of $ZnCl_2$ solution was added and stirred for 10 mins with a clean glass rod

↓

- All the particles were filtered with 300 μm custom sieve and collected into a beaker

↓

- 20 ml of $FeSO_4$ solution & 20 ml of 30% H_2O_2 were poured into each beaker and waited for 5 mins

↓

- The mixtures were placed on a hotplate until gas bubbles were appeared

↓

- 6 g of NaCl for each 20 ml sample was added

↓

- WPO solution was transferred into the DS using 700 ml of $ZnCl_2$ solution & left for 24 hrs.

↓

- Filtration was done by using cellulose nitrate filter (5 μm pore)

⇩

- Filters were viewed below an Olympus CX41 microscope with a camera from 4X to 100X magnification

⇩

- MPs data were calculated in Kg

India has been a land of MPs since long but studies on the various sites are limited. A total of 74 research articles were found, where most publications are from 2018, with 81% of all literature being published during the last five years (Fig. 1) and most studies were conducted in the Northern Hemisphere compared to the Southern Hemisphere, as already documented by Gago et al. [8]. Figure 1 clearly depicts those studies on MPs that have not been carried out in ISBR.

3 Results and Discussion

The data sets generated from the ground-zero observation confirm relatively more vulnerability of microplastic pollution in the western sector of ISBR as compared to the central sector. The western sector is exposed to effluents discharged from several industries in and around the region (Table 2). In addition to industrial effluents, waste generated from fish landing stations, brick kilns, shrimp farms, pilgrims, and tourism units also contribute to appreciable amounts of MPs along the Hooghly estuary flowing through Kakdwip and Sagar South. All these anthropogenic activities contribute to the discard of nets, breakdown of fibreglass from vessels and the discard of plastic bottles, wrappers and other plastic non-biodegradable debris, which eventually get deposited in the sediments and lead to a greater load of microplastics in these stations.

On the other hand, Bali and Chotomollakhali are least populated and have low fishing pressure and tourist load. Hence, the level of microplastics in sediments is also lesser than the stations in the Western sector of the ISBR. We presume that tourism and fishing pressure is directly proportional to the microplastic load in the aquatic phase and sediments. These have significant negative impact on the biotic community by ways of choking the gills of aquatic animals preferably fishes that are consumed by the people in and around the region as a part of their daily diet. In the food chain, bio-magnification to the higher tropic levels including human species causes toxic effects and reduced immunity thereby putting human health at risk to new infectious pathogens.

The overload of microplastics directly leads to the destruction of natural ecological niches of the flora and fauna and increases pathogenic pressure on the aquatic species due to decreased immunity and choking of gills.

Microplastic-related pollution in water and sediment disrupts the food chain and poses negative impact on the members of higher trophic level in the aquatic ecosystem, thereby causing a threat to sustainable goals. Sustainable fishery, being a subset of sustainable goal includes several agendas like better integration of fisheries management in the coastal zone, control of land-based activities that degrade the coastal and marine environment, increased awareness of factors beyond the conventional realm of fisheries management and stronger control of access to co-resources. To cover these agendas, we analyzed the scenario of the two sectors of Indian Sundarbans to trace the source of microplastics and observed more sources of MPs in the western sectors due to high frequency of anthropogenic footfall compared to the central sector (Table 1).

In the above figures (Figs. 2, 3, 4, 5 and 6) it is clearly obvious that MPs have become a part and parcel of human civilization. Much of these larger fragments are degraded and get washed away into the oceans and ISBR region due to tidal influx. This along with low levels of awareness among tourists and pilgrims have led to the deposition of these microplastics in the surface water and sediments, which in turn are causing a negative impact on the environment (Table 3).

From the graph it can be inferred that Kakdwip has the maximum load of microplastics in the soil sediments followed by sagar islands. Kakdwip and Sagar islands are known for their fishing activities, trawlers, and high load of tourism (Table 4).

The current figure depicts the load of microplastics in the surface water of the sampling stations. The MPs load in the study area followed the order Sagar Island > Kakdwip > Bali Island > Chotomollakhali.

Table 2 Spatial variations of different anthropogenic activities that generate MPs in Indian Sundarbans

Indian Sundarban Sector	Activity									
	Fish landing	Movement of Fishing vessels and trawlers	Passenger vessel related activity	Conditioning of boats, vessels, and trawlers	Shrimp farming	Industrial activity	Brick kilns	Tourism	Pilgrim related activity	Navigation
Western Sector										
Central Sector										

INDEX

	Maximum		Least
	Medium		Nil
	Less		No information

Fig. 2 Fencing using plastics in Sundarbans

Fig. 3 Plastic boats harbored at docks

Fig. 4 Pilgrimage and anthropogenic pressure at tourist hotspots

Fig. 5 Fiberglass being used rampantly

Fig. 6 Fishing nets being used by local fisherfolks and trawlers

Table 3 Distribution of microplastics in the sediments of the selected stations

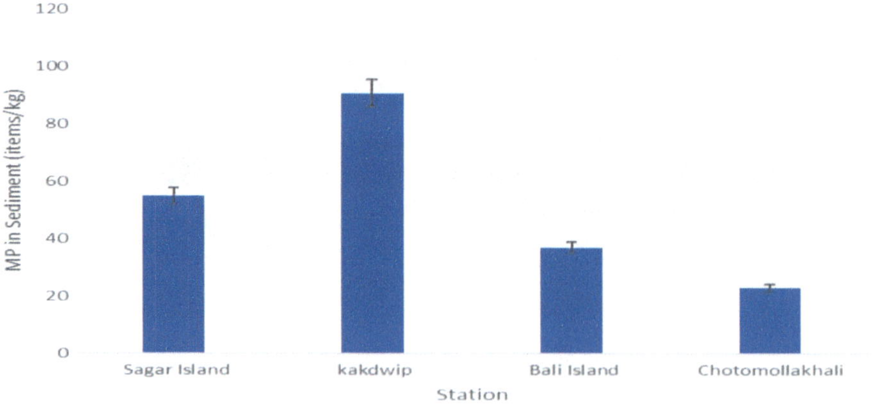

Table 4 Distribution of microplastics in the surface waters of the selected stations

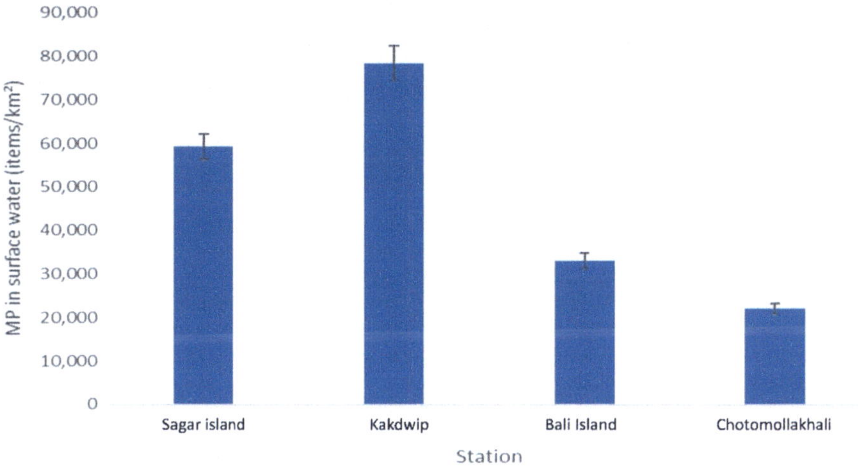

4 Conclusion

Microplastics are now known as a major health concern for human life and also aquatic and domestic animals. Uncontrolled use of plastic bottles, personal care products, polystyrene, polypropylene, and other fibre pollutants along with uncontrollable footfalls in the deltaic Ganges region is leading to a mass deposition of MPs in the surface waters and sediment of Indian Sundarbans. There is no major policy to restrict the usage of MPs in this region which ultimately reaches the islands of the delta complex due to tidal influx. Mangroves, being a very vulnerable ecosystem are now under a threat due to these microplastic debris. The present documentation serves as the baseline data bank on microplastic pollution in the ISBR. Implementation of strict policy coupled with frequent awareness programs can be an effective roadmap to retard the pace of MPs pollution in this mangrove-dominated delta complex, which is in alignment with sustainable development goals.

References

1. Barnes, D. K. A., Galgani, F., Thompson, R. C., & Barlaz, M. (2009). Accumulation and fragmentation of plastic debris in global environments. *Philosophical Transactions of the Royal Society B, 364*, 1985–1998. https://doi.org/10.1098/rstb.2008.0205
2. Thompson, R. C., Olsen, Y., Mitchell, R. P., Davis, A., Rowland, S. J., John, A. W. G., McGonigle, D., & Russell, A. E. (2004). Lostat sea: Where is all the plastic? *Science, 304*, 838. https://doi.org/10.1126/science.1094559
3. Duis, K., & Coors, A. (2016). Microplastics in the aquatic and terrestrial envi- ronment: Sources (with a specific focus on personal care products), fate and effects. *Environmental Sciences Europe, 28*, 2. https://doi.org/10.1186/s12302-015-0069-y

4. Cincinelli, A., Scopetani, C., Chelazzi, D., Lombardini, E., Martellini, T., Katsoyiannis, A., Fossi, M. C., & Corsolini, S. (2017). Microplastic in the surface waters of Ross Sea (Antarctica): Occurrence, distribution and characterization by FTIR. *Chemosphere, 175*, 391–400. https://doi.org/10.1016/j.chemosphere.2017.02.024
5. Derraik, J. G. B. (2002). The pollution of the marine environment by plastic debris: A review. *Marine Pollution Bulletin, 44*, 842–852.
6. Ivar do Sul, J. A., & Costa, M. F. (2014). The present and future of microplastic pollution in the marine environment. *Environmental Pollution, 185*, 352–364. https://doi.org/10.1016/j.envpol.2013.10.036
7. Lusher, A. L., Tirelli, V., O'Connor, I., & Officer, R. (2015). Microplastics in Arctic polar waters: The first reported values of particles in surface and sub-surface samples. *Scientific Reports-UK, 5*, 14947. https://doi.org/10.1038/srep14947
8. Gago, J., Carretero, O., Filgueiras, A. V., & Viñas, L. (2018). Synthetic microfibers in the marine environment: A review on their occurrence in seawater and sediments. *Marine Pollution Bulletin, 127*, 365–376. https://doi.org/10.1016/j.marpolbul.2017.11.070
9. Karbalaei, S., Hanachi, P., Walker, T. R., & Cole, M. (2018). Occurrence, sources, human health impacts and mitigation of microplastic pollution. *Environmental Science and Pollution Research, 25*, 36046–36063. https://doi.org/10.1007/s11356-018-3508-7
10. Rist, S., Carney Almroth, B., Hartmann, N. B., & Karlsson, T. M. (2018). A critical perspective on early communications concerning human health aspects of microplastics. *Science of the Total Environment, 626*, 720–726. https://doi.org/10.1016/j.scitotenv.2018.01.092
11. Mitra, A. (2013). *Sensitivity of Mangrove ecosystem to changing climate*. Springer. ISBN 978-81-322-1509-7 (eBook).
12. Mitra, A., & Zaman, S. (2015). *Blue carbon reservoir of the blue planet* (Vol. XII, pp. 299). Springer. ISBN 978-81-322-2106-7. https://doi.org/10.1007/978-81-322-2107-4
13. Mitra, A., & Zaman, S. (2016). *Basics of marine and estuarine ecology* (pp. 1–481). Springer. ISBN 978-81-322-2707-6.
14. Mitra, A. (2020). *Mangrove forests in India exploring ecosystem services* (Vol. XV, p. 361). Springer, Chem. ISBN 978-3-030-20595-9. https://doi.org/10.1007/978-3-030-20595-9
15. Mitra, A., & Zaman, S. (2021). *Estuarine acidification: Exploring the situation of Mangrove dominated Indian Sundarban Estuaries* (Vol. XII, pp. 402). Springer. e-Book. ISBN 978-3-030-84792-0. https://doi.org/10.1007/978-3-030-84792-0
16. Ahmed, S., Trivedi, S., Pal, N., Pramanick, P., Zaman, S., & Mitra, A. (2023). Study on floral carbon storage potential in the megacity of Kolkata: A roadmap towards net-zero carbon emission. *Journal of Environment and Sociobiology, 19*(2), 113–123.

Green Solutions and Technologies

Overview of Plastics, Usage, Pollution and Mitigation

Gautam Sen

1 Introduction

Plastic is a loose term for materials that are pliable and can be moulded into different shapes under heat and pressure. Natural plastics like horns, tortoise shells, rubber, amber and shellac have been in use, even during pre-industrial era. John Wesley Hyatt [1], in 1869 treated cellulose derived from cotton fibre with camphor and created plastics. This helped in overcoming manufacturing constraints set by nature. Leo Baekeland [2] in 1907 discovered Bakelite a polymer made out of monomers of phenol and formaldehyde. Bakelite was made without using any naturally occurring substance and hence is termed as the first total synthetic plastic. Bakelite is a good insulator, durable, heat resistant, and ideally suited for mechanical mass production. Bakelite could be shaped or moulded into almost anything, providing endless possibilities. Chemical companies started investing in the research and development of new polymers, following Hyatt's and Baekeland's successes. This gained momentum especially during second world war and post-war period.

Modern synthetic plastic is made through a process called polymerization of monomers such as ethylene, propylene, or even higher hydrocarbons in the presence of a catalyst and other additives. These polymers are called plastics. Utility of plastics has far exceeded the expectations when plastics were first discovered. Plastics have saturated our world, replacing traditional materials like glass in crockery, steel in cars and cutlery, paper and glass in packaging, and wood in furniture. Plastics are also used in computers, cell phones, and in modern medicine. Modern cars use plastics instead of steel, for plastics are light weight and help in saving fossil fuel. Plastics have created a utopian vision of a future with abundant material wealth.

G. Sen (✉)
Ex, RIL, Mumbai, India
e-mail: gautamsenindia@yahoo.com

Ex, ONGC, New Delhi, India

M. Goel and N. G. Tripathi (eds.), *Plastic Pollution*,
https://doi.org/10.1007/978-981-97-5528-8_8

Though plastics are extremely convenient to use because of its flexibility, light weight, strength, low costs and their easy disposability, they have generated around six billion tons of cumulative plastic waste over the years. The world is reeling under plastic waste for synthetic plastics are not bio-degradable. Single-use plastics used as grocery and retail store bags are worse and account for a significant component of the plastic waste. Plastics have polluted our landscapes, oceans and we breathe microplastics. Landfill plastics are a common sight and in spite of the usefulness of plastics, it also symbolizes as a cheap, and a flimsy product. Plastic debris in the oceans was first observed in the 1960s [3]. Today, there are ample evidences of sea animals dying because they ingested too much of plastics. The Great Pacific Garbage Patch [4], a swirl of plastic garbage the size of Texas, floating in the Pacific Ocean is the ultimate symbol of the problem of plastic waste.

Earth's rotation around its axis has created five gyres [5], namely North Pacific gyre which stretches from North America to Japan, South Pacific gyre, North Atlantic gyre, South Atlantic gyre and Indian Ocean gyre. Plastics which find its way to oceans through rivers, wind, storm water, beaches and other means ultimately end up in these gyres. They disintegrate due to prolonged exposure to ultraviolet rays and ocean currents and winds breaking up into microparticles. These microparticles are often not visible and are ingested by marine species including fishes. Human beings too are affected through fishes and other sea animals. Additives like Bisphenol A and a class of chemicals called Phthalates [6], which are used during the manufacturing process of some of the types of plastics, are harmful to human beings especially children.

Recycling of plastics, as a solution to plastic waste, has been tried, since the 1980s. Municipalities were encouraged to collect and process recyclable materials as part of their waste-management systems. However, globally recycling of plastics is still a mere 9–10% [7]. Recycling degrades the quality of plastics; hence, it is far from being the panacea. Incineration produces toxic chemicals hence it needs to be done under controlled environment. Globally only 11–12% of plastics are incinerated [7]. Most plastics therefore end up in landfills or in the environment. Since it's clear that plastics have a valuable place in our lives, scientists are attempting to make plastics safer and more sustainable. Research to find new plastic eating bacteria in the presence of enzymes offers hope for mankind to survive the menace of plastic pollution. Bioplastics are also being developed to produce plastics from plant crops instead of fossil fuels, for they are environmentally friendly. Conversion of plastics back to fossil fuels and using plastics in road constructions are other ways to minimize plastic pollution. Today, there is an acceptance of the reality, that plastics are not perfect yet are a necessary part of our future. Summing up, though plastics as a product, is nothing short of a miracle, with its low cost, durability, low weight, convertibility to required shapes, capability to withstand rough usage, aesthetical appeal, and recyclabslitty, yet the fact that they are non-biodegradable takes away all its advantages. Plastic accumulates as garbage and has polluted the earth so much, that we eat and breathe microplastics.

2 Chemistry and Usage of Synthetic Plastics

Modern plastics are entirely made of petroleum products. Naphtha is one of the ingredients produced during fractional distillation of crude oil. Plastics are made by cracking Naphtha, as shown Fig. 1.

Synthetic Plastics are polymers of monomers like ethylene, propylene or even higher hydrocarbons, etc., derived from cracking naphtha, which is obtained from fractional distillation of crude oil. Modern plastics are then synthetically prepared by combining a large number of monomers ethylene or propylene or higher hydrocarbons, as the case may be, using high pressure and temperature in the presence of a catalyst. Additives like oxygen, nitrogen, chlorine and sulphur in different proportions are used to modify the properties of the final product, lending variety to plastics. This process is known as polymerization and this creates long chains of atoms, arranged in repeating units. It is the length of these chains and the patterns in which they are arrayed that make polymers strong, lightweight, and flexible, and makes them exceptionally useful. Millions of polymer chains, called resins are formed at the same time. Polyethylene resins, differing in density and weight are sold to plastics factories, as powder, tiny granules, or pellets. Plastic granules or resins are moulded under high temperature and pressure. After cooling the mould is opened and plastic is formed.

There are two types of plastics.

Thermoplastics: They melt when heated, and harden again when cooled and can be remoulded.

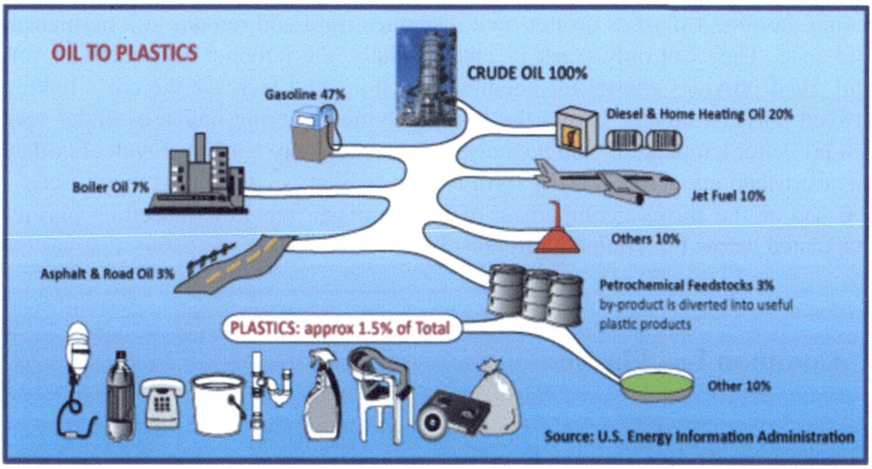

Fig. 1 Fractional distillation of crude oil to produce various ingredients including petrochemicals

Table 1 Compiled by author

Thermo-plastics	Thermosetting plastics
Covalent bond between monomers, weak Van Der Wall interaction between molecular chains	Strong cross-links and a 3D network of covalently bonded atoms
Processed by injection moulding extrusion process, blow moulding thermoforming process and rotation moulding	Processed by compression moulding, reaction injection moulding
Low molecular weight as compared to thermosetting plastics and low melting point, low tensile strength	High molecular weight and high melting point, high tensile strength
Flexible	Are stiff for stiffness of plastics increases with the number of cross links in the structure

Thermosetting Plastics: They do not melt and only at very high temperatures cracks or char. They are therefore ideal for high-heat applications such as electronics and appliances. Eventually, they also end up in creating e-waste.

Thermoplastics, i.e., Polyethylene, Polypropylene, Polystyrene and Poly Vinyl Chloride (PVC) constitute around 4/5th of the plastics produced. Thermoplastics have long, linear polymer chains that are weakly chemically bonded (covalent bonds), to each other. When a thermoplastic object is heated, these bonds are broken. They can then again be remoulded and reshaped into other products. They also have low melting points.

On the other hand, in Thermosets the linear chains are cross linked—strongly chemically bonded in three dimensions. These bonds cannot be broken easily. On heating thermoset plastics do not melt, becomes rigid and remain in a permanent solid state. They can only crack or char. Initially, the polymer is a liquid or soft solid. Heat provides energy for chemical reactions that increase the cross-linking between polymer chains, curing the linear polymers. Curing increases under very high pressure, temperature and a catalyst. Thermosets are hard to recycle, but they can be crushed into a fine powder form for use as fillers in reinforced thermosets.

Based on the above a comparison of Thermoplastics and thermosetting plastics is tabulated below for a quick reference (Table 1).

3 Common Use Plastics

3.1 Thermoplastics

Thermoplastics can be categorized as either semi-crystalline or as amorphous plastics depending on their molecular structure. In the former the long polymer chains are arranged in a regular and in an ordered manner while in the latter they are arranged in a haphazard manner within the material structure. As a result of the orderly arrangement, semi-crystalline thermoplastics have the following properties.

- A clearly defined melting point.
- High rigidity and toughness.
- Excellent chemical resistance.
- Hard-wearing materials that are tough and resistant to abrasion.

Polyethylene (PE), Polypropylene (PP), Polybutylene (PB), Polyethylene Terephthalate (PET), Polyoxymethylene (POM), Polyamide (PA), Polyketone (POK) are typical examples of semicrystalline thermoplastics.

Amorphous thermoplastics do not have clearly defined melting temperatures, but rather begin to soften while being heated. This occurs because different parts of the polymer structure melt over a range of temperatures.

Polymethyl methacrylate (PMMA). Polysulfone PSU), Polyphenylsulfone (PPSU), Polyethersulfone (PESU), Cyclic Olefin Copolymer (COC), Polycarbonate (PC), Acrylonitrile Butadiene Styrene (ABS) and a blend of PC/ABS are common examples of amorphous Thermoplastics. Common Thermoplastics and their usages are enunciated below [8].

Polyethylene (PE), is a polymer of the monomer ethylene C_2H_4. It is the most common plastics in the world. High-density Polyethylene is strong and resistant to moisture and chemicals, which makes it ideal for cartons, containers, pipes and other building materials. Milk cartons, detergent bottles, cereal box liners, toys, buckets, park benches and rigid pipes are other examples of high-density polyethylene, and they are in everyday use. Low-density polyethylene is used as a liner inside beverage cartons, and in corrosion-resistant work surfaces and other products like sandwich and bread bags, bubble wrap, garbage bags, grocery bags and beverage cups.

Polyethylene Terephthalate (PET or PETE), is a polymer of the monomer $C_{10}H_8O_4$, is lightweight, strong, typically transparent and is often used in food packaging and fabrics like polyester, beverage bottles, food bottles/jars and polyester clothing or rope.

Polypropylene (PP) are polymers of monomers C_3H_6, used in food packaging and food storage. It's flexible enough to allow for mild bending, but it retains its shape and strength for a long time and is used in making Straws, bottle caps, prescription bottles, hot food containers, packaging tape, disposable diapers and DVD/CD boxes.

Polymethyl Methacrylate (PMMA/Acrylic) are polymers of the monomer $C_5H_8O_2$. These are transparent plastics used in lenses, acrylic nails, paint, security barriers, windows, medical devices, enclosures around exhibits, etc. Acrylic does not contain the harmful bisphenol A. It is inexpensive, readily available, scratch resistant, and is a good alternative to polycarbonates when strength is not required.

Acrylonitrile–Butadiene–Styrene (ABS) is a common amorphous polymer made by polymerizing styrene and acrylonitrile in the presence of polybutadiene. Its chemical formular is $(C_8H_8)_x \cdot (C_4H_6)_y \cdot (C_3H_3N)_z$. Its properties can be tuned by varying the ratio of its monomer units and can be prepared for different grades. The proportions can vary from 15 to 35% of acrylonitrile, 5 to 30% butadiene and 40 to 60% styrene depending on the end usage. ABS has impact resistance, toughness and rigidity. ABS is used in computer keyboards, LEGO bricks, face guards for wall sockets, automobile industry, plastic alloys, construction industry, plastic tubing, corrugated plastic structure, helmets, printers, vacuum cleaners, kitchen utensils, musical instruments, toys, 3D printing, etc. It is inexpensive aesthetically pleasing, and can be tailor-made. Few disadvantages are that it is inappropriate for high temperatures, less tolerant to UV radiation, and pollutes when incinerated.

3.2 *Thermosetting Plastics*

Common examples of thermoset plastics include the following; Vulcanized rubber, Fibre glass (a fibre-reinforced polymer composite), Polyester resin, Melamine, Bakelite Epoxy, Silicone, Polyurethane and phenolic are other examples of plastics [9] in this category. Epoxy resin is a resin ideal for repairing items such as vases, ornaments, figurines, etc. Silicone is used as sealants, adhesive, lubricants, medicine, cooking utensils, thermal and electrical insulation. Polyurethane is used as insulation of refrigerators and freezers, as building insulation. Phenolic resins are used in brake pads, billiard balls, coatings for metals to strengthen, etc.

Summarizing, plastics are a sought-after commodity by the industry with its use in varied sectors like building and construction, packaging, consumer industry, electronics, automobiles, etc. Packaging alone is responsible for 36% of plastic usage globally, but contributes around 50% of global plastic wastage [10]. In India plastics are mostly used in packaging industry. Flexible packaging accounts for 42% of plastic consumption in India and building construction accounts for 13%. India is one of the highest plastic polluters and plastic pollution will only increase with the growth in Indian economy. A pie chart showing the different sectors where plastic is used is shown in Fig. 2.

3.3 *Bioplastics*

Three common sources of Bioplastics are corn starch, cellulose and protein [10]. Fifty percent of the bioplastics are produced from corn starch and are used for the production of drug capsules, loose fill foams for transport packaging and in service ware like cups, plates and cutlery. Starch-based bioplastics exhibit high strength, thermal stability, water solvent resistant, self-healing ability and are degradable. Arrowroot starch, tapioca starch, corn starch, potato starch, rice flour and potato flour are common sources of starch-based plastics. Cellulose-based bioplastics also called cellulose acetates are produced from cotton pulp or wood liners. They are used for manufacturing of films, moulding and extrusion compounds, fibers and lacquers. This bioplastic can be reused, recycled and is environment friendly. Protein-based bioplastics are made from wheat gluten, casein, milk and soya. Soya proteins have been used in plastic production for over one hundred years. Even the body panels of an

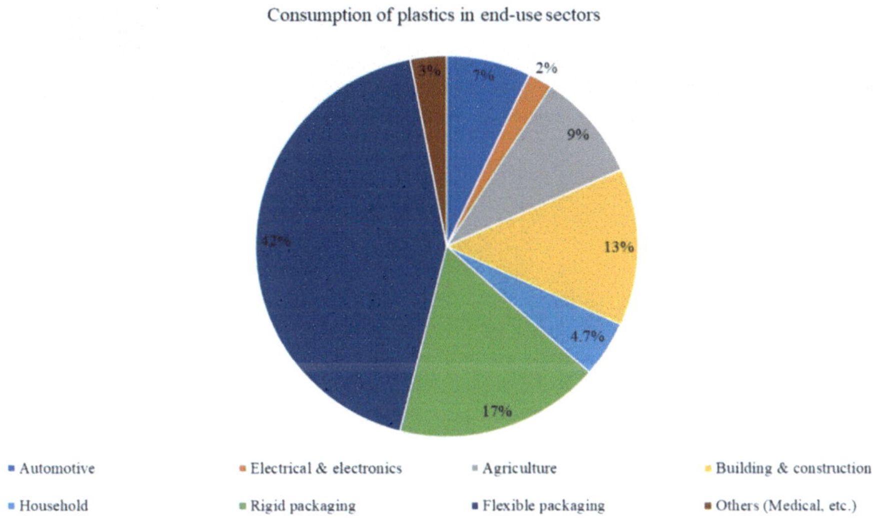

Fig. 2 Distribution of plastic use sector wise in India 2018 [11]

original Ford automobile were made of soya-based plastic. Proteins have amino acids and through crosslinking of amino acids by a certain enzyme, protein bioplastic can be created which are biodegradable, have nutritional value, is water resistant, edible and flexible and are primarily used in food industry. China, US, Germany, Brazil and Canada are the five largest producers of bioplastics. Indian bioplastic production is still in its initial days however its market is projected to grow at a rate of 24% in the next 5 years. Microorganisms present in fertile soil rich in bacteria can disintegrate most of the bioplastics into carbon dioxide and water.

3.4 Plastic Production

The use of plastics across a plethora of industries has resulted in an exponential increase in demand for plastics, especially after the second world war. Synthetic plastics are produced from fossil fuel, removing any natural constraints, thereby supply of plastics could keep pace with the demand. As a consequence, global plastic production has increased from 2 million metric tons in 1950 to 458 million metric tons in 2022 [12]. The major spurt in growth in plastic production happened post-1980 when Asia particularly China and then India, which together makes up 33 pc of the World's population, started developing. Even single-use plastic production annually is around 150 million metric tons. Cumulative plastic production since 1950 is over 10 billion tons and is projected to go up to 34 billion tons by 2050. The growth of global plastic production is shown schematically below in Fig. 3.

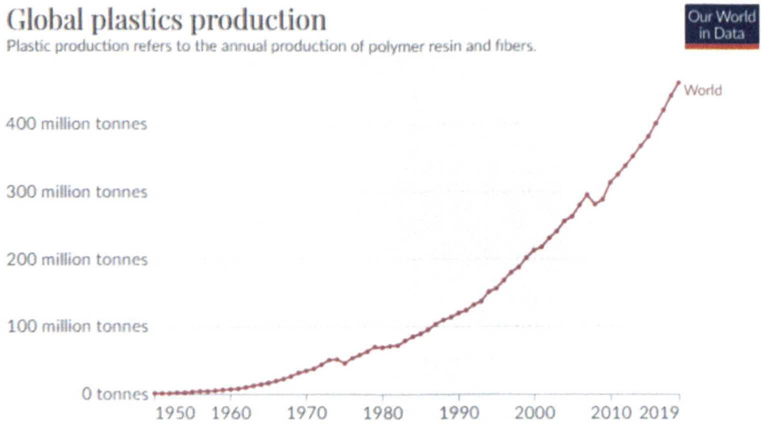

Fig. 3 Exponential growth in plastic production [12]

China accounted for 32% of global plastic materials production in 2021, making it the world's largest plastic producer. North America was the second-largest plastics producing region that year, accounting for almost 20% of global production. As per the latest data India consumes over 21 million metric tons annually and produces 9.3 million metric tons. Plastic production region wise is shown in Fig. 4.

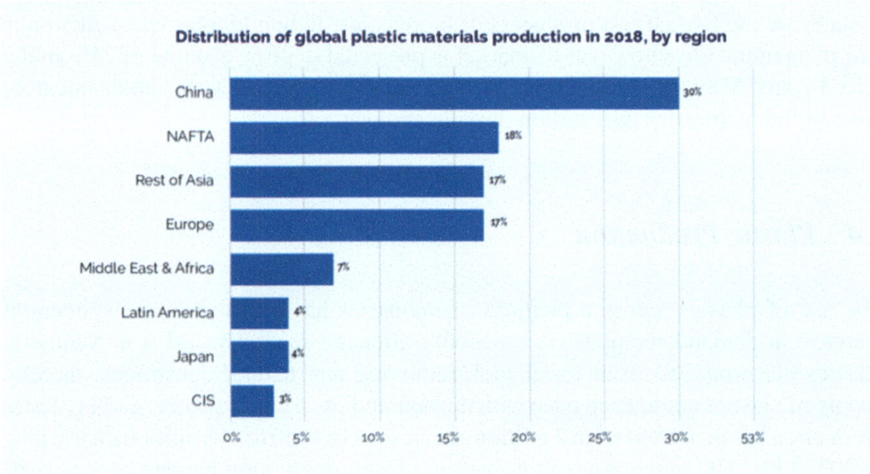

Fig. 4 Region wise plastic production [13]

4 Plastic Waste Generation

Demand driven, exponential growth in synthetic plastic production including single-use plastics has resulted in the accumulation of huge plastic wastage (Fig. 5). Plastics have an average life cycle of around ten years while it takes up to 500 years for it to decompose depending on their composition and disposal. This has created a total imbalance. OECD (Organization for Economic Cooperation and Development) countries, with a very high per capita consumption of plastics also trade plastic waste to less developed countries, aggravating plastic waste generation. As per UNEP (United Nations Environment Program) report 2021 [14], out of 9.2 billion tons accumulated production till 2017, 5.3 billion tons have accumulated as plastic waste. 2.9 billion tons are in use and only 700 million tons could be recycled. Cumulative plastic production is projected to go up to 34 billion tons by 2050. In a business-as-usual scenario this could easily generate over 20 billion tons of plastic waste. This is schematically shown below.

As per OECD (organization of developed countries) global plastic output report 2022 [7] in the year 2019, out of 460 million tons of plastics available globally only 29 million tons were recycled plastic. 353 Mt of plastic waste was generated of which 174 million tons went into landfill and 79 million tons were mismanaged. Even a large share of the plastics in landfill will also end up in being mismanaged. This is shown in Fig. 6.

In an article authored by Bruna Alves [15], in Statista dated August 2023, Plastic production worldwide today is 390.7 million tons and global plastic waste generation is 353.5 Mt and plastic waste collected for recycling is around 15%. Recycled plastics are around 9–10% of total plastic waste generated accounting for the losses.

The World today is therefore battling with around 6 billion tons of accumulated plastic waste which can go up to 20 billion tons by 2050, in business-as-usual scenario. This plastic waste is ubiquitous in our planet, in landfills, in water drains, in tap waters, in the atmosphere, in rivers and other water bodies, in beaches, in mountains and even in remote corners of this planet like in Arctic, in Antarctic, in uninhabited islands. A large proportion, around 10 million tons, of plastic waste finds its way to the oceans annually. At least 80–90% of the plastic waste found in oceans are from land.

Plastic waste has been found even in the deepest trenches, in oceanic sediments and in sea beds. There are approximately 50–75 trillion pieces of plastic and microplastics currently in the ocean [16]. Approximately 593,043,485 pounds of plastic pollution float on the ocean's surface. Only around 1% of all plastic pollution in the oceans is on the surface. The other 99% stays below the surface. As per United Nations report if this trend continues, there will be more plastics than fish in oceans by 2050. Plastics have become a source of great danger to all marine lives. Large marine animals like whales ingest large quantities of plastic together with the water they use to collect their food. Turtles mistake plastic bags for their favourite food, jelly fish. Sea birds feed their chicks with plastic that they find on the ocean surface. Many marine animals

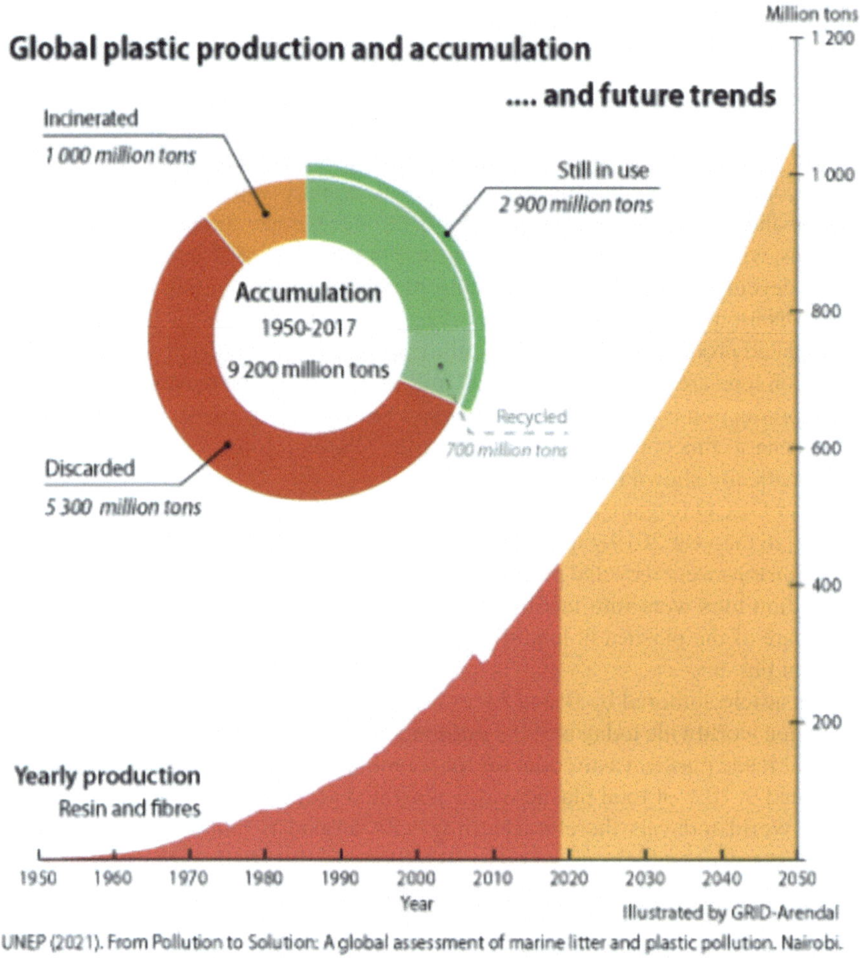

Fig. 5 Global plastic production, accumulation and future trends [14]

die because of starvation as the accumulation of plastic in their stomachs and their digestive tracts gives them a false sense of being full.

Most of the plastic waste in oceans eventually lands up in the five gyres mentioned above with the North Pacific gyre being the most polluted. The ocean gyres are circulating systems of currents where the water moves sluggishly. The slow currents create a vortex that collects all the plastic from the whole ocean and locks it in. Once trapped there, no plastic leaves those gyres again. Under the influence of solar UV radiation, wind, currents and other natural factors, plastic breaks down into small particles called microplastics (particles smaller than 5 mm) or nanoplastics (particles smaller than 100 nm). Microparticles are ubiquitous today and are found in sea beds (higher concentration), in tap water, in beer, salt and in sea ice. Melting of ice caps

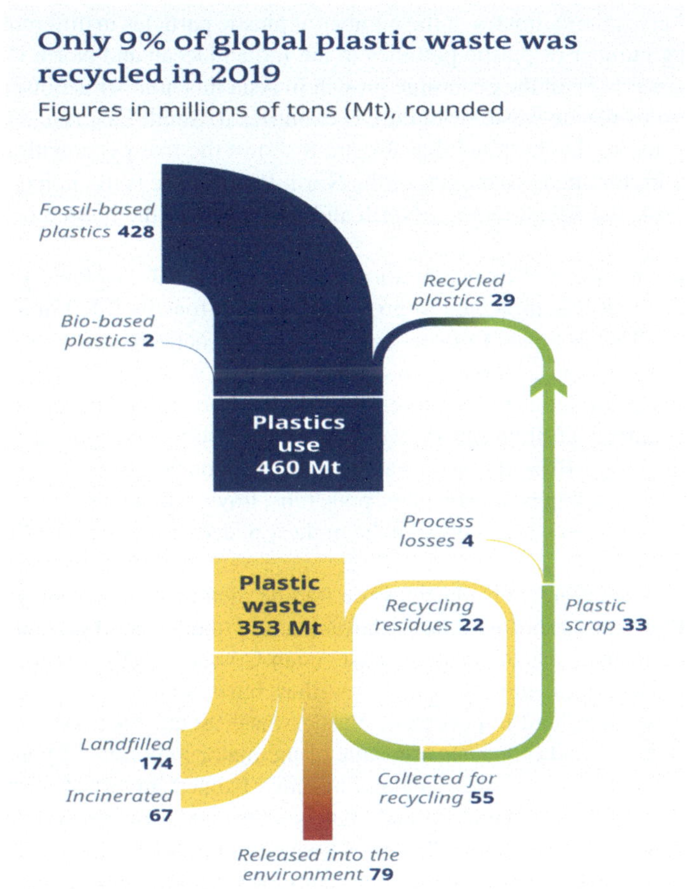

Fig. 6 Plastic Cycle a schematic diagram [7]

will further increase their concentration. They form a kind of a thin plastic soup within the five gyres. North Pacific gyre has attracted headlines because of its sheer size. It's nearly impossible to clean up the gyres. First, because of their distance from mainland. Second, because most of the debris floats below the surface or sinks to the ocean floor. And third, because a big percentage of the debris is microplastic particles too small to be caught in nets.

Microplastics affect marine ecology, their presence alters the ecological equilibrium. These small particles are released into the ocean and often contain toxic chemicals like Bisphenol A, Phthalates, DDT, and polychlorinated biphenyls [6]. Marine animals consume and breathe in these microplastics. These substances are then consumed by humans through seafood. They are known to cause problems with the immune systems, hormonal imbalances, liver disease, and issues with childhood development.

Figure 7 gives the estimates of the number of plastic particles in different Oceans. The growing number of plastic particles in the Indian Ocean and North Pacific is a matter of concern. With the economic growth in Asia this number is poised to grow more. Export of plastic waste from OECD countries to South East Asian Countries also is reflected in this high number. Figure 8 shows the five gyres with the North Pacific gyre highlighted. To the left of the North Pacific gyre is the Indian ocean, to the right of it is the Atlantic gyres, North and South, and to the bottom is the South Pacific gyre.

Plastic production and its incineration contribute to 3.5% of global greenhouse gas emissions and it is projected to go up to 3 billion tons by 2050 in a business-as-usual case. This is a matter of concern for achieving net zero emission. Although China is the world's largest plastic-producing country with a 32% share of global production, it is the United States which produces more annual plastic waste than any other country.50 billion plastic water bottles are sold each year in the United States. A staggering **46%** of plastic waste comes from **packaging** comprising items such as bottles, wrappings, plastic bags, pots, tubs, trays, bubble wrap, rubbish bags, etc. Top few countries which generate the highest plastic waste are listed in Table 2 [18]

Another area of concern is plastic waste trading mentioned above. In spite of the fact that OECD countries have a better infrastructure to manage plastics they export a large share of their plastic waste to East Asian Countries. Till recent past China was one of the biggest importers and now they have banned it. Figure 9, shows the leading exporters and importers of plastic waste. South East Asia particularly Vietnam, Malaysia and Indonesia are some of the major importers of plastic waste. India too imports plastic waste. Thailand has now banned import of plastic waste; Vietnam is also going to ban imports and Malaysia has issued temporary halts. Almost all the world nations have signed the Basel convention treaty restricting the trade in hard-to-recycle waste plastics. Some of the imported plastic waste is recycled some are used in road construction, etc., but a large portion is not managed well, or land

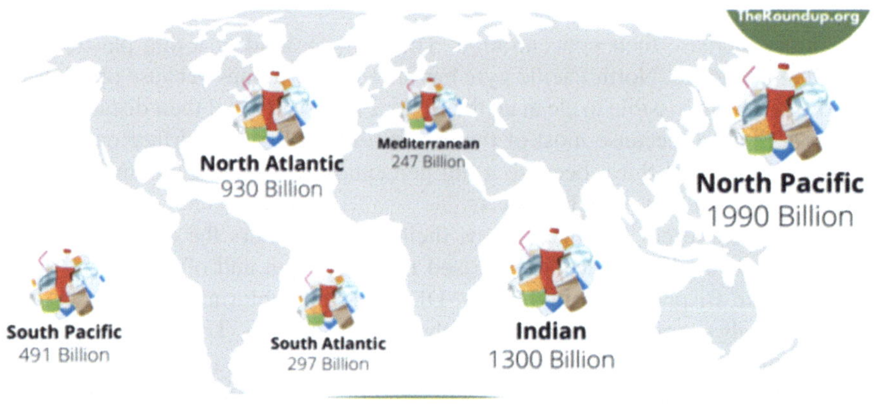

Fig. 7 Pieces of plastic pollution in our ocean [17]

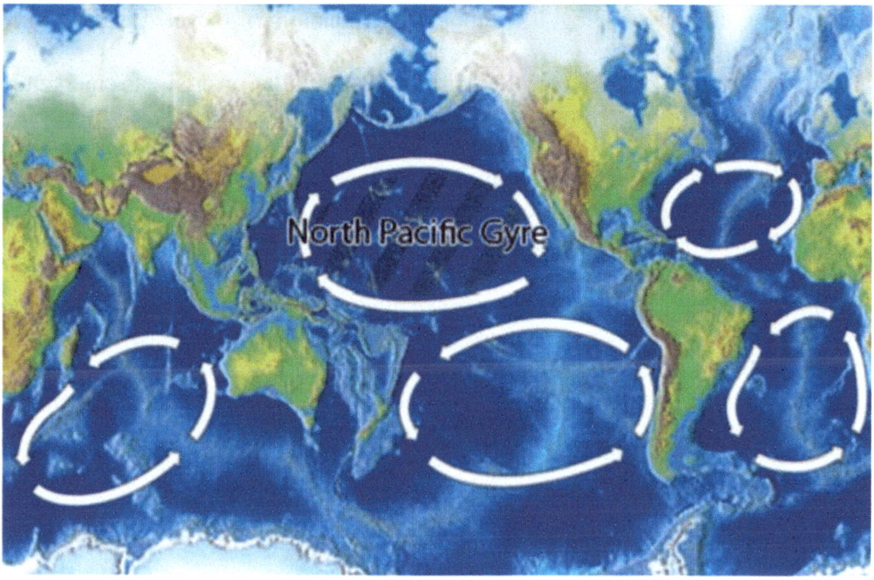

Fig. 8 The five ocean gyres with North Pacific highlighted [5]

Table 2 Top few plastics wastes generator

United States	34.02
India	26.33
China	21.60
Brazil	10.68
Indonesia	9.13
Russia	8.47
Germany	6.68
United Kingdom	6.47

up in landfills or even incinerated. Incineration of plastics can produce toxic gas and therefore ideally, it needs to be carried out under control conditions. Figure 10 depicts how OECD and non-OECD countries are managing their plastic wastage. Developed countries incinerate their plastic waste far more and therefore there is less quantum of mismanaged plastics per total plastic waste. Yet, it is the OECD countries which generates half of the global plastic waste., Plastic pollution per person is by an order of magnitude more in OECD countries as compared to non-OECD countries and today the world is producing far more plastic than what they can control, with waste management and recycling falling far short of requirement, causing exponential increase in plastic pollution. A study was also conducted to estimate the global plastic waste in future [21] in a business-as-usual scenario. Results of the study show that the plastic waste will triple by 2060, though the quantum of recycled plastics may go up marginally (Fig. 11).

United Nations in 2023 has accorded global plastic pollution crisis, an emergency status. They are still hopeful that pollution could be drastically reduced by cutting down the demand on plastics, increase reuse and recycling and wherever feasible replace plastics with greener alternatives. Recycling also has its own hazards, unless we can upscale. A global Plastic Waste Partnership (PWP) has been established under the Basel Convention in 2017 to promote environmentally sound management (ESM) of plastic waste at the global, regional and national levels and to prevent and minimize its generation. It will also act as a regulatory body for trade of plastic waste. It is pertinent to mention here, that there are 13,000 known chemicals associated with plastics of which 3200 are of major concern to all living beings. A financial estimate of the costs incurred due to plastic pollution is around 300–600 billion dollars annually.

Fig. 9 Import and export of plastic waste in tons 2020 data [19]

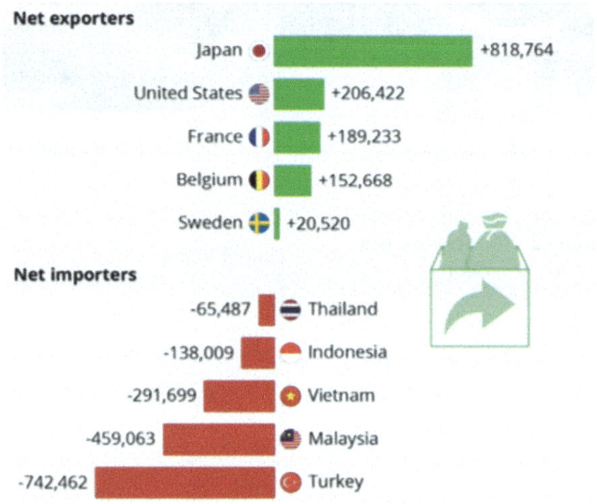

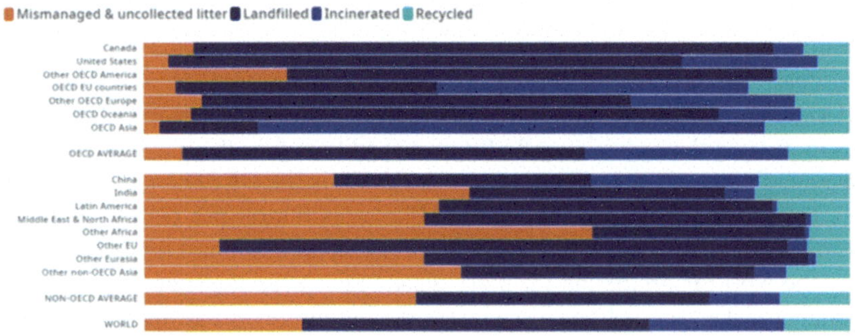

Fig. 10 Global plastic waste management scenario [20]

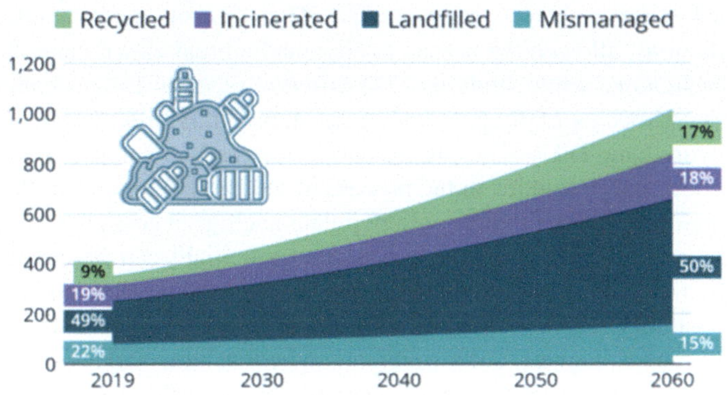

Fig. 11 Estimated global plastic waste in future [21]

Thus, controlling the menace of plastic pollution like global warming has become the greatest challenge to the very existence of living beings.

5 Scientific Solutions to Mitigate Plastic Pollution

Nature has its own solutions for preservation. Different combination of enzymes secreted by bacteria, wax worms, and fungi, can ingest certain types of plastics. Globally extensive research is happening in discovering such combination of enzymes. Some of the success stories have been described below.

Polyethylene terephthalate (PET) is the most commonly used plastic in drinking bottles and polyester clothing. They are being recycled but each such cycle degrades the quality of plastic. In 2016, a Japanese scientist [22] discovered an enzyme PETase in 2016 that was capable of breaking down Polyethylene terephthalate (PET). PET consists of long strands made from monomers of ethylene glycol and terephthalic acid. Ideonella Sakaiensis 201-F6, is a bacteria that can digest plastic by secreting PETase. As the enzyme breaks down plastic the bacteria produce another enzyme MHETase which further breaks the pieces the combination of these two enzymes is called super enzyme. A combination of DNA of PETase and MHETase makes the breakdown process faster. The challenge is to use such combination of enzymes or its DNA to break down the pieces into their constituent's ethylene glycol and terephthalic acid which could be combined to make new PET.

Researchers have discovered that wax worms, or moth larvae [23] that live in the honeycombs of beehives, have two enzymes in their saliva and these enzymes can breakdown polyethylene, even at room temperatures within few hours. The chance discovery came about as an accident in 2017. Federica Bertocchini, noticed that beehives were plagued with wax worms. When she put these worms in a plastic bag, she noticed lots of holes in the bag because worms were actually breaking down the plastics. The key was in the wax worms' saliva. Their enzymes, or proteins that speed chemical reactions, allowed the insects to chemically break down the polyethylene.

Samantha Jenkins, while studying [24] different types of fungus came across a dark-pigmented species of fungus, known as *Aspergillus Tubingensis*, which can degrade polyurethane (PU). The plastic-eating fungus had eaten a plastic sponge that was used to seal it. Jenkins in the process of testing this fungus on PET and PU plastic also discovered the fungus could populate itself, as it consumes more plastic.

Scientists have created tiny magnetic coils, that potentially can target microplastics in the ocean. This experimental nanotechnology is able to break down microplastic in the water without causing any harm to marine life. Xiaoguang Duan, a co-author of this study created magnetic coils, thinner than human hair, made out of carbon and coated with manganese which are magnetic and nitrogen [25]. These coils resemble bed springs under a microscope. These chemicals produce reactive oxygen molecules under water which oxidizes plastic. This study is still in the research stage.

Australian company Licella Holdings has developed a new patented technology, known as the Catalytic Hydrothermal Reactor (Cat-HTR), that can convert unrecyclable plastic into oil [26]. There is no need to separate the different plastic types and it can include end-of-life plastic. It allows plastic waste to be recycled over and over again and on a commercial scale. This company could convert 20,000 tons of plastic waste annually. However, critics have labelled the technology as an environmental trade-off as the process may produce further carbon emissions.

Indonesian start-up company Evoware [27] has been researching ways of converting seaweed into a bioplastic. They work with local seaweed farmers to create a range of different types of packaging such as sandwich wraps, burger wraps, sachets for spices, and soap, which can be dissolved in hot water, and in some cases, edible. Indonesia produces 10 million tons of seaweed each year and could reach up to 19 million tons by 2020, which could help supply Evoware's expanding efforts.

Plastic waste can also be converted into other useful products like construction materials including bricks, furniture, carpets, feedstock chemicals, etc. Plastic waste is being utilized in road construction. Plastics are more durable than asphalt and require less heavy equipment and time to install, which makes its carbon footprint smaller.

6 Users' Responsibility

Apart from various scientific solution to reduce the existing plastic pollution global users also have their share of responsibility for at least minimizing the menace of plastic pollution. Users can mitigate the problem incase the following suggestions are accepted;

- Reducing the use of single-use plastics includes plastic bags, water bottles, straws, cups, utensils, dry cleaning bags, take-out containers, microbeads, etc.
- Replacing single-use plastic by bioplastic to the extent possible.

- Supporting Legislation to Curb Plastic Production and Waste, Awareness.
- Adopting circular economy, recycling, reusing.
- Beach and river cleaning. Maximum weathering of plastics happens in beaches.

7 Plastic Waste in India

India s plastic demand in 2021–22 was around 21 million tons and is projected to go up exponentially. Plastic waste has consequently quadrupled in India in the last 5 years though the industry is expected to adopt a circular economy for it can open new opportunities for growth and innovation as well as minimize waste. Increased urbanization, growth in retail outlets with grocery packages, food packages besides consumables have all added to the plastic growth. India has an ineffective waste management due to lack of sorting at source, leading to improper collection of waste. This makes managing, recycling and disposing waste difficult. As per Central Pollution Control Board there are around 5000 registered units that are engaged with plastic management besides a large number of informal sectors, many of whom lack basic disposal techniques and eventually most of the plastic waste lands up in landfills. Also, a substantial amount of plastic waste is contaminated that cannot be recovered and eventually disposed in landfills. 20% of plastic waste is either incinerated or used in road construction and the rest is either recycled or is disposed in landfills. Plastics often clog water bodies and drainage systems in urban India resulting in major floods after heavy rainfall.

According to UNEP, more than 15000 Mt (most of it is plastic) of waste are dumped on south Asian seas generated from 60 major Indian Cities. India's coastline spans over 7500 km with a population of 420 million living in this part, marine plastic waste is set to grow bigger, most of which eventually lands in the gyre in Indian Ocean and gets broken into microplastics.

8 Recommendation and Conclusions

Plastics have been a revolutionary discovery and have changed the lifestyle of human beings. Their use in multifaceted activities starting from the simple straw to real estate, transport, infrastructure, consumer industry, furniture and fixtures, health, food and packaging industries have resulted in an exponential growth in plastic production after second world war. Most of these plastics are synthetic and are generated from fossil fuels. As a result, there is no natural material constraint in their growth. Plastics can be broadly classified into two types, both being polymers. Within each type there are variations, and different plastics have different chemical compositions and different usages. The bad news is that synthetic plastics do not biodegrade and as a result today the earth has accumulated around six billion of tons of plastic waste and they are everywhere starting from mountain summits to the deepest trenches, in polar regions within permafrost in air, land and in oceans. Today all living beings breathe and eat microplastics. This can potentially have a very harmful effect on the biodiversity and ecology. The biggest challenge is to mitigate this nuisance of plastic pollution and together with effects of climate change, this can potentially make this planet uninhabitable. Recycling of plastics can only be accomplished in a limited scale for quality of plastics deteriorate with each recycle and fresh plastics are often cheaper. Researchers have discovered enzymes and their combinations, secreted by bacteria, wax worms, fungi, etc., which can potentially eat plastics. Nanotechnology in the form of thin magnetic coils can potentially reduce plastic pollution in oceans. Much more research is essential to identify the right combination of enzymes for different types of plastics for there is no one single solution for all plastics. Usage of artificial intelligence and a detailed database could help in identifying such super enzymes which could break different types of plastic and their source of these enzymes. Global users also have a big role to play in restricting the demand for plastics especially the single user plastic. Bioplastics which are degradable could replace them. International organizations have realized the havoc plastic waste has created, and the importance of regulating and mitigating plastic waste. Amongst the different usages packaging potentially creates the highest plastic waste. India needs to reduce usage of plastics in packaging.

Acknowledgements Thanks are due to Dr. Ruchira Sen, Assoc. Prof. in Jindal Global University, for carefully editing this chapter. Thanks, are also due to Dr. Malti Goel for meticulously going through the chapter and suggesting few corrections.

References

1. Britannica, The editors of Encyclopedia. "John Wesley Hyatt". Encyclopedia Britannica, revised edition (2023).
2. Britannica, The editors of Encyclopedia. "Leo Baekeland". Encyclopedia Britannica, revised edition (2023).
3. National Academy of Sciences, USA, Plastic waste and its management (2023).
4. NOAA, National Oceanic and Atmospheric administration USA, Marine Debris-Demystifying The great Pacific garbage Patch, revised (2023).
5. NOAA, National Oceanic and Atmospheric administration USA, what is a Gyre, revised (2023).
6. Plastic Soup Foundation USA, BPA and Plastic Additives: What health effects (2023).
7. OECD (organization of economically developed countries), Global Plastic Outlook (2022).
8. TWI (Training within industry), Granta Park, Cambridge, UK; What is a Thermoplastic (Definition and examples), updated (2023).
9. Helmenstine. Anne, Science notes .org, what is a thermosetting plastic? Definition and examples (2021).
10. Tsakona Maria and Rucevska Leva, Grid-Arendal, Report as per request by Secretariat Basel waste convention for Plastic Waste partnership (2020).
11. Plastic Waste generated in India, Published by Statista Research Department, Mar 24, 2023.
12. Our World in Data. (2020). *Global Plastic production.* University of Oxford.
13. Statista 2020b and Plastics Europe (PEMRG), Consultic, 2019b, Distribution of global plastic produced by region (2018).

14. UNEP 2021, from source to solution; A global assessment of marine litter and plastic pollution.
15. Bruna Alves, Plastic waste worldwide -Statistics and facts, Statista, chemicals and resources, August 2023.
16. Marta Fava, Ocean Plastic pollution an overview data and statistics, ocean literacy portal, UNESCO.
17. The roundup.org USA Plastic pollution Statistics.
18. Upkanah Inemesit, Top 10 Countries producing Most Plastic Waste, Green Match, UK (2023).
19. UNCOMTRADE Database, Import and export of plastics, selected countries (2020).
20. OECD (Organization of economically developed countries), Plastic pollution is growing relentlessly as waste management and recycling falls short, Global plastic outlook database, Feb 2022
21. World trade Organization, Global Plastic outlook, Policy Scenario to 2060, September 2022
22. Yoshida, S., Hiraga, K., Takehana, T., et al. (2016). A bacterium that degrades and assimilates Polyethylene Terephthalate, Science, Vol 351, issue 6278, PP1196-99, Mar 2016, American Association for the Advancement of Science, USA.
23. Sanlus-Verdes, A., Colomer-Vidal, P., Rodriguez-Ventura, et al. (2022). Wax worm Saliva and the enzymes therein are the key to polyethylene degradation by Galleria mellonella. *Nature Communication, 13*, 5568.
24. Mackenzie. Lois, the fungus and bacteria tackling plastic waste, UK, Manufacturer, July 2021.
25. Fermin Koop, Magnetic coils, the new way to deal with microplastics, ZME Science, Romania, August 2019.
26. Licella Holdings limited, Cat-HTR Technology Core, North Sydney, Australia.
27. Evoware, Replacing plastics with seaweed, Jakarta, Indonesia.

Role of Digitization in Achieving Sustainable Waste Management of Plastic and Other Materials

Sourabh Manuja

Abbreviations

APIs	Application Programming Interface
CCP-ME	Cities Combating Plastic Entering the Marine Environment
GIZ	Gesellschaft für Internationale Zusammenarbeit
GPS	Global Positioning System
IoT	Internet of Things
MoHUA	Ministry of Housing and Urban Affairs
MRVs	Monitoring Review and Verification
MSW	Municipal Solid Waste
PIBOs	Producers Importers Brand Owners
PWPs	Plastic Waste Processors
RFID	Radio Frequency Identification
SLRM	Solid Liquid Resource Management

1 Introduction

1.1 Introduction to Waste Management Industry

Waste generation is a consequence of development in cities. Increasing urbanization, population growth, and changing lifestyles with economic growth and development in India evidently imply that waste generation will keep on growing further [1]. India

S. Manuja (✉)
CCP-ME Project, Deutsche Gesellschaft für Internationale Zusammenarbeit (GIZ) GmbH, New Delhi, India
e-mail: sourabh.manuja@giz.de

M. Goel and N. G. Tripathi (eds.), *Plastic Pollution*,
https://doi.org/10.1007/978-981-97-5528-8_9

is expected to generate around 165 million tonnes of municipal solid waste (MSW) by the year 2050 [2]. While India strives to become a developed nation by 2047 [3], the challenges that remain in the management of MSW in cities will be still enormous.

Broadly, waste management is linked to two major global issues: marine litter and climate change. Globally, over 1.5 million and 8 million tonnes of micro- and macro-plastics, respectively, enter the ocean every year [4], adding to the 170 million tonnes of plastics already circulating in the marine environment [5]. The continued growth of plastic consumption and inappropriate disposal are key environmental concerns because of the long time it takes plastic products to break down in the environment. Consequently, approximately 100,000 marine mammals every year are killed by plastic pollution. Nearly 81 out of 123 marine mammal species are reported to have eaten or become entangled in plastic, and all seven sea turtle species are reported to be affected [6].

In India, we consume around 13.6 kg c^{-1} y^{-1} plastics and generate around 9.5 million tonnes of plastic waste every year, of which around 60–70% is estimated to get recycled [7]. This plastic consumption may be way less than the global average of 30 kg^{-1} y^{-1}, [7] with 9% recycling rates [8]. Yet, India caters to one-sixth of the world's population, a point to the amplification of issues due to high population in the country. Two of the ten world's most plastic-carrying rivers (>90% of plastic pollution load)—*Ganges and Brahmaputra,* also lie in India [9]. Overall, this indicates a strong need for Indians to act and combat plastic from entering marine environments.

Hot spots of plastic waste generation, population growth, and proximity to marine and freshwater ecosystems mean that cities are at the front line of the plastic pollution crisis. It has been cited by many studies that around 80% of the marine plastic litter in seas and oceans originates from land-based sources, and most of this litter is associated with inappropriate waste management in cities [10]. Making it clear, to reduce plastics from entering the marine environment, we need to target and help cities manage waste appropriately.

On the other side, inappropriate disposal of biodegradable waste leads to methane emissions from disposal sites which were estimated to 11% of the global methane emissions [11] and around 15,832 Gg CO_2 equivalent in 2016 from Indian solid waste disposal sites [12]. Clearly indicating the extent of global climate impact, we can have with inappropriate management of waste in cities.

On date in August 2023, India reported generating 1,54,090 tonnes per day of municipal solid waste (MSW) of which 76% gets processed [13]. This MSW comprised plastic, paper, glass, metal, other non-biodegradable, biodegradable waste, domestic hazardous waste, and sanitary waste. The composition of waste varies from year to year as well as in seasons of a year [14]. Waste generation rates also change from city to city due to various factors including and not limited to climate, precipitation, geography, and geology. There is limited knowledge about the different components of waste generated in cities, processed, or diverted for recycling or recovery in India. Largely, a city administrator neither has an idea of what all types of non-biodegradable wastes (subcategories of plastics, for example) are not getting catered in the city waste processing facilities, nor how much materials the city diverts for recycling each day. The problems in waste management are complex but are mostly linked with a need to have appropriate data and management strategy [15]. It would be very unlikely for a city to accurately track how much quantity of dry waste is available at the city MRFs for recycling every day. Contrary, a city may track the amount of dry waste getting diverted for recycling/energy recovery/disposed/available for recycling to help understand the need for interventions, strategies, and policies. This will even help a city understand why all low-value recyclables are not catered via MRF operators in a city.

A possible approach for improving waste management is to identify the gaps and weaknesses linked to garbage vulnerable points and hotspots. Thereafter, strategizing services, and creating forward linkages for municipal waste components for recycling and recovery. This also indicates a strong role for having digitalization in the waste sector to bring appropriate monitoring, review, and verification (MRVs) mechanisms in place. Thus, helping transform socio-economic and environmental surroundings of a city, with a transformation towards a circular economy.

Solid Waste Management Rules 2016 Rule 15(h) specify the duties and responsibilities of local authorities and village panchayats of census towns and urban agglomerations to "setup material recovery facilities or secondary storage facilities with sufficient space for sorting of recyclable materials" [16]. However, the need of the day is also to track and monitor the flow of materials and performance of these MRFs in cities, probably through digital systems and innovations.

1.2 Industrial Evolution Towards Digitalization

Since 1784, the first Industrial Revolution started in Great Britain, which was focussed on mechanization and adoption of factory systems. From 1870 onwards, the second IndustrialRrevolution started, which focussed on electrification and petroleum as sources of energy. The standard of living increased in this phase and water supply systems, sewerage systems, public health, and sanitation came into focus. The Third Industrial Revolution started in 1969, focussing on automation. This was an era of the digital revolution, where we all saw the transitions to solar, personal computers, cellular phones, internet, e-commerce, and many more. Thereafter, since 2011, the Fourth Industrial Revolution started [17], also known as Industry 4.0 focussing on digitalization with a convergence of digital, physical, and biological innovations. Advance robotics, smart machines, 3D printing, self-diagnostics, and artificial intelligence are getting introduced. We are seeing increased application of science in industry and significant growth in the transportation and communication sectors. This revolution has also brought together the focus of governments towards sustainable development, with balanced growth and environmental conservation. Focus has also raised issues like climate change, marine litter, and carbon footprints. Thus, reinforcing the role of digitalization in the management of waste.

Industry 4.0 environment basically has three pillars—Digitalization with integration of vertical and horizontal value chains; Digitalization of products and services

Table 1 Design principles of industry 4.0

Principles	Remarks
Interoperability	Ability of things, devices, softwares, and other equipments to interact with each other, i.e. the ability for multiple Internet of things (IoT) platforms from different vendors to work together [20]
Virtualization	This is useful for the simulation of programmes and services, and allows the sharing of a single resource or app among multiple users at a time [21]
Decentralization	For ultimate scalability—Agility including customization of products and services for undertaking self-diagnosis and optimization
Modularity	Enables recombination of software components for reusability and interoperability
Scientific orientation	Enabling integration of digital interface with physical and biological systems
Real-time capability	For important data-enabled decision-making by recognizing micro trends

offerings; Digital business models and customer access [18]. These pillars have data analytics as core capabilities. Looking at waste management, where monitoring segregation levels, collection of segregated waste from generators, and further tracking transportation to appropriate processing units are closely linked to the first pillar, marking attendance of sanitation staff and performance of processing units for quantities handled and processed are strongly linked with the second pillar. The third pillar is linked to the offtake of prepared or offered products and its linkages with the market (Fig. 1). There are broadly six design principles on which Industry 4.0 has been progressing—Interoperability, virtualization, decentralization, modularity, scientific orientation, and real-time capability [19]. Table 1 provides a brief description of each principle of design under Industry 4.0.

1.3 Waste Management and Digitalization in Solid Waste Management

Digitization facilitates easy replicability, uniformity, and transparency. Governments can enhance their services, save funds, and even make cities more liveable by digitizing processes and making organizational changes [22]. Digitalization is slowly progressing into municipal systems, particularly waste management. We could see collection and transportation vehicles getting tracked with a global positioning system (GPS) for real-time monitoring in cities, Radio Frequency Identification (RFID) enabled smart bins enabling the city control center to take appropriate actions and provide collection services appropriately. In fact, our city waste grievance system for citizens is also through a digital app "Swachhata App", which the city uses to facilitate citizens. However, there are many developments and roles of digitalization to facilitate the recycling Industry as well as protect our environment. Studies

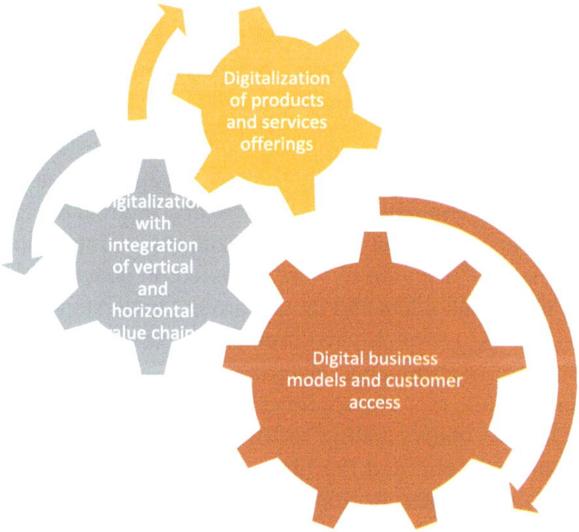

Fig. 1 Pillars of digitalization

suggest that, with the growth of digitalization and shifting of the waste sector towards a circular economy, Urban Local Bodies could save between 30 and 35% of the municipal waste management budget [23].

If MSW from cities is appropriately monitored via digitalized mechanisms for segregation, collection, transportation, treatment, and further offtake, the cities' waste management will be entirely different. Global issues related to waste management will also be much more impactfully addressed.

Efficiency of existing waste management systems and processes could be further increased with rigorous MRV mechanisms in place, which is only possible through digitalization and the capacities of stakeholders to use digital systems. The era of Industry 4.0 provides us with ample opportunities to improvise and build circular economy principles in our day-to-day activities.

2 Methodology

2.1 Baseline Study for Understanding Plastics Waste Management

To understand the flow of plastics from cities and the monitoring mechanism adopted by ULBs in the country, a baseline study was undertaken. This study helped in understanding the lacuna in data availability and material linkages.

Targeting digitalization with the third pillar in focus (digital business models and customer access), the study targeted to enhance digital monitoring and capabilities of the waste sector. The requisite was to monitor dry waste management through MRFs and provide materials available for reuse and recycling visible to recyclers and energy recovery agencies in the market.

As part of the study, a detailed understanding of the current recycling scenario of plastic waste and other non-biodegradable waste was studied. Detailed discussions and consultations were undertaken with different categories of stakeholders including recycling associations, bulk recyclers, aggregators and scrap dealers, government agencies, non-government organizations, and start-ups, at the national, state, and city levels. The discussions were focussed on understanding the needs and expectations of the stakeholders and mostly around understanding the current recycling scenario with respect to quality requirements, ease of access to waste materials, specific waste material requirements, materials with unfeasible recycling, database management, and specific material challenges.

While understanding the flow of dry waste management from pilot cities of Kanpur, Kochi, and Portblair, it was learned that cities are not effectively tracing the flow of materials from MRFs in an appropriate manner, nor do MRFs make these materials visible for recyclers to be attracted, for market linkages. There was a need felt to come up with a framework, which enables transparency, ease of replicability, and the ability to connect with markets—A digital waste exchange platform.

The 85+ interactions with recyclers, traders, and city administrators helped the development team understand that recyclers for various materials look for particular sub-materials with specific colours, conditions, and packagings. For example, one recycler recycling green glass would not like to get amber colour glass mixed with this. Thus, the project came out with a need tohave 17 categories and 52 subcategories of materials with varying colours, conditions (mixed, soiled, clean, and sorted), and packagings (shredded, loose, and baled). As a pilot test, physical entries of these materials with specific colours, conditions, and packagings were also undertaken at 06 MRFs in Kerala and Portblair, allowing to verify pragmatism and the vetting of the findings of the study. A detailed waste characterization study was also undertaken at Kanpur. The list of sub-materials was thus increased to 68, including categories of organic materials which are often also sold via integrated Solid Liquid Resource Management (SLRM) centres. The intention was to offer a wish list to MRFs for reporting and monitoring, which is appropriate and useful for establishing linkages with the recycling industry. A detailed data input format was developed for feeding data onto the digital waste exchange platform and reporting.

2.2 Digital Waste Exchange Platform

The digital waste exchange platform "Sansaadhan" was developed, with abilities for integrating with the existing monthly inspection system (Swachhatam Portal) for ULBs by the Government of India. Using the principles of digitalization, the digital portal Sansaadhan facilitated visualizing data from within city MRFs to the

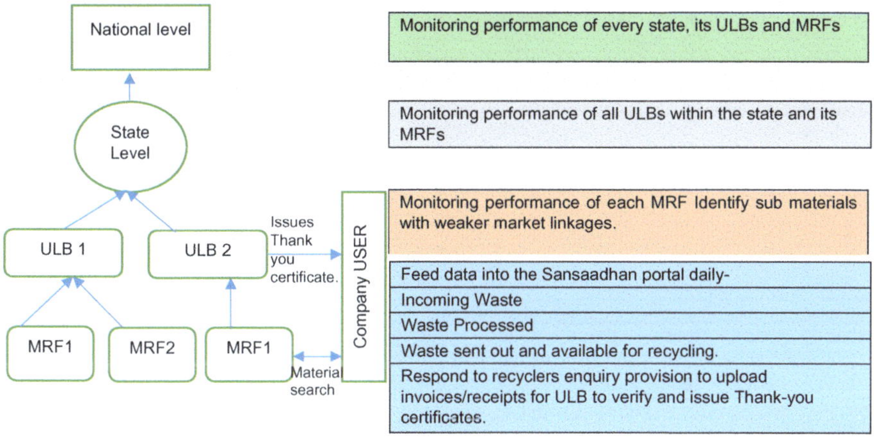

Fig. 2 Role of each stakeholder on Sansaadhan portal

city administration as well as state-level administration for making informed decisions. The intention of this portal was also to help cities reduce material leakages into the marine environment and help combat marine litter. This required visibility of the resources available from MRFs to recycling markets and enabling transactions by providing access and communication mechanisms. Thus, a national portal "Sansaadhan" under Swachh Bharat Mission 2.0 of the Government of India was planned.

The portal enables cities to monitor MRF operations in terms of materials coming in, getting processed, available for reuse and recycling, and further diverted for recycling, energy recovery, disposal, or utilization in city and road construction in a unified format. Thus, enabling a city administrator to understand the challenging materials, and material flow trends and identify possible linkages. States are enabled to monitor the performance of cities at a glance and likewise national-level agencies can monitor the performance of states, and respective cities (Fig. 2).

The portal allows MRF to make entries in two specific registers—the input material register and the output material register. The input material register takes entries for incoming waste—dry, wet, mixed, domestic hazardous, and sanitary waste. In case an MRF is working with a model, which allows operators to buy specific dry waste items, the portal allows them to enter for specific dry waste items they have coming in the MRF. The other entry in this register of materials processed on that particular day. This allows to monitor the flow of materials coming in and the operational processing capacities of MRF. Output material register allows an MRF to showcase stocks of processed materials made available for reuse and recycling as well as materials going out for recycling/energy recovery/disposal/road construction. Thus helping monitor the flow patterns and stocks.

Figure 3 demonstrates the material entry form of the output register for MRF. This enabled MRF to feed data on a daily basis and also upload the latest picture of the material, which could be made available to recyclers looking for specific material

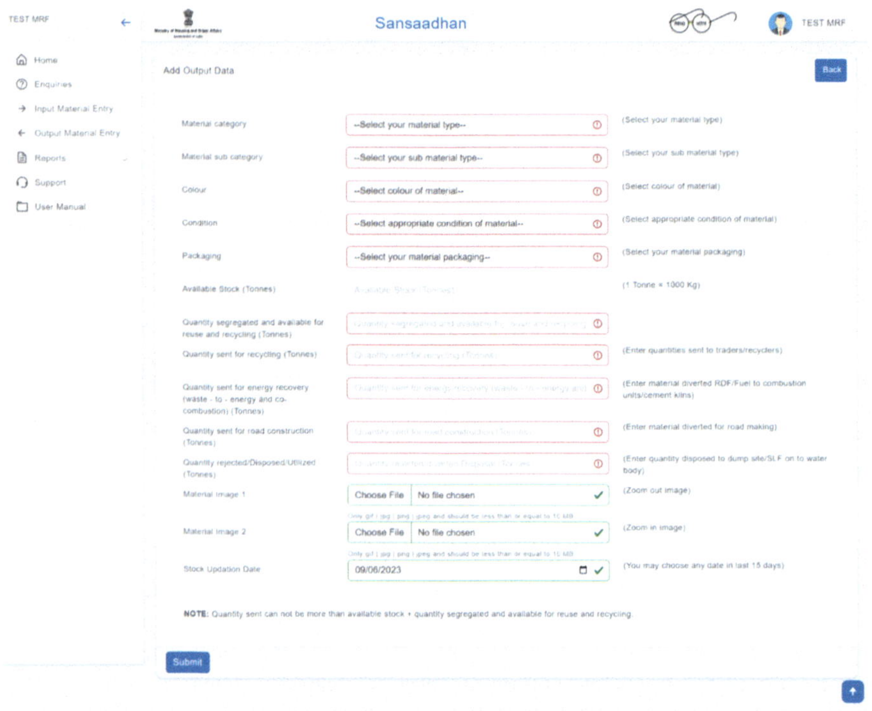

Fig. 3 Snapshot from Sansaadhan portal of material entry in input register (Sansaadhan Portal, 2024)

on the portal. MRFs had the flexibility to feed data/edit data up to the last 15 days, after which data editing was not allowed.

Digital tool Sansaadhan enabled easy replication: any number of MRFs in a city could be added, any number of cities in a state can be operational, and further, all data in clustered and decluttered manner can be analysed with the help of bar graphs, pie charts, and line graphs, demonstrating the performance of specific materials and sub-materials over months.

Figure 4 shows a snapshot of recyclers who can search for materials available from MRFs for reuse and recycling. The recyclers can go to specific states of interest having materials, narrow down to specific MRF, and send an enquiry for specific material demand. Once the enquiry is registered, the contact information of MRF to whom the enquiry is floated is made available to the recycler. Simultaneously, MRF also gets the contact information of the recycler sending the enquiry.

The portal is conceptualized in a manner that data can flow from the Sansaadhan portal to the existing Swachhatam portal (MIS portal for ULBs), which cities use to report. Data from the Sansaadhan portal capturing the performance of MRFs (dry waste management) may be pushed through the application programming interface (APIs) to the Swachhatam portal between 01st and 05th of every month as this data

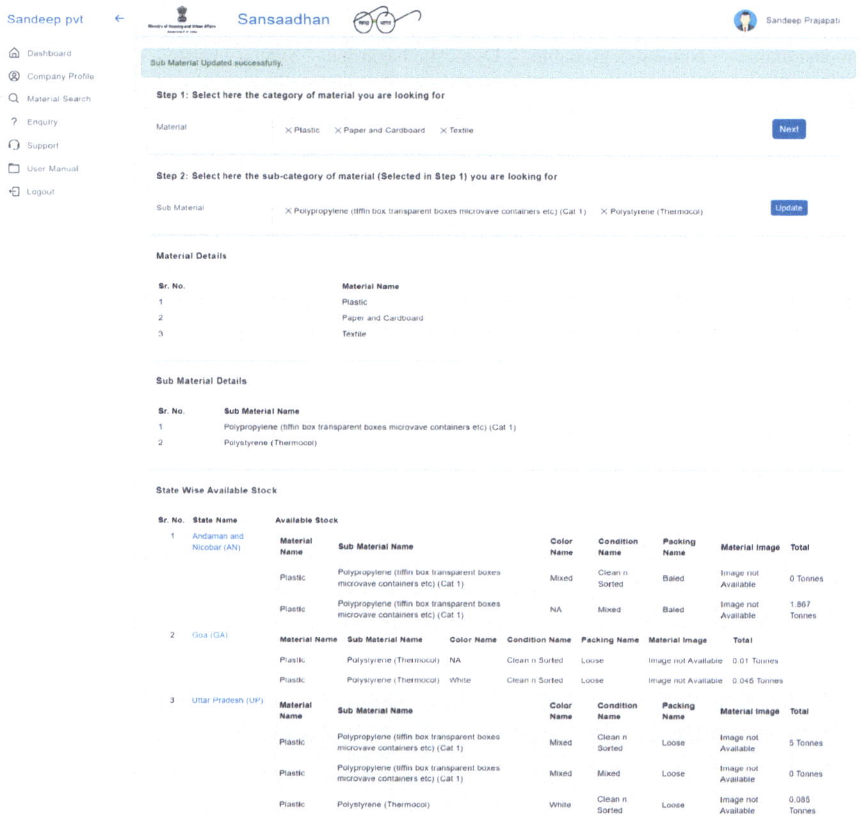

Fig. 4 Material search template for companies searching materials via Sansaadhan portal (Sansaadhan Portal, 2024)

is fed directly via MRFs in a city. Thus, helping reduce workload for a city and enabling first-hand data to flow to the MIS portal—contributing to the efforts of cities to become zero waste to landfill.

The market linkage with Sansaadhan is expected to make all different materials, around 17 broad categories as shown in Table 2, with varying material colours, conditions (soiled, mixed, clean, and sorted), and packagings (baled, loose, and shredded) to recyclers community around the country, as shown in Fig. 4. For example, materials made available by Panaji Municipal Corporation MRFs can be seen by a recycler sitting anywhere in the country to send an enquiry.

Each enquiry at the Sansaadhan portal will be registered with a unique enquiry number, indicating the purpose for demanding material, category, subcategory of material, and quantity required. Once the enquiry is submitted, both the company and the MRF to which the enquiry is raised will be able to see each other's contact information, as well as chat with each other. The company also has the privilege to

Table 2 Broad category of materials identified for reporting at Sansaadhan portal

Biodegradable	Non-biodegradable	Domestic hazardous waste	Sanitary waste	Others
Leather	Plastic	Bio-medical waste (generated from households)	Sanitary waste—diapers	E-waste
Coconut shell	Paper and cardboard	Other domestic hazardous waste	Sanitary waste—Sanitary pads	C&D waste
Wood (engineered)	Metal			Others
Hair	Glass			
Food scrap waste	Rubber and tyre			
Green/garden waste	Textile			
	Used beverage cartons			
	Ceramic			

mark an enquiry inactive, in case the demand no longer exists or has been fulfilled. No transactions happen through the portal. This was intentionally kept for markets to be driven by demand and supply and other business factors.

Certificate to Thank You

This is to certify that __First Company__ (Company Name)

has successfully transacted ___2____ tonnes of waste materials Plastic (Materia sub category)

Pet bottle (Cold drink bottle etc) (Cat 1) from MRF DummyMRF on the date 17-Aug-2023.

Details : Enquiry number.: 202308174225. Purchase order (PO).: 1262623649448.

Certificate Issued by: ____KANPUR______

Certificate No.: RH-2023090525

Issuing Date: 05-Sep-2023 12-11-13 PM

This Certificate is issued via Sansaadhan Portal (https//:sansaadhan.sbmurban.org)

Once the transaction is made between the company and city MRF (separately and not on the portal), the company has the option to get an acknowledgement certificate of the transaction from the city, called "Thank you certificate" at Sansaadhan Portal. This certificate has a unique number and is also verifiable by any of the parties through the Sansaadhan website (https://sansaadhan.sbmurban.org). Recycling companies may upload unpriced purchase orders against the generated enquiry number. MRF thereafter will have to upload the lorry receipt and unpriced invoice for material supply to the portal, against this same enquiry number. Once all these three documents are uploaded, an option will come to ULB, to verify the uploaded documents and issue the "Thank you Certificate". In case the material transaction is for energy recovery, an additional document is needed from the company, "Appropriate disposal certificate". The "Thank you certificate" is issued to the company and is released to both the MRF operator as well as the recycler.

3 Result and Analysis

Digital portal has a perspective to provide transparent operations and data flow for ease of monitoring for cities and states. MRF operation trends will be visible not only to MRF operators, but also to the city. National level and States get a holistic view of operations and can narrow down the monitoring lens even to particular levels below them, dragging down to the lower level of MRF in a city.

- Trends of particular material and sub-material management through MRFs can be monitored.
- Provides a chance to the states and cities for taking informed decisions and supporting MRFs in identifying material linkages.
- Connects with markets by showcasing materials available for reuse and recycling to companies.
- Trends for material flows over months can be seen for understanding the status and impact of policies.

The Sansaadhan portal provides outputs in the form of a snapshot, depicting broad outputs from MRF.

Figures 5, 6, 7, 8, 9 and 10 showcase the status of material coming in MRF, and the status of material available in stock or sent out of MRF. These are the outputs from two registers, which MRF maintains at the portal—Input register and Output register.

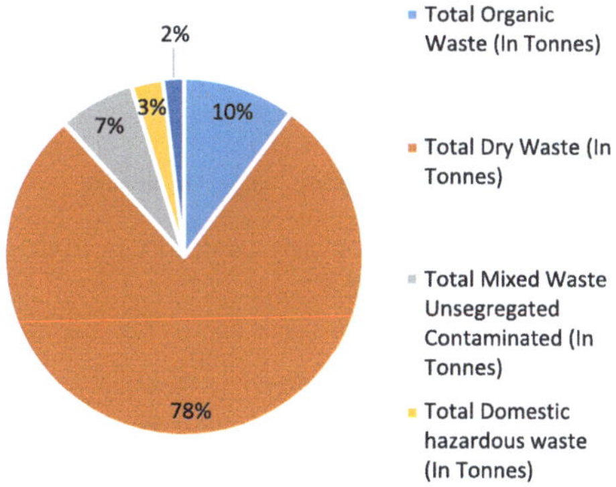

Fig. 5 Current status of materials in stock and going out of MRF as shown on Sansaadhan portal (Sansaadhan Portal, 2024)

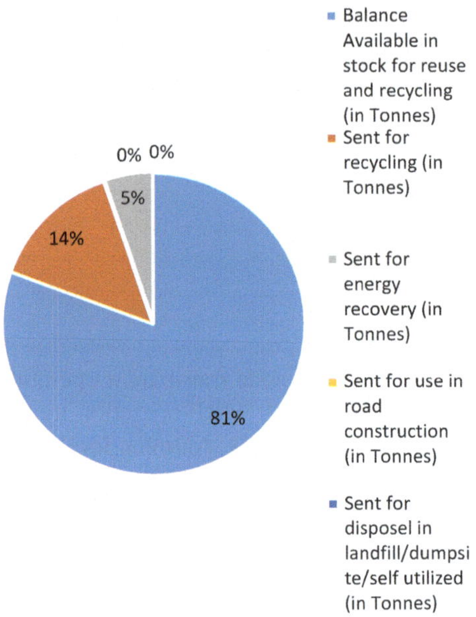

Fig. 6 Current status of materials coming In MRF as shown on Sansaadhan portal (Sansaadhan Portal, 2024)

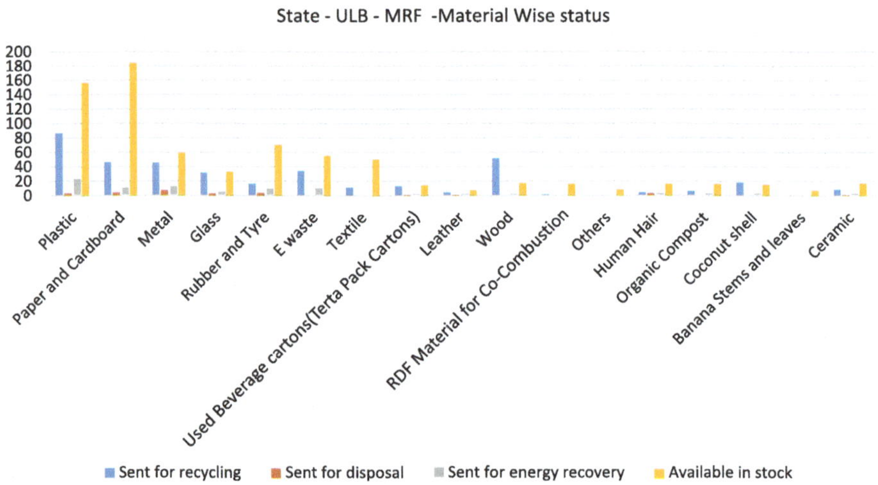

Fig. 7 Status of each material category in stock or diverted from MRF from Sansaadhan portal (Sansaadhan Portal, 2024)

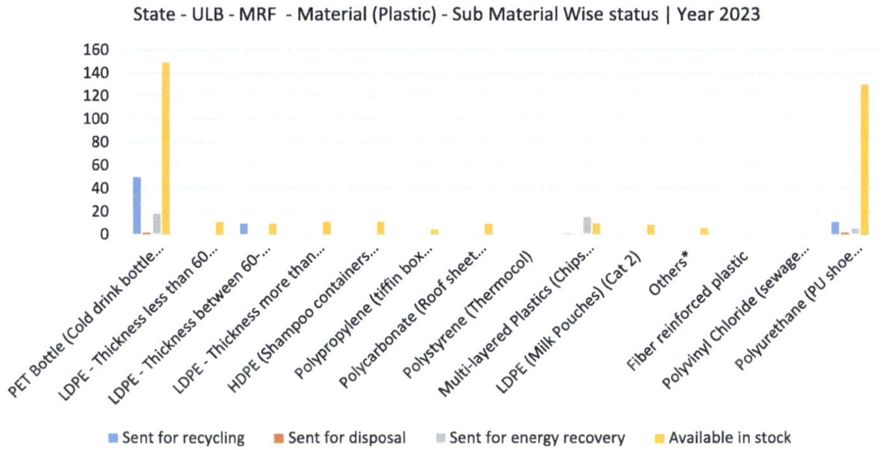

Fig. 8 Status of each plastic material sub-category in stock or diverted from MRF from Sansaadhan portal (Sansaadhan Portal, 2024)

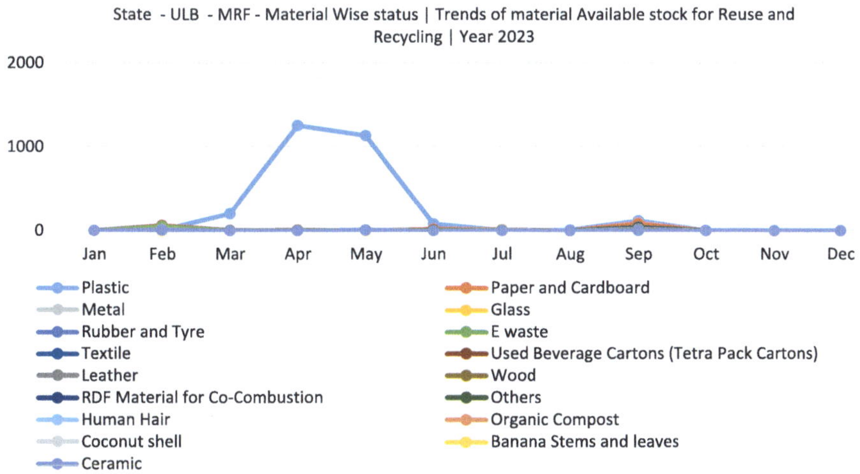

Fig. 9 Trend of plastic materials in stock for reuse and recycling (Sansaadhan Portal, 2024)

Once we click on the pie chart showing the output register, the portal diverts us to details of each material category, giving a showcase of the performance of MRF over various material categories. Each of the material categories can be further explored for its subcategories, as per the interest of the viewer.

Figure 4 shows the performance of each of the 17 materials in MRF and Fig. 5 shows the performance of each subcategory of plastics. Likewise, there are 68 subcategories of different materials, which the portal is capable of displaying and help city and state agencies understand the trends and problem materials, which either are not

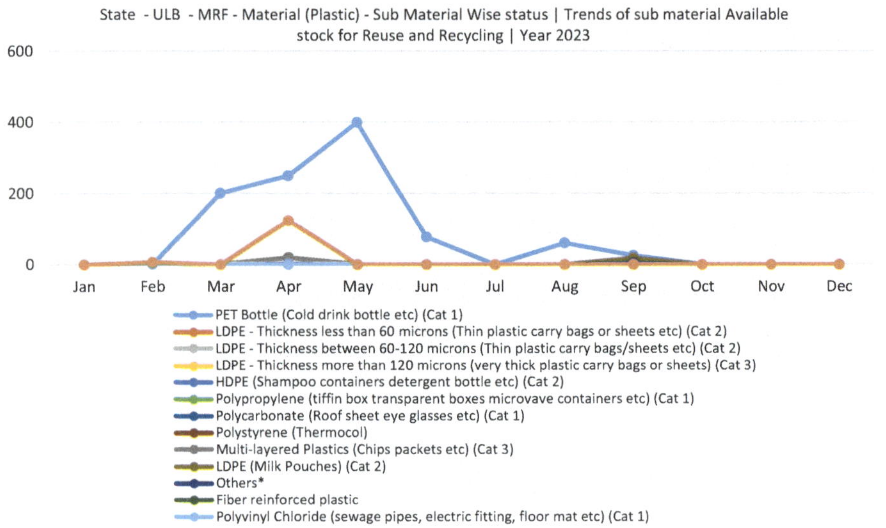

Fig. 10 Trend of each plastic material sub-category available for reuse and recycling (Sansaadhan Portal, 2024)

coming to MRF or not going out due to weak linkages. This also gives an opportunity to policymakers to arise at appropriate policies to help the city move more waste materials into the circular economy loop.

In case a city administrator is willing to understand the trend of specific materials and sub-materials over months in a year, there is an option to view the line graph, which shows the performance trend of material and sub-material as shown in Figs. 9 and 10. The line graphs can be generated for materials in stock, materials sent for recycling, materials sent for energy recovery, materials diverted for disposal, or materials sent for utilization in road construction. All these graphs are visible to an individual MRF as his performance, to the city as a performance of the city, or a particular MRF, to the state as a performance of the whole state, or a particular city, or even a particular MRF in a city, and this pattern continues up to the national level.

Once the company searches for available materials, the portal displays all the states in which the materials in question are available. Company users can trickle down to the specific cities and MRF where the materials are shown as available and raise an enquiry for further communication. Company users are also enabled to mark their enquiries as inactive, to close the enquiries for MRF, in case transactions were concluded/fulfilled through other MRFs or other sources. The portal allows contact information between company users and MRF users to be shared but neither interfere with pricing nor transactions are allowed through the portal. This is intentionally kept to let the trades happen freely without any interference.

Thank you certificate acknowledges the transaction for post-consumer materials being traded from city MRFs. This will further benefit ULBs in establishing linkage with plastic waste processors (PWPs) and get plastic credits for further trading with producers, Importers, and Brand Owners (PIBOs). City MRFs can also establish contractual arrangements with plastic recyclers to share the profits gained from the sale of plastic credits (or any future material credits) of post-consumer waste material provided by the city. These will be acknowledgements for PWPs and PIBOs for transacting post-consumer waste materials with city MRFs. Going ahead, if targets for post-consumer materials are established under EPR regulation for PIBOs to meet, the demand for materials from MRFs will see a rise, as well as a better price.

4 Conclusion and Way Forward

Digital transformation is the way forward. It can transform the way we see the waste management industry. Of course, this will definitely demand more security for data and capacities in cities at all levels to manage digitalization at workspaces.

The trends and reports we can get from the created waste exchange platform will help cities and states not only to compare the performance of MRFs across their boundaries, but also to compare trends and materials getting picked up for recycling. This will open various fronts for discussion and peer learning among cities and states and also establish a basis for taking informed decisions. Further, it will establish baselines for cities to set targets for better waste management from MRFs under the clean India mission.

The acknowledgements generated through the digital portal, are verifiable and authentic proof for helping PIBOs and PWPs obtain certificates for sourcing post-consumer waste materials from cities. It is foreseen that if the near future will require EPR targets to be binding with weightages for post-consumer products, such acknowledgements will help gain more importance for sourcing materials from city MRFs as well as help cities fetch funds for the collection of targeted waste streams like plastics.

Further, the need shall arise for establishing standards for operating MRFs and model contracts among states, which can help cities engage contractors/PPP arrangements for operating medium and large MRFs. The role of digitalization will also expand, looping in the performance of all processing facilities for dry as well as wet waste in the future, facilitating easy monitoring at all levels.

Acknowledgements We acknowledge the support from the Federal Ministry for the Environment, Nature Conservation, Nuclear Safety and Consumer Protection (BMUV) for the project "Cities Combating Plastic Entering the Marine Environment (CCP-ME)" anchored with the Ministry of Housing and Urban Affairs (MoHUA), Government of India, which helped executing the digital portal. Authors also acknowledge the work undertaken by contractor Sakura Solutions Private Limited under contract from GIZ for "Design, development and maintenance of plastic waste exchange platform on SBM portal at National, State and City Level" under contract 83381317. We also thank all the authors whose work has been cited here.

References

1. Tahir, M., Hussain, T., & Behaylu, A. (2015). Scenario of present and future of solid waste generation in India: A case study of Delhi Mega City. *Journal of Environment and Earth Science, 5*(8), 83–91. Retrieved September 11, 2023, from https://core.ac.uk/download/pdf/234664208.pdf#:~:text=Solid%20waste%20generation%20is%20the%20by-product%20of%20the,and%20consumption%20pattern%20and%20lavish%20lifestyle%20of%20urbanites.
2. Cheela, V. S., Ranjan, V. P., Goel, S., John, M., & Dubey, B. (2021). Pathways to sustainable waste management in Indian Smart Cities. *Journal of Urban Management, 10*(4), 419–429. https://doi.org/10.1016/j.jum.2021.05.002
3. Press Trust of India. (2023, August 04). India to be a developed nation by 2047, modernising infrastructure at rapid pace, says R K Singh. *Business News*. Retrieved September 11, 2023, from https://www.financialexpress.com/business/infrastructure-india-to-be-a-developed-nation-by-2047-modernising-infrastructure-at-rapid-pace-says-r-k-singh-3199984/
4. Lau, W. W., Shiran, Y., Bailey, R. M., Cook, E., Stuchtey, M. R., Koskela, J., et al. (2020). Evaluating scenarios toward zero plastic pollution. *Science, 369*, 1455–1461. https://doi.org/10.1126/science.aba9475
5. Horton , H., & Carrington, D. (2023, March 08). *The Guartian*. Retrieved from https://www.theguardian.com/environment/2023/mar/08/plastic-particles-oceans-marine-pollution-production
6. World Wide Fund. (2023, June 27). Plastic in our oceans is killing marine mammals. Retrieved September 12, 2023, from https://wwf.org.au/blogs/plastic-in-our-oceans-is-killing-marine-mammals/
7. The Energy and Resources Institute. (2020). *Plastic waste management-turning challenges into opportunities*. The Energy and Resources Institute.
8. OECD. (2022). *Global plastics outlook: Economic drivers, environmental impacts and policy options*. OECD Publishing. https://doi.org/10.1787/de747aef-en
9. Pendharkar, V. (2018, January 12). Indus, Brahmaputra and Ganga among the top 10 plastic waste carrying rivers. *Mongabay*. Retrieved September 12, 2023, from https://india.mongabay.com/2018/01/indus-brahmaputra-and-ganga-among-the-top-10-plastic-waste-carrying-rivers/#:~:text=Out%20of%20the%20ten%20rivers,together%20carry%20the%20sixth%20highest
10. Gallo, F., Fossi, C., Weber, R., Santillo, D., Sousa, J., Ingram, I., Romano, D., et al. (2018). Marine litter plastics and microplastics and their toxic chemicals components: The need for urgent preventive measures. *Environmental Sciences Europe, 30*(13). https://doi.org/10.1186/s12302-018-0139-z
11. Singh, C., Kumar, A., & Roy, S. (2018, February 13). Quantitative analysis of the methane gas emissions from municipal solid waste in India. *Scientific Reports, 8*(2913). https://doi.org/10.1038/s41598-018-21326-9
12. MoEFCC. (2021). *India - Third Biennial Update Report to the United Nations framework convention on climate change*. Ministry of Environment, Forest and Climate Change, Government of India. Retrieved September 21, 2023, from file:///C:/Users/manuja_sou/Downloads/INDIA_%2520BUR-3_20.02.2021_High%20(1).pdf
13. Swachh Bharat Mission Urban 2.0. (2023, September 12). *SBM mission progress*. http://sbmurban.org/swachh-bharat-mission-progess
14. Agarwal, A., Pandey, R., & Agarwal, M. (2013). Seasonal variation in composition and characteristics of Indian municipal solid waste—a case study. *Science and Technology*, 40–43.
15. Kumar, S., Stephen, S. R., Fowier, G., Velis, C., Kumar, S., Arya, S., Cheeseman, C., et al. (2017). Challenges and opportunities associated with waste management in India. *Royal Society Open Science, 4*. https://doi.org/10.1098/rsos.160764
16. MoEFCC. (2016). *Solid waste management rules 2016*. The Gazette of India. Retrieved September 12, 2023, from https://cpcb.nic.in/uploads/MSW/SWM_2016.pdf

17. Luenendonk, M. (2019, September 23). Industry 4.0: Definition, design principles, challenges, and the future of employment. *Cleverism*. Retrieved September 21, 2023, from https://www.cleverism.com/industry-4-0/
18. Geissbauer, R., Vedso, J., & Schrauf, S. (2016). *Industry 4.0: Building the digital enterprise*. PWC. Retrieved September 21, 2023, from https://www.pwc.com/sk/sk/odvetvia/assets/global-industry4-0.pdf
19. Beier, G., Ullrich, A., Niehoff, S., Reißig, M., & Habich, M. (2020). Industry 4.0: How it is defined from a sociotechnical perspective and how much sustainability it includes—A literature review. *Journal of Cleaner Production, 259*(120856). https://doi.org/10.1016/j.jclepro.2020.120856
20. Noura, M., Gaedke, M., & Atiquzzaman, M. (2019). Interoperability in internet of things: Taxonomies and open challenges. *Mobile Networks and Applications, 24*, 796–809. https://doi.org/10.1007/s11036-018-1089-9
21. Bhist, N. (n.d.). *Virtualization in cloud computing and types*. Retrieved from https://www.geeksforgeeks.org/virtualization-cloud-computing-types/
22. McKinsey&Company. (2016). *Digital by default: A guide to transform government*. McKinsey&Company Center for Government. Retrieved September 12, 2023, from https://www.mckinsey.com/~/media/mckinsey/industries/public%20and%20social%20sector/our%20insights/transforming%20government%20through%20digitization/digital-by-default-a-guide-to-transforming-government-final.pdf#:~:text=By%20digitizing%20processes%20and
23. Kurniawan, T. A., Othman, M. D., Xue , L., Goh, H., Gikas, P., Kusworo, T., Chew, K., et al. (2023). Decarbonization in waste recycling industry using digitalization to promote net-zero emissions and its implications on sustainability. *Journal of Environmental Management, 338* (ISSN 0301-4797). https://doi.org/10.1016/j.jenvman.2023.117765
24. Sansaadhan Portal. (2024, February 12). *Sansaadhan*. Retrieved from Swachh Bharat Mission Urban 2.0: https://sansaadhan.sbmurban.org/company_home/search_material
25. Sansaadhan Portal. (2024, February 12). *Sansaadhan Portal*. Retrieved from Swachh Bharat Mission Urban 2.0: https://sansaadhan.sbmurban.org/adddata/createinputformat
26. Sansaadhan Portal. (2024, February 12). *Swachh Bharat Mission Urban 2.0*. Retrieved from Sansaadhan: https://sansaadhan.sbmurban.org/
27. Sansaadhan Portal. (2024, February 12). *Swachh Bharat Mission Urban 2.0*. Retrieved from Sansaadhan: https://sansaadhan.sbmurban.org/home/material_wise_bar_graphp/MzM=/MTEzMQ==/MTI3

Plastic Waste Crisis: How Smart Cities Can Lead the Change

Neha G. Tripathi

Abbreviations

AI	Artificial Intelligence
CCP	Corporation of the City of Panaji
CSR	Corporate Social Responsibility
EPR	Extended Producer Responsibility
IoT	Internet-of-Things
LDPE	Low-Density Polyethylene
MT	Metric Tons
MLP	Multi Layer Plastic
MRF	Material Recovery Facility
NEA	National Environment Agency
PPP	Poly P-Phenylene
RFID	Radio Frequency Identification

1 Introduction

Plastic waste generation in urban areas has been rapidly increasing globally. Around 2.01 billion metric tonnes of municipal solid waste was produced globally in 2016, with at least 12% of the total waste consisting of plastics. This is expected to grow to 3.40 billion tonnes by 2050 [1]. The most prevalent forms of plastic pollution in cities are single-use packaging, plastic bottles, wrappers, bags, and foam containers, along with microfibers from textiles and microbeads from cosmetics [2].

N. G. Tripathi (✉)
School of Planning and Architecture, 4, Block-B, Indraprastha Estate, New Delhi 110002, India
e-mail: nehagtripathi@gmail.com

M. Goel and N. G. Tripathi (eds.), *Plastic Pollution*,
https://doi.org/10.1007/978-981-97-5528-8_10

In Indian cities, plastic waste generation ranged from 0.10 to 0.34 kg per capita daily in 2021 [3]. The types of plastic waste trends in cities are similar—packaging films constitute the major share at about 54%, followed by single-use items at 25% on average. Over 26,000 tonnes per day of plastic waste is generated from major Indian cities [4].

Consequently, plastic pollution poses a critical challenge for rapidly expanding urban areas. As cities grow, they need to implement comprehensive plastic waste management to reduce environmental and health impacts from uncontrolled accumulation. By transitioning to a circular economy, cities can not only reduce plastic pollution but also create economic opportunities, and decrease their reliance on the production of new plastics.

2 Circular Economy Principles and Plastic Pollution

The concept of a circular economy has gained significant traction globally. As per the analysis by Kirchherr et al., 2017, a circular economy aims to radically transform current economic systems that are largely linear take-make-dispose and rely heavily on finite resources [5]. Circular economy instead envisions the regeneration of materials flows in a way that allows reusing products and recycling resources continuously. So materials like plastics remain economically productive and circulate within the system instead of polluting the environment.

Implementing circular economy principles involves designing systems that keep plastics in circulation for longer periods by supporting initiatives such as recycling programmes, extended producer responsibility, and encouraging the use of reusable materials.

Circular economy principles involve [5]:

1. Eliminating waste and pollution: Designing products and systems that prevent waste generation in the first place through strategies like dematerialisation, utilising bio-based materials, and enabling extended life cycles via durability, easy disassembly, modularity, etc. For example, packaging could be designed with mono-materials that can easily be recycled instead of multilayer composites that end up as waste.
2. Circulating products and materials: Keeping materials in use for as long as possible through reuse, remanufacturing, refurbishment, repair, and high-quality recycling to retain their inherent value [6]. For instance, using recycled plastic instead of virgin plastic for manufacturing new goods. This extends the lifespan of products while minimising the extraction of finite virgin resources.
3. Regenerating nature: Restoring and replenishing natural systems impacted by resource extraction and waste generation [7]. Strategies include composting organic waste to regenerate soils, using recycled materials like plastic for land restoration instead of bitumen, and supporting biodiversity through the creation of value from waste.

By adopting these it is imperative for cities to sustainably manage plastic waste flows and pollution. Scholars project that by transitioning to a circular economy; cities can reduce plastic leakage into oceans by over 80% and double their plastic recycling rates [8]. According to estimates, the yearly economic benefit for India from a circular transition could range from ₹14 to ₹40 lakh crore over the next decades [9].

3 Role of Smart Cities in Plastic Pollution Management

One of the most pressing environmental issues that cities face is dealing with waste effectively. The Internet of Things (IoT) offers promising solutions to tackle this problem head-on. Smart cities are deploying Internet-of-Things (IoT) sensors and RFID[1] tags across waste bins to enable real-time monitoring of waste levels as well as provide visibility into waste hotspots. For example, in smart cities, smart garbage bins have sensors that send alerts to authorities when they are full so waste can be collected faster before trash overflows onto the streets. Furthermore, authorities can also track the amount and locations where waste, especially plastic, is thrown using tools like RFID tags on bins. This information can then be used in making effective policies that promote less plastic usage by analysing the data collected [10].

Smart cities are increasingly installing advanced automated units for identifying and segregating various recyclable waste streams like plastics, paper, glass, and metals. They utilise near-infrared spectroscopy, X-ray fluorescence sensors, laser-induced breakdown spectroscopy along with integrated robotics [11]. For example, the use of RFID tags in packaging has been considered to enable the identification and classification of individually packaged items. The concept envisages embedding RFID tags in individual packaging items, which can then be read either at pick up or in the sorting system to enable the precise sorting of different types of plastic [12].

Additionally, smart cities use mobile apps, online dashboards, and other IT mediums to actively spread awareness among communities on the growing issue of plastic pollution. Mobile apps targeted at citizens provide localised information on plastic pollution statistics, single-use item consumption, waste segregation guidance, and responsible disposal mechanisms in their neighbourhood.

Since plastics have low density, they float on the surface of water bodies, often accumulating into clusters, which can be transported over long distances. Several smart cities have begun leveraging the Internet of Things (IoT)-based sensors and satellite imagery to identify the build-up of plastic waste in urban rivers and coastal areas [13]. Sensor floats deployed in river beds act like probes continuously transmitting key water quality indicators and plastic concentration levels to an integrated

---

[1] RFID [Radio Frequency Identification] systems allow assigning unique identifying codes to each waste bin. Networked sensors attached to these tagged bins can then detect and transmit data on aspects like bin capacity utilisation, fill levels, pick-up frequency, etc., to city authorities.

geospatial database. This allows dynamically mapping of plastic hotspots along city waterways.

Another upcoming technology is to use of drones fitted with infrared cameras that can capture thousands of aerial photos. These photos are then used to train an AI algorithm to recognise images of plastic trash and distinguish between shells, jellyfish or plastic bags, or bottle tops [14].

The 2023 research by Szpilko et al. analyses current and developing waste management trends across smart cities, covering topics ranging from applied technologies to community participation to policy implications. Findings reveal that data-driven solutions around machine learning, mobile computing, blockchain, IoT sensors, and LPWAN networks have significantly enhanced the efficiency, costs, and sustainability of municipal waste systems. The systematically reviewed literature concluded that smart city efforts are indispensable and underutilised to leverage technology capabilities and transform urban waste management onto a circular track in line with global sustainability goals [15].

4 Global Initiatives in Smart Cities for Circular Economy and Waste Management

While cities are increasingly embracing smart waste innovations, the application of these advanced technologies specifically for plastic pollution management is still an emerging area. As such, there are currently limited well-documented examples on this front, but momentum is gaining as the potential benefits become clearer. Here are some examples from smart cities across the world that harness technology to manage their waste including plastic waste holistically.

4.1 Amsterdam, the Netherlands

Amsterdam has been at the forefront of smart city initiatives, including those related to waste management. The city utilises technology to monitor and manage waste collection, with a focus on increasing recycling rates. Smart bins and sensors help optimise waste collection routes, reducing the environmental impact and contributing to circular economy goals.

Amsterdam was also one of the first cities in the world to implement an EPR scheme for single-use coffee cups. Under the scheme, coffee cup producers must pay a fee based on the number of cups they put on the market in Amsterdam. This fee is used to fund the collection and recycling of the cups [16]. It is estimated; this policy intervention resulted in a 70% reduction in cup litter within two years demonstrating that regulatory measures combined with physical infrastructure improvements enabled by smart technologies can accelerate circular resource flows in cities [17].

4.2 Singapore

Singapore has launched a "Zero Waste Nation" initiative that utilises emerging technologies like artificial intelligence (AI) and blockchain to optimise waste collection routes and incentivise responsible waste disposal. Recycle Nation is a nationwide digital recycling initiative launched by the Singapore's National Environment Agency (NEA) in 2021. Under this initiative there are installed high-tech reverse vending machines called "eco-stations" in public locations like grocery stores, schools, libraries, and parks. These eco-stations use intelligent scanning and sorting technology to identify and separate different recyclable materials including plastic, glass, aluminium, and paper deposited by residents. Citizens can create an online Recycle Nation account connected to the eco-stations. When they deposit eligible recyclable items, their account is instantly credited with points based on the material type and weight. The points can then be used to claim vouchers and discounts for partner grocery stores, restaurants, shops, sustainable products, and other rewards [18].

Singapore also utilises advanced optical sorting facilities with near-infrared scanners to automate identification of plastic resins and purities to allocate waste to appropriate end processing—recycling, incineration, or landfilling. It can be seen that the combination of data-based optimisation, involving citizens through incentives and automation across the waste management value chain is successful and aligns with sustainability goals.

4.3 Shanghai, China

On 1 July 2019, Shanghai became the largest Chinese city to implement mandatory waste sorting. Under the new regulations, waste is divided into four groups: "wet" waste (organic waste, including food scraps and yard trimmings), recyclables, hazardous waste, and "dry" waste.

To encourage public participation, Shanghai has implemented mobile applications that provide citizens with information on proper plastic disposal, recycling locations, and real-time updates on waste management initiatives. Interactive displays, social media campaigns, and educational apps, video games are used to inform citizens about the importance of reducing plastic consumption, properly segregating waste, and supporting recycling initiatives [19].

Furthermore, the city has installed smart waste bins equipped with sensors and IoT connectivity in high-traffic areas. These bins monitor fill levels in real time and send alerts to waste collection teams when they reach capacity. This helps in optimising collection routes (Fig. 1).

Advanced technology is now allowing for "smart recycling" in Shanghai. Artificial Intelligence has been adapted to allow residents in Jing'an district to simply deposit

Fig. 1 Smart waste bins in Shanghai

their recycling in a specially designed sorting machine available online at https://www.chinadaily.com.cn/.

Shanghai is also piloting advanced automated sorting mechanisms utilising AI-enabled robotic arms and computer vision techniques to accurately segregate waste items. These intelligent sorting systems have shown improvement in capturing plastics such as packaging film and PET bottles into their dedicated recycling streams versus manual sorting [20].

4.4 Phoenix, USA

The city of Phoenix, Arizona has deployed smart waste management solutions like RFID-enabled smart recycling bins across the city to optimise collection routes and improve recycling rates. These sensors monitor real-time capacity allowing route optimisation that reduced miles driven for collection by 15% leading to lower carbon emissions [26]. Over the last 10 years, Phoenix's smart recycling infrastructure has helped double recycling participation rates. It has also doubled diversion rates—how much material is diverted from landfills for recycling [21].

Phoenix is testing new technologies to improve waste management. The city implemented a pilot programme using artificial intelligence (AI) on city cameras to detect illegal dumping sites. This digital monitoring discourages dumping and allows faster responses [22]. Additionally, Phoenix is exploring automated plastic sorting at its recycling facility using near-infrared and optical sensors. This technology diverts over 30% more plastics into recycling streams instead of landfills [22].

5 Indian Smart Cities and Waste Management

The Indian Government launched the Smart Cities Mission in 2015 with the aim of developing 100 futuristic and sustainable cities across the country. A key feature of smart cities is the use of emerging technologies like the Internet of Things (IoT), big data analytics, and artificial intelligence to optimise infrastructure and service delivery [23].

The Smart Cities Mission has also provided cities opportunities to integrate innovative solutions for urban challenges like waste management through strategic use of technology. While the adoption of smart technology for urban waste management is still relatively low overall in India. Some cities are beginning to implement smart solutions specifically for managing plastic waste:

- Indore has deployed GPS tracking on plastic waste collection vehicles to optimise routes and ensure efficiency.
- Pune and Bengaluru are using RFID tags on bins to track waste generation patterns and collect user fees digitally.
- Hyderabad is trying out AI-enabled automated waste segregation to reduce plastic waste sent to landfills.
- Delhi is piloting an IoT sensor network and mobile app to map plastic pollution hotspots for targeted cleaning.
- Panjim has launched a blockchain rewards scheme for waste pickers to collect and submit plastic waste.
- Mumbai is exploring drone-based monitoring of plastic waste in storm drains and nullahs

6 Smart City Indore, India

Indore, the largest city in Madhya Pradesh State, is home to over 2 million residents. Indore is divided into 19 zones and 85 wards, each containing approximately 6,000 households and 600 commercial establishments [24]. In 2015, the city embarked on its smart city journey as part of India's Smart Cities programme. This initiative aimed to enhance infrastructure, utilities, transportation, and waste management. Sanitation, including solid waste management, is a core infrastructure element in the city.

6.1 Indore Waste Management

Presently, Indore generates more than 1,115 metric tonnes (MT) of garbage daily. Out of all waste, 58.25% is wet or natural waste, 41.75% is dry waste and 0.5% is household hazardous and sanitary waste. From the dry waste, 14% of it is plastic waste [24]. In Indore, door-to-door waste collection commenced in 2016 covering all households and commercial establishments. The mandate for segregation at source was also introduced in 2016, requiring residents to separate waste into categories such as plastic, e-waste, hazardous, and wet waste [25]. Currently, Indore segregates waste into six distinct categories at the collection point.

The success of Indore collection lies in the residents' compliance with the municipal body's mandated six-category waste separation system. With the help of NGOs and safai karamcharies, this is possible. One bin is for wet waste, a second for paper, a third for plastics, a fourth for domestic hazardous material (like tablets or chemicals), a fifth for sanitary waste (menstrual waste, condoms, syringes, diapers, and cotton), and the sixth bin is for electronics like TVs, laptops, printers, and their components [25] (Fig. 2).

The Indore Municipal Corporation has installed India's first 300 MT Fully Automatic Material Recovery Facility-I at Devguradia Trenching Ground, which is spread over approximately 3 acres. This high-tech facility is equipped with an automated sorting system that separates all kinds of dry waste, such as plastic, metal, cloth, paper, and other substances, more efficiently than manual sorting.

Furthermore, Indore has established Asia's largest 550-tonnes/day waste-to-bioCNG plant, producing 17,000 kg of bio-CNG and 100 tonnes of compost daily [26]. Additional infrastructure includes a 15 MW waste-to-energy plant and a 200 TPD composting plant. As a result, there has been a 70% reduction in landfill burden through composting and recycling treatments [24].

Fig. 2 Waste collection in Indore city

6.2 Use of Technology for Waste Management

Indore ensures 100% coverage of wards through its door-to-door collection system. This is possible due to the use of mapping and technology. The years 2015–2016 were spent mapping each of Indore's 86 city wards to plan the collecting, transporting, segregating, and recycling of over 1000 tonnes of municipal waste that is generated daily across the city. It was only after mapping and calculating the capability of different vehicles to travel door-to-door and manoeuvre the narrowest of streets that a fleet of 800 mini-tippers, dumpers, compactors, containerised tippers, hook loaders, and even CNG rickshaws was ordered [24].

A fleet of over 600 GPS-enabled vehicles was deployed to collect segregated waste daily from all households and commercial establishments. Indore has introduced a system featuring 600 GPS-tracked waste collection vehicles for real-time routing optimisation and monitoring. An RFID-enabled tracking system traces waste at the household level during pickup using tag sensors [26]. A command centre hub closely monitors operations through data-driven analytics and decisions for improved efficiency [27].

Along with door-to-door inspections technology like CCTV cameras help verify whether the six-bin segregation is being followed in the city.

An "Indore 311" helpline app was launched to promote communication with the public. This mobile app aids grievance redressal, information access, and tracking for citizens regarding waste management services. The app has a feature that relies on the before-after pictures uploaded by citizens on the 311 app to help in monitoring and ensuring transparency.

Automated "Smart Bags" vending machines, sponsored via Corporate Social Responsibility funds, dispense affordable reusable cloth bags are installed across crowded public areas like markets as a deterrent to plastic bag use [24].

6.3 Plastic Management Initiatives in Indore City

Indore has implemented a diligent plastic waste collection system. All households under the Municipal Corporation jurisdiction have received two large bags to separately store plastic waste [27]. The waste is collected through partitioned vehicles known as 'Tippers' and transported to the Garbage Transfer Stations for secondary collection. Once a bag is full, interested residents notify the authorised recycling agency, which purchases the plastic at Rs. 1.5–5 per kg based on quality [26]. This incentivises segregation while enabling agency collection for treatment.

The collected plastic from households is processed at a treatment plant, where 8 tonnes daily are converted into 3,500 L of diesel. The remaining plastic is sent to a Neemuch cement plant facility, where it is transformed into bricks [32]. Additionally, non-recyclable plastic is sent to the M.P. Rural Road Development Corporation for constructing sidewalks and roads, with 10 projects utilising plastic waste [28].

Thus, Indore's system promotes plastic circularity through household separation, collection, and treatment for recycling or energy recovery, significantly reducing landfilling.

7 Smart City Panaji, Goa

With a population of 114,759 in the metropolitan area, Panaji stands as Goa's largest urban agglomeration. The Corporation of the City of Panaji (CCP), declared a Portuguese city in 1843, serves as the world's smallest municipal corporation, governing Panaji's capital, Panjim [29]. The Panaji Smart City initiative was inaugurated on October 14, 2015, in Panaji, Goa.

7.1 Panaji Waste Management

Panaji has successfully achieved the status of a zero-landfill city through effective waste segregation and recycling, ensuring nothing is sent to landfills as of 2022 [30]. In 2017, Panjim generated an average of 47 tonnes of garbage every day, deposited at Patto Plaza (two treatment plants) and various composting units. According to CCP statistics, a total of 130 composting units are present across the city.

Approximately 40% of housing colonies in Panaji have onsite waste sorting stations [30]. On average, 14 tonnes per day of dry waste are collected from homes and establishments.

In October 2020, CCP mandated citywide 16-way source segregation through bylaws, facilitating effective recycling. CCP provides color-coded bins for this multi-way separation. Waste is segregated into paper, plastics, e-waste, tetra packs, coconut shells, and even ceramics [31] (Fig. 3).

The Material Recovery Facility (Swachhta Kendra), opened in 2014, enables efficient sorting and recycling. Six waste management facilities in Panjim were initiated by the CCP (Corporation of the City of Panaji) in March 2021. The facility, also known as the Sorting Station or Mini MRF (Material Recovery Facility), was established to manage waste at a decentralised level. Here, dry waste can be sorted into sixteen categories. About 700–1,000 kg of unsorted dry waste from different wards of the city comes to their respective MRFs. The "safaisathins" working at the facility segregate it into various fractions like PET, Hard Plastics, Glass, Paper, and so on. Each category of waste is then sent to the concerned recyclers, reducing the load on the main MRF (Fig. 4).

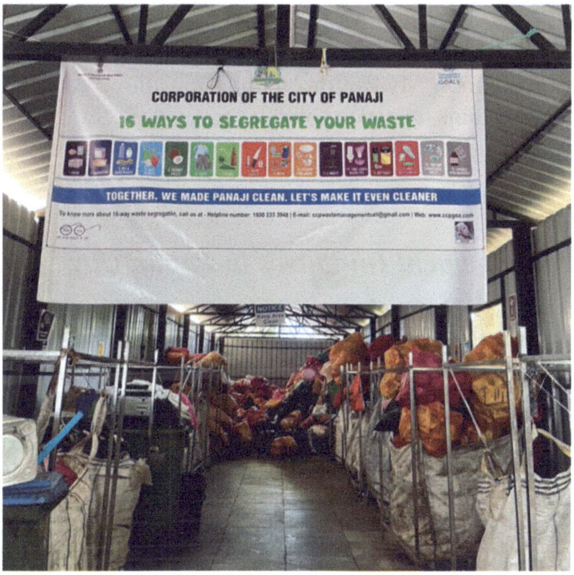

Fig. 3 Ayya waste management in Panaji (published online GoaGomantak Times). *Source* https://www.gomantaktimes.com/my-goa/art-culture/wondering-how-to-manage-your-waste-better

Fig. 4 "Safaisathins" engaged in waste segregation in Panaji. Conveyor belts keep the waste moving at the Panaji material recovery facility. *Source* Chryselle D'Silva Dias

7.2 Use of Technology for Waste Management

GPS-enabled vehicles and RFID tags: The waste collection vehicles in Panaji are fitted with GPS devices to track locations and optimise routes in real time. RFID (Radio Frequency Identification) tags are placed in household bins which get scanned during pick up. This generates digital records of waste collected from each household [32].

Command and Control Center: An integrated command centre monitors Panaji's waste management operations using the latest surveillance and tracking technologies.

Mobile apps and hotline: "My Panjim" mobile app allows convenient payment of bills and filing of complaints related to solid waste services. Swachhata-MoHUA app has features for registering waste-related grievances and monitoring redressal [33].

7.3 Plastic Management Initiatives in Panaji City

Segregation at Source Program: Households and establishments have been given separate bins for wet waste, dry waste, and rejected waste to enable proper segregation of garbage. This allows for more effective waste processing at the material recovery facility. The waste is further manually sorted by sanitation workers. Plastic and paper are further segregated into different fractions based on quality and colour. After sorting, bales of paper, plastic, tetra packs, and cloth are made.

Recyclable waste, including plastic, is sold to registered local vendors, generating revenue that is fed back into the facility. A local plastic recycling industry utilises Panaji's sorted plastic waste as raw material.

The Corporation of City of Panaji (CCP) also plans to start manufacturing benches, chairs, and other public park furniture out of plastic waste. The city has also used the corporate social responsibility (CSR) funds towards buying a shredder that can help plastic break into smaller pieces that makes it easier for recycling (Fig. 5 and Table 1).

8 Discussions

According to SWM Rules, 2016, waste needs to be separated at the household level into three categories: biodegradable, non-biodegradable, and hazardous waste. Indore and Panaji have gone beyond the standard segregation by separating waste into 6 different streams in the case of Indore. Similarly, Panaji mandates separating waste into 16 categories. This allows for more efficient waste collection and processing.

Indore and Panaji showcase innovative circular economy approaches to plastic waste management enabled under the smart city initiatives. Indore has implemented an advanced source segregation system that separates waste into six specialised streams at the household level: dry, wet, plastic, sanitary, domestic hazardous, and others. This sorting in categories allows for more targeted material recovery and processing through localised recycling ecosystems and waste-to-energy plants. Formal integration of waste pickers into collection provides inclusive livelihoods as well.

Similarly, Panaji follows a highly localised strategy with tailored ward-level plans spanning customised awareness campaigns, training workshops, and robust 16-way

Fig. 5 A rate card placed outside the Ayya waste management unit. Gomantak Times. https://www.gomantaktimes.com/my-goa/art-culture/wondering-how-to-manage-your-waste-better

source separation mandates. This decentralised capacity building empowers residents while allowing the recovery of diverse waste types. Integration with the local recycling industry unlocks circularity across entire plastic lifecycles.

In both cities, leveraging data and analytics allows for optimised waste collection routes and vehicle tracking. Given Indore's size, its waste management system is significantly strengthened by IoT-enabled vehicle tracking, RFID tracing, command centres, mobile apps, and other smart technologies. This helps in operational efficiency and enables superior monitoring, decision-making, and accountability.

Furthermore, Indore enforces its six-way waste segregation system through technology with CCTV cameras visually verifying compliance across the city, while door-to-door municipal inspections directly check strict adherence by households and establishments.

The success of smart waste management in both cities relies on collaboration between proactive stakeholders including community associations, citizen volunteers, NGOs, and local government officials. Technologies like the 311 App in Indore and "my panjim" app in Panaji enable more transparent and responsive waste management by allowing residents to actively participate.

In summary, smart waste management involves the utilisation of technologies to enhance the efficiency, productivity, and sustainability of waste management operations. These technologies draw upon various computing disciplines, including artificial intelligence, IoT (Internet of Things), cloud computing, fog computing, and

Table 1 Summarising waste management initiatives in Indore and Panaji

Waste management initiatives	Indore	Panaji
Waste segregation and collection	Indore promotes plastic waste recycling through authorised centres for sorting, cleaning, and conversion (Times of India, 2022). Non-recyclable plastic heads to waste-to-energy plants as alternative fuel in cement kilns, allowing up to 80% plastic to stay in the circular economy	Sorting station or mini MRF (material recovery facility. dry waste like white paper, coloured paper, newspaper, PET plastic, hard plastic, LDPE (low-density polyethylene) soft plastic, MLP (multi layer plastic) soft plastic, and PPP (poly p-phenylene) soft plastic, commonly found, can be dropped thereafter separation
Circular economy initiatives Extended producer responsibility (EPR)	Indore municipal corporation has become the first urban body in the country to receive the extended producers responsibility credit by recycling 8 tonnes of banned single-use plastic items	EPR (extended producer responsibility) authorisation registration is mandatory for producers belonging to the electronic sector in Panaji
Consumer awareness and education	Various campaigns to eliminate single-use plastic have contributed to Indore city's successful model. To encourage public participation in the drive to eradicate single-use plastic, the Indore Municipal Corporation has run several awareness campaigns, including "Main Hoon Jholadhari Indori", "Halla Bol Drive", "Plastic Heist Gang", and "Plastic Premier League", among others Citizens of Indore played a significant role in making the city neat and clean. The cleanliness scenario of Indore was changed by improved habits of its people. While certain IEC media, such as nukkad nataks, rallies, and swachhata samitis connected citizens with service providers	CCP's "Shop with Your Waste" campaign motivates residents to recycle dry waste items to barter for goods

Source Multiple reports and documents

big data. They are strategically applied across various phases of the waste management process in the specific city context. This typically includes interconnected stages such as bin disposal, source separation, collection, waste treatment, and final disposal transportation.

9 Conclusions

In conclusion, cities as major contributors of plastic waste must proactively adopt solutions to tackle pollution at the source. The current collection and segregation in cities has been made smart and efficient as shown in the selected case studies. Proactive measures taken by smart cities like Indore and Panaji in India showcase effective pathways to manage post-consumer plastic waste through circular principles, technology integration, and localised strategies. As India continues rapid urbanisation, replicating and scaling such smart data-driven models will be instrumental in steering other cities onto sustainable trajectories.

References

1. Kaza, S., Yao, L., Bhada-Tata, P., & Van Woerden, F. (2018). *What a waste 2.0: A global snapshot of solid waste management to 2050*. World Bank Publications.
2. UNEP. (2015). Biodegradable plastics and marine litter. Misconceptions, concerns and impacts on marine environments. United Nations Environment Programme (UNEP), Nairobi.
3. MoEFCC. (2022). Report of the committee to identify quantification of plastic waste generation. Ministry of Environment, Forest and Climate Change. Government of India.
4. CPCB. (2019). Inventory of plastic waste generation and management in India. Control of Pollution in Environmental Pollution.
5. Kirchherr, J., Reike, D., & Hekkert, M. (2017). Conceptualizing the circular economy: An analysis of 114 definitions. *Resources, Conservation & Recycling, 127*.
6. Geissdoerfer, M., Savaget, P., Bocken, N. M. P., & Hultink, E. J. (2017). The circular economy – A new sustainability paradigm? *Journal of Cleaner Production, 143*, 757–768.
7. Merli, R., Preziosi, M., & Acampora, A. (2018). How do scholars approach the circular economy? A systematic literature review. *Journal of Cleaner Production, 178*.
8. Lau, W. W., Shiran, Y., Bailey, R. M., Cook, E., Stuchtey, M. R., Koskella, J., Velis, C. A., Godfrey, L., Boucher, J., Murphy, M. B., & Thompson, R. C. (2020). Evaluating scenarios toward zero plastic pollution. *Science, 369*(6510).
9. Ellen MacArthur Foundation. (2021). The circular economy opportunity for urban and industrial innovation in India. Ellen MacArthur Foundation. https://www.ellenmacarthurfoundation.org/
10. Smart City Waste Management. Spectrum Smart Cities. https://spectrumsmartcities.com/articles/smart-city-waste-management.html
11. Wilts, H., Garcia, B. R., Garlito, R., Gómez, S., & Prieto, E. (2022). Artificial intelligence in the sorting of municipal waste as an enabler of the circular economy. *Sustainability, 10*(4), 28. https://www.mdpi.com/2079-9276/10/4/28
12. Abdoli, S. (2009). RFID application in municipal solid waste management system. *International Journal of Environmental Research, 3*, 447–454.
13. Themistocleous, K., Papoutsa, C., Michaelides, S., & Hadjimitsis, D. G. (2020). Investigating detection of floating plastic litter from space using Sentinel-2 imagery. *Remote Sensing, 12*(16), 2648. https://doi.org/10.3390/rs12162648
14. ITU. (2020). How AI-powered drones are helping fight ocean plastic. From International Telecommunication Union Hub. https://www.itu.int/hub/2020/04/how-ai-powered-drones-are-helping-fight-ocean-plastic/
15. Szpilko, D., de la Torre Gallegos, A., Jimenez Naharro, F., Rzepka, A., & Remiszewska, A. (2023). Waste management in the smart city: Current practices and future directions. https://www.mdpi.com/2079-9276/12/10/115

16. Hartley, K., van Santen, R., & Kirchherr, J. (2020). Policies for transitioning towards a circular economy: Expectations from the European Union (EU). *Resources, Conservation and Recycling, 155*, 104663.
17. Buren, N. V., Demmers, M., Heijden V. D. R., & Witlox, F. (2016). Towards a circular economy: The role of Dutch logistics industries and governments. *Sustainability, 8*(7), 647.
18. Empowering Singaporeans. (n.d.). *RecycleNSave.sg*. https://recyclensave.sg
19. Yuan, F., Shi, K., & Hongzhi, W. (2019). *How does technology assist in waste sorting and Smart City development in China?, AllTechAsia*. https://alltechasia.com/2019/07/30/how-does-technology-assist-in-waste-sorting-and-smart-city-development-in-chin
20. Partnership with Lovere. (n.d.). Alliance to end plastic waste. https://endplasticwaste.org/en/our-work/innovation-through-collaboration/lovere
21. Phoenix Environmental Sustainability Goals - waste (n.d.). City of Phoenix. https://www.phoenix.gov/sustainability/waste
22. Duda, J. (2024). Phoenix plastic recycling facility touted as the first of its kind, Axios Phoenix. https://www.axios.com/local/phoenix/2024/02/07/microfactory-phx-plastic-recycling
23. Smart Cities Mission Government of India (Ministry of Housing and Urban Affairs). (2023). Smart Cities Mission. https://smartcities.gov.in
24. Indore Smart City Development Limited. (2023). Solid waste management. https://www.smartcityindore.org/solid-waste/
25. Frontline. (2022). How does Indore do it? India's cleanest city 7 years in a row strategy. The Hindu. https://frontline.thehindu.com/the-nation/how-does-indore-do-it-india-cleanest-city-7-years-in-a-row-strategy/article67737621.ece
26. Perinchery, A. (2022). *Here's why the new Indore bio-CNG plant is likely to succeed – though others like it haven't*. https://thewire.in/environment/heres-why-the-new-indore-bio-cng-plant-is-likely-to-succeed-though-others-like-it-havent
27. Outlook India. (2022). How Indore became the cleanest city in India for the sixth time. Outlook India. https://www.outlookindia.com/national/how-indore-became-the-cleanest-city-in-india-for-the-sixth-time-news-227284
28. Singh, R. (2021). Municipal solid waste management in the City of Indore- A case study. *Journal of Civil Engineering and Environmental Sciences, 7*(1), 008–017. https://doi.org/10.17352/2455-488X.000039
29. CCP. (n.d.). About CCP. Corporation of the City of Panaji. https://ccpgoa.com/about-panjim/
30. Scroll. (2023). Could Panaji's waste revolution be a blueprint for other Indian cities? *Scroll*. https://scroll.in/article/1048071/could-panajis-waste-revolution-be-a-blueprint-for-other-indian-cities
31. Times News Network. (2020). Soon, segregate waste into 16 categories in Panaji. https://timesofindia.indiatimes.com/city/goa/soon-segregate-waste-into-16-categories-in-state-capital/articleshow/78711336.cms
32. IMAGINE Panaji Smart City Development Limited. (2023). Projects Portfolio. https://imaginepanaji.com/portfolio
33. Smart Cities Mission. (n.d.). Panaji. Ministry of housing and urban affairs. https://smartcities.gov.in/smartcities/panaji.html

Plastic Waste Management in Fertilizer Sector—A Case Study

Manish Goswami and S. Nand

Abbreviations

BOPP	Biaxially Oriented Polypropylene
CPCB	Central Pollution Control Board
EC	Essential Commodity
EPR	Extended Producer Responsibility
FAI	Fertiliser Association of India
HDPE	High-Density Polyethylene
PET	Polyethylene Terephthalate
PP	Polypropylene
PROs	Producer Responsibility Organizations
PWM	Plastic Waste Management
SOP	Standard Operating Procedures
WMAs	Waste Management Agencies

1 Introduction

India is the second largest consumer of fertilizers in the world next, only to China. The total fertilizer consumption in terms of products was 64 million tonnes in fertilizer year 2022–23 (April 2022–March 2023) [1]. Fertilizers are sold in large HDPE bags of 45 and 50 kg to the farmers. Minor quantities of specialty fertilizers are also sold in smaller sizes of 5, 10 and 20 kg bags. The fertilizers are sold to the farmers through well-established channels of public, cooperative and private sectors and made

M. Goswami (✉) · S. Nand
The Fertiliser Association of India, FAI House, 10 Shaheed Jit Singh Marg, New Delhi 110067, India
e-mail: tech@faidelhi.org

M. Goel and N. G. Tripathi (eds.), *Plastic Pollution*,
https://doi.org/10.1007/978-981-97-5528-8_11

available to the farmers in every corner of the country. There are more than 200,000 retail outlets in the country. The bags are transported by rail (80%), road and small quantities through inland waterways. Fertilizer bags are handled at multiple points of supply chain and stored in warehouses of wholesalers and retailers for a long period of 4–12 months. This is because of sale/consumption of fertilizers is seasonal but production and imports take place continuously throughout the year. Fertilizer bags have to be strong enough to withstand multiple handling and be impervious to the moisture during storage.

Fertilizers are classified as essential commodity under the Essential Commodity (EC) Act 1955. A Fertiliser (Control) Order has been notified under the EC Act which specifies the quality parameters of each product, methods of analysis, procedures for the collection of samples, etc. Therefore, preservation of the quality of fertilizers right upto the retail outlets is an essential requirement. Therefore, fertilizer bags are of high strength and have to meet Indian Standard IS 9755:2021.

2 Logistics of Fertilizer Distribution

Sale of fertilizers takes place mainly in two crop seasons for a period of about 2 months in each season. Of the 64 million tonnes of fertilizers sold in fertilizer year April 2022 to March 2023, 48.7 million tonnes or 76% was produced within the country and 15.3 million tonnes was imported. Domestic production is packed in the factory itself and at most times being despatched continuously by rail or road. If there is insufficient availability of rail rakes or trucks, the bulk material is stored in factory warehouses. It is reclaimed and packed when it can be despatched. The imported material is stored in bulk within ports where it is packed before despatch by rail or road. Road transport is economical upto a distance of about 300 km. Fertilizer distribution and freight policy of the Government of India permits road transport up to 500 km. From the rail heads, fertilizer bags are transported to the field warehouses by road. Figure 1 shows the channels for distribution and sale of fertilizers. It can be seen that fertilizer bags undergo multiple handling and storage before they reach 125 million farmers.

3 Fertilizer Packaging

The fertilizer industry contributes about 1.6% to the total plastic wastes generated in the country. The entire packaging material used by the fertilizer industry is mainly flexible and consists primarily of High-Density Polyethylene (HDPE) material with inner lining of Biaxially Oriented Polypropylene (BOPP) and polypropylene (PP). Water soluble fertilizers including micronutrient solutions are packaged in Polyethylene Terephthalate (PET) bottles. In terms of share of material used, HDPE accounts

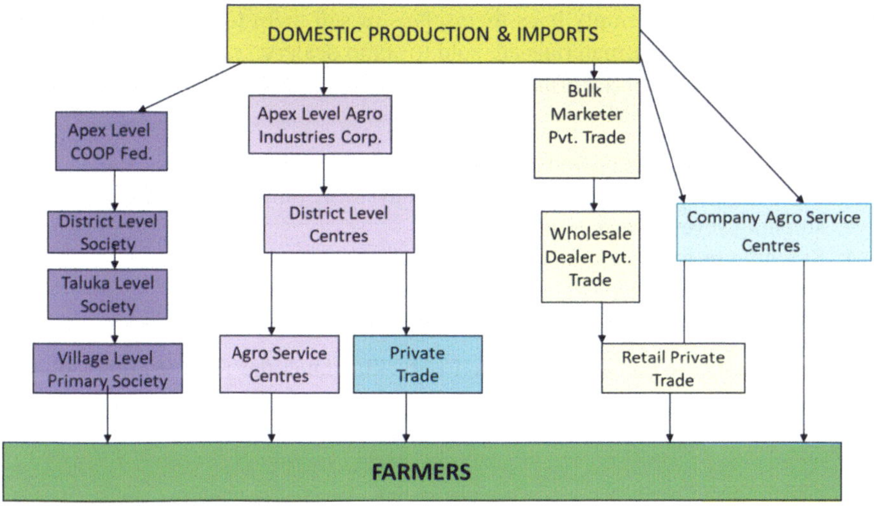

Fig. 1 Distribution channels for fertilizers

for 90%, polypropylene (PP) for 7% and the balance consists of PET and other materials.

Large bags of 45 and 50 kg account for 97% of packaging material. The large empty bags (50 kg) weigh about 130 g. The total plastic packaging material is estimated to be more than 150,000 metric tonnes per annum [2].

4 Life Cycle of Fertilizer Bags

Fertiliser Association of India (FAI) got a field study done on the life cycle of fertilizer bags. It is necessary to mention at this stage that small and marginal farmers cultivating less than 2 ha of land constitute 86% of farming community in India. The average size of land holding in India is 1.1 ha [3]. This has bearing on the resume and disposal of empty fertilizer bags.

A field survey across eight locations in equal number of states in India was carried out to study the lifecycle of fertilizer packaging.

4.1 Uses of Empty Bags

It was found from the study that large bags are invariably used till their active life. This is because these bags have high strength, light weight, flexible for multiple types of usage and are durable. These bags do not degrade in hot and cold temperatures.

Farmers either use the empty bags themselves or sell them to small traders. Small bags for packaging 5 kg materials are sold to scrap dealers.

In a way, fertilizer bags form an essential part of the rural and semi-urban economy. The empty bags are used for:

(i) Storage of grains, coconuts, household items, building materials, etc.
(ii) Providing cover for cattle, vehicles, haystack, animal dung cakes, etc. by stitching several bags together.
(iii) Stitching to form shopping bags and mats.
(iv) Making ropes.
(v) Miscellaneous uses like bedroll, curtains, sun shade, crop protection.

The above list is not exhaustive and only illustrates the role of empty plastic fertilizer bags in rural and even in urban economies. Some of these uses are illustrated in Figs. 2, 3, 4, 5, 6, 7 and 8.

Fig. 2 For covering haystacks/farm produce to protect them from rain

Fig. 3 Fertilizer bags to store grains

Fig. 4 Charpoy strings

Fig. 5 A popular use of empty fertilizer bags is the multi-utility shopping bag

Fig. 6 Bags stitched into a multi-utility mat

4.2 Final Disposal/Recycling of Empty Bags

The life of a large fertilizer bag is 2–3 years if it is used in an external environment and exposed to elements of nature. The life is longer of 4–5 years for reuse indoors. The ropes made of empty plastic fertilizer bags can last for more than 10 years. Used bags are freely traded between sellers (farmers) and users. After the active life of empty bags, large bags are sold by farmers to either recyclers or scrap dealers who

Fig. 7 Ropes—An innovative twist to bags

Fig. 8 Curtain cover for cattle

also in turn sell them to recyclers. These bags are then melt by recyclers and made into granules. Figure 9 [2] shows a picture of granules obtained from used plastic bags. These granules are sent to processing units which make hard plastic spares, bumpers, buckets and other articles from these granules. The flow chart for the life cycle of fertilizer bags is shown in Fig. 10.

5 Regulatory Environment

Plastic Waste (Management and Handling) Rules were notified in 2011 under the Environment (Protection) Act 1986. These rules have been amended from time to time. The latest amendments were notified in April 2023. These rules are called Plastic Waste Management (PWM) Rules, 2023 [4]. Producers and Marketers of

Fig. 9 Granules made by recycling of plastic bags

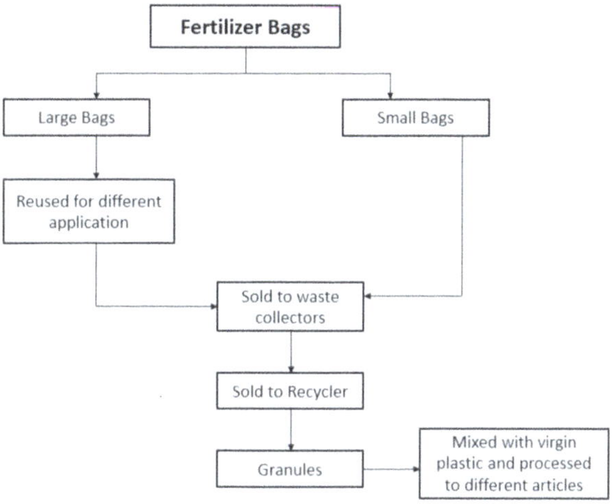

Fig. 10 Lifecycle of fertilizer bags

fertilizers are called "Brand Owners". The rules define a brand owner as "Brand Owner means a person or company who sells any commodity under a registered brand label/trademark". They are also covered under the clause of extended producer's responsibility which means "the responsibility of producers for the environmentally sound management of the product until the end of its life" [4] under Plastic Waste Management (PWM) Rules, 2016.

The plastic packaging has been defined under four categories viz. Category I: Rigid plastic packaging, Category II: Flexible plastic packaging of single layer or multilayer (more than one layer with different types of plastic), plastic sheets or like and covers made of plastic sheet, carry bags, plastic sachet or pouches.

Category III: Multilayered plastic packaging (at least one layer of plastic and at least one layer of material other than plastic) and Category IV: Plastic sheet or like used for packaging as well as carry bags made of compostable plastics. The fertilizer bags largely fall under the category II, i.e. flexible packaging. Further, the compliance of PWM Rules is for post-consumer plastic wastes. This is defined as the plastic packaging waste generated by the end-use consumer after the intended use of packaging is completed and is no longer being used for its intended purpose.

The purpose of compliance with the PWM Rules, the reuse of plastic bags by farmers and other stakeholders will be difficult to quantify. The regulator has made the compliance of recycling as product and locational agnostic. The target for recycling plastic packaging may be carried out within the same category and anywhere in the country. The MOEFCC notified the Extended Producers Responsibility Guidelines along with PWM Rules, 2022 [4].

5.1 Extended Producers Responsibility

Under Plastic Waste Management (Amendment) Rules 2022, guidelines for extended Producers Responsibility (EPR) were introduced. The EPR guidelines cover the reuse, recycling, use of recycled plastic content and end of the life disposal of plastic packaging. According to guidelines, any entity falling under more than one sub-category viz. producer, importer or brand owner has to register under each sub-category separately and entity having operating units under a sub-category in different states have to register separately. The Central Pollution Control Board (CPCB) shall register entity under each category. The registration is granted initially for 1 year and subsequently granted for 3 years. In addition, CPCB has also prescribed standard operating procedures (SOP) for the registration and filing of annual return on an online system. Brand owners and EPR were given graded targets for recycle of quantity of plastic used reaching 100% quantity to be recycled in 2023–24. Any entity recycling more than used plastic quantity used in a year can use it to (i) offset previous year shortfall (ii) carry forward for use in succeeding year and (iii) sell to other producers, importers and brand owners.

EPR target will be applicable for virgin plastic packaging material which means the quantum of recycled plastic used in the packaging material will be excluded from the obligation. There is also provision to utilize recycled plastic in packaging material starting from the year 2025–26. These targets will not be applicable for plastic packaging which is 100% biodegradable in the ambient environment.

6 Compliance by Fertilizer Producers

Fertilizer industry started implementing Extended Producer Responsibility (EPR) in 2021–22 with a collection and recycle/reuse target of 25% in 2021–22, 70% in 2022–23 and 100% in 2023–24. As per the EPR portal of CPCB [5], the total EPR targets of 3.07 million tonnes was given for the year 2022–23 for all industries. The flexible packaging accounts for 54.4% of total EPR target. Considering the 70% recycle requirement for 2022–23, fertilizer industry was responsible for 5% of total EPR requirement in the country.

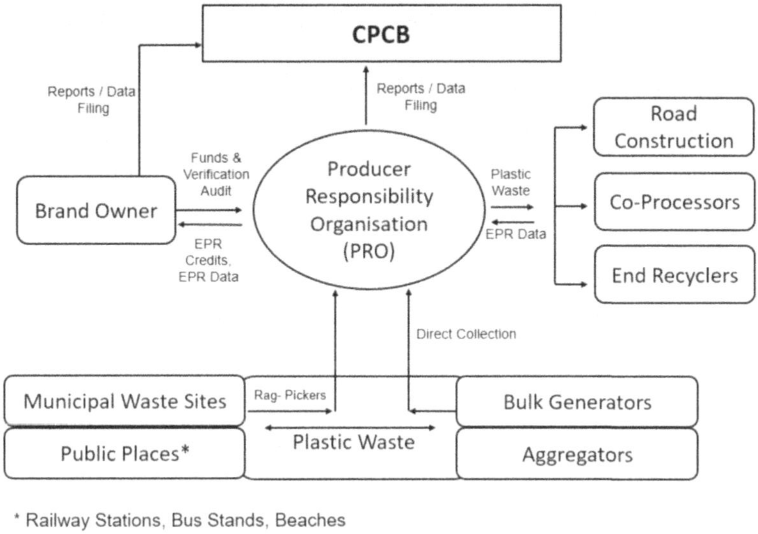

Fig. 11 Plastic waste management system

All fertilizer producers have registered with the CPCB. The compliance is being done through Producer Responsibility Organizations (PROs)/Waste Management Agencies (WMAs) which are also registered with the CPCB. PROs/WMAs are responsible for collection/recycling of post-consumed plastic waste for the brand owners with mutually agreed charges per kg of plastic waste. In another alternatives collectors and recyclers of plastic waste may be different entities. Figure 11 shows the flow of work for compliance. Initially, there were problems that some bag producers and recyclers were either not registered or not filing returns with CPCB. The compliance with EPR can be verified by CPCB only when all the stakeholders viz. bag producers/suppliers, fertilizer producers, collectors and recyclers file the complete return in time. These teething problems are still being ironed out to verify the EPR obligations. In the meantime, fertilizer producers have by and large complied by recycling the equivalent plastic waste materials.

7 Conclusions

Fertilizers are sold to 125 million farmers through more than 200 thousand retailers across India. Fertilizer bags are transported through rail and road and stored in warehouses of distributers, wholesalers and retailers. To ensure the quality of fertilizers, the plastic bags with lining are strong enough to withstand multiple handling and storage. Consumption of plastic packaging in fertilizer sector is about 150 thousand tonnes which is about 1.6% of total plastic waste in the country. Large fertilizer bags for 45 and 50 kg packing, account for almost 97% of the total plastic packaging used in the fertilizer sector.

Empty fertilizer bags are used both by farmers' and non-farmers' rural and semi-urban families for a variety of applications. Used fertilizer bags after their useful life are given to waste collectors who in turn sell them to recyclers. Fertilizer producers as brand owners have to recycle the waste plastic bags or equivalent plastic waste under Plastic Waste Management Rules 2016. Fertilizer producers, bag producers, plastic recyclers have to register with the Central Pollution Control Board and file annual returns. Fertilizer companies in India are ensuring compliance through Producer Responsibility Organizations for the recycle of plastic waste equivalent to the plastic used for packing the fertilizers.

References

1. Fertiliser Association of India—Internal Communication 2023.
2. Report on Life Cycle Analysis of Fertilizer Bags 2020: Commissioned by Fertiliser Association of India.
3. Fertilizer Statistics 2021–22, Vol. 67, P II 17 Fertiliser Association of India, New Delhi.
4. Plastic Waste Management Rules 2011, 2016, 2022 and 2023, Ministry of Environment, Forest and Climate Change, Government of India, New Delhi.
5. https://eprplastic.cpcb.gov.in

Plastics and the Sustainability Paradox

Swapan Ray

1 Preamble

From time immemorial, humanity has been on a constant search for materials to help meet the basic needs of hunting and gathering for food, making shelters, and fashioning clothes to protect from the vagaries of nature. Materials like stones, wood, leaves, and skins were *gathered* from around human habitats to meet these needs. From this stage of *gathering*, humans moved on to *extracting* material ushering in the copper and bronze age, followed by the iron age. Phenomenal progress was made during the *extraction* phase in understanding the science of *metallurgy*. Over time, sophisticated processes enabled production of the steel and aluminum. The range of metals obtained through *extraction*, overcame the deficiencies of materials that were earlier *gathered* from nature.

With advances in science and technology, the stage was set to *synthesize* materials at the dawn of the twentieth century by applying the knowledge of *chemistry*. Thus, the quest for materials progressed from *gathering* to *extracting* and then to *synthesizing* materials with properties and functionalities that were hitherto not available. This ushered in the era of plastics, a family of materials, like metal, eventually occupying center stage in the material domain. What is noteworthy is that with every transition, the domain of materials became more sophisticated. Plastics enabled making a range of products with improved functionalities which were earlier not possible with traditional materials. This platform of *synthesized material* also enabled the design of new products that were not possible in past. An example is the range of digital instruments and accessories made with plastics. The digital revolution ushered in this century has been built on the platform of synthetic material—plastics.

S. Ray (✉)
Indian Center for Plastics in the Environment (ICPE), 401, Choksey Mansion, 303, Shahid Bhagat Singh Road, Fort, Mumbai 400001, India
e-mail: secretary@icpe.in

Somaiya Vidyavihar University, Vidyavihar, Mumbai 400077, India

M. Goel and N. G. Tripathi (eds.), *Plastic Pollution*,
https://doi.org/10.1007/978-981-97-5528-8_12

2 Plastics Materials

Plastics, as natural polymers, are present in nature for millennia. These are shellac, ivory, animal horns, and casein, to name a few. However, their limited availability was a major hurdle for any large-scale deployment. Manmade plastics arrived in the field during the middle of nineteenth century, initially in the form of cellulose acetate, commonly known as Celluloid. This replaced ivory in billiard balls, an early step to save the tuskers! Credit for the major commercial success of Celluloid goes to John W. Hyatt, an American inventor without much formal scientific training. Celluloid started being used to make decorative handles, artifacts, and cinematographic films.

The first truly synthetic plastic, Phenol Formaldehyde (PF), was invented in 1907 by Leo Hendrik Baekeland, a Belgian chemist [1]. This ushered in the "Plastics Era." Material scientists and inventors were now less constrained by limitations imposed by nature in meeting the need for materials. Global demand for plastics accelerated and crossed 2.0 million tons in 1950, reaching 460 million tons now. Per capita use of plastics which was less than 1 kg in 1950 is currently at 58 kgs; more than a 50-fold increase in slightly over seven decades. Plastics have now become ubiquitous in all areas of human activities. The primary reasons for such rapid growth are the versatility of plastic materials and the affordability of plastic products.

Now plastic products are indispensable in every field of human activities. Plastics are used in the food sector, healthcare, consumer goods packaging, transportation, building and construction, electrical and electronic gadgets, and household utility items (Fig. 1) just to name a few. In most of these areas, plastics have replaced or supplemented use of traditional materials offering superior properties, improved functionalities and greater affordability.

2.1 Plastics and SDGs

In 2015, United Nation in it's General Assembly adopted 17 Sustainable Development Goals (SDGs) [2] to be realized by 2030. These include SDG-2 (Zero Hunger), SDG-3 (Good Health and Wellbeing), SDG-6 (Clean Water and Sanitation), SDG-7 (Affordable and Clean Energy), SDG-8 (Decent Work and Economic Growth), SDG-9 (Industry Innovation and Infrastructure), SDG-11 (Sustainable Cities and Communities), SDG-13 (Climate Action), and SDG-15 (Life on Land). Plastic products make a positive contribution in the pursuit of each of these 9 SDGs out of 17 SDGs (Fig. 2) adopted by the global community.

In agriculture, plastic products are used for growing, storing, handling, packaging, and distributing food and other commodities. Greenhouses for the production of horticultural products, by creating a controlled environment, allow the growing of food in harsh climatic conditions. Precision farming using drip irrigation and mulch films results in saving of 50–70% water and nutrients. During storage and distribution, plastic crates are used to reduce wastage. Bulk bags and sacks are used for storage

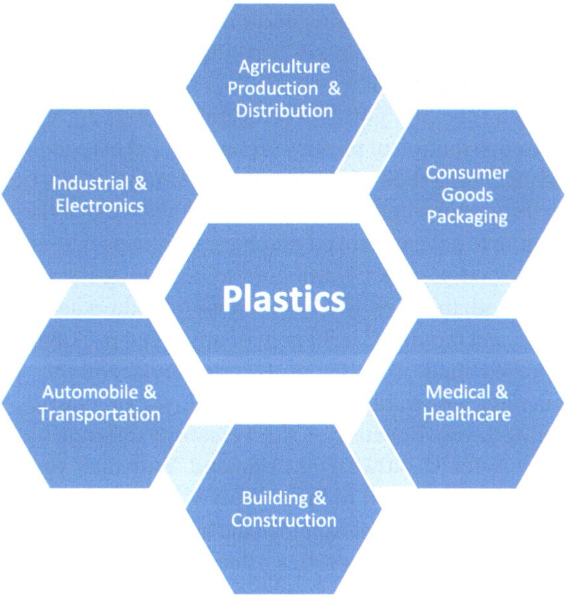

Fig. 1 Use Spectrum of Plastic Products. *Source:* Indian Center for Plastics in the Environment (ICPE)

Fig. 2 Sustainable Development Goals

and distribution of grains and other agricultural commodities. These substantially reduce wastages and losses, making food available to consumers in an efficient and cost-effective manner.

Both food and non-food consumer goods are widely packed in plastic products. Here contribution of plastics is phenomenal in reducing losses, protecting the content, and extending their shelf life. The versatility of plastics is highly expressed in design flexibility, substantially reducing package weight vis-à-vis content. These also offer a medium to have improved graphics. Plastic packaging has brought enormous benefits to the food sectors by reducing losses along the entire value chain. For perishable fresh food, the shelf life has increased by days and weeks and for packaged food by weeks and months. The benefit of advanced packaging of food with plastics had been unprecedented. To offer protection, the seemingly paper or multilayer packaging structures invariably incorporate a layer of plastics for barrier properties or as a sealing layer. Because of the versatility offered by plastics, the packaging sector accounts for 35–40% of total demand. With their versatility and affordability, plastics offer critical support to the pursuit of SDG-2.

Plastics are indispensable in the healthcare sector. Most of the products used in this sector are single use plastics offering protection to patients and healthcare workers. Plastic products are also used for dispensing medicines in a safe and efficient manner. These are indispensable at critical care centers for infusing blood and IV fluids as well as in preventing infectious diseases through vaccination using disposable syringes made of plastics. The non-reactive nature of plastics makes them safe for administering medicines and vaccines. Surgical masks and PPEs, critical accessories for doctors and health care workers, are made from plastic materials. Most of the diagnostic equipments and patient care accessories are made with plastic parts. Plastic products thus make a significant contribution to SDG-3.

Building and Construction sector uses plastic pipes for water distribution and sanitation areas. Plastic-insulated wires and cables are used for power distribution. PVC double-glazed windows, sidings, rotomolded plastic tanks made from PE for storage of water, and electrical switches are made of plastic materials. Distribution of clean water and wider coverage of sanitation infrastructure depends on plastics to help realize SDG-6.

To combat greenhouse gas (GHG) build-up, the global community is striving to increase the share of renewables in the energy mix. The present share of approximately 20% of renewable energy in the mix is likely to go up to 27% by 2030. Major part of this is solar and wind energy, which accounts for 13% of current share, where plastics play a critical both in making solar panels as well as windmill blades. With an aggressive push to decarbonize the energy sector, the share of both solar and wind energy would rapidly increase. Thus, plastics would play a significant role in our energy transition. This would help pursue SDG-7 on the plastic material platform.

The use of plastic parts in automobiles and airplanes has brought multiple benefits. The high strength-to-weight ratio and design flexibility have not only made these parts lighter but also more aerodynamic thereby substantially enhancing fuel efficiency and passenger safety. The higher fuel efficiency of internal combustion engines in cars translates to lower emissions and hence lower GHG generation. Supplementing these developments is rapid adoption of electric vehicles to further reduce GHG emissions wherein the contribution of plastic parts remains critical.

The electrical and electronic industries have been a major beneficiary of plastics due to their insulating properties. Plastics combine the insulating property with other wide-ranging physical properties of being extremely flexible or highly rigid depending on the types of plastic and/or the compounding technology deployed. All electrical and electronic gadgets extensively use plastic parts. The digital revolution is fully built on a plastic material platform.

Plastics are widely used in households as utility products for storage, leisure goods, cleaning, and grooming accessories. Modern-day living is dependent on plastic products. Since plastics have become integral to human society, it is necessary to evaluate their impact on the environment.

While plastic plays an enabling role in realizing the SDGs listed above, special attention should be paid to analyze its role in SDG-13, Climate Action.

3 Climate Change and Plastic

Climate on this planet had never been static. It had been oscillating between extreme hot and extreme cold periods. Present *Holocene Epoch* is the most benign period for life forms to thrive on this planet. The concern, however, is changes triggered by human activities. This has prompted scientific community to consider the proposal to mark the beginning of a new period, the *Anthropocene Epoch.* Prevailing scientific evidence attributes the GHGs build-up due to human activities as the primary contributor to global warming and climate change.

At an aggregate level, growing population and increasing affluence are implicated as primary drivers. These lead to increasing demand for food, energy, goods, and services emitting higher quantities of GHGs. It is estimated that the planet has reached 411 ppm of CO_2e in the atmosphere from the pre-industrial level of 280 ppm. The Intergovernmental Panel on Climate Change (IPCC) has set a target on global warming to a maximum of 2.0 °C but preferably limiting it close to 1.5 °C. This appears to be a daunting task since there are indications that the global average temperature rise has already breached 1.1 °C threshold. Paradoxically, the role of plastics in this complex phenomenon is beneficial. A simplistic view of the pathway to climate change is presented (Fig. 3).

It has been observed that more affluent households, societies, and countries are relatively more resource intensive. This leads to higher GHG emissions resulting in global warming and hence changes in climate. While the global community is striving to reduce GHG emissions by reducing energy production based on fossil fuel, the role of material in this endeavor is not adequately focused on and researched.

Nearly, 95% of plastics produced now are derived from fossil fuels. At the stage of production, plastics are highly energy intensive. However, as it moves forward along the value chain, the equation significantly changes to finally leaving a plastic products with a lower energy footprint. This happens due to three critical attributes:

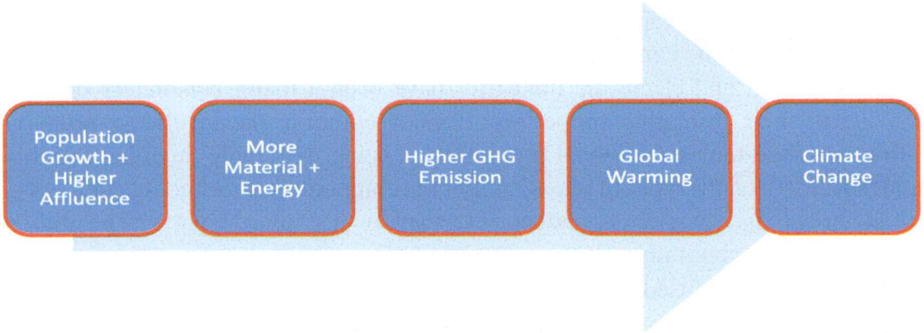

Fig. 3 Pathway to Climate Change. *Source* Indian Center for Plastics in the Environment (ICPE)

lower weight, enormous design flexibility, and lower energy needs for conversion as well as for recycling.

Weight Advantage: The density of most widely used plastics ranges from 0.90 to 1.38 g/cm^3 with an average value of less than 1.00 g/cm^3, as bulk of the commodity plastics (PE, PP, and PS) have densities in the lower range. Thus, a part made from plastics would typically weigh 1/8th, of steel, and approximately 1/3rd, of aluminum or glass which are often the alternative materials. The other alternative, paper, has similar densities as plastics but is deficient in physical properties and susceptible to moisture as well as insect infestation.

Conversion Energy Advantage: Bulk of the widely used plastic products are converted at temperatures ranging from 180 to 230 °C. While steel has three stages of high energy uses; during the smelting of ore, conversion to iron ingot, and eventually to steel. All these processes are carried out at very high temperatures ranging from 1400 to 2000 °C. Aluminum also uses a similar temperature range in the smelting process. Conversion of aluminum ingot to products is carried out at a lower temperature close to 700 °C. Glass bottles are formed. As well as recycled at 1500–1600 °C. These differences in conversion and recycling temperatures substantially compensates for the initial energy handicaps plastic material.

Energy Advantage in Use: Since plastic products typically weigh a fraction of identical products made from alternative materials like metal or glass, there is substantial savings in energy spent during transportation and distribution of plastic products as well as products distributed in plastic packaging. The design flexibility of plastics offers an additional advantage in weight reduction by moving away from a rigid structure to a flexible one. A unique example is a supply of fresh milk in pouches replacing the previous packaging medium, glass by Indian dairies (Fig. 4).

For packing and delivering one liter of milk, the weight of the packaging medium in past (glass bottle) was 650 g. The same quantity of milk is now packed by dairies in 5 g of flexible polyethylene films (Fig. 4). This shift in material and package design combines the benefits of density differences and design flexibility of plastics.

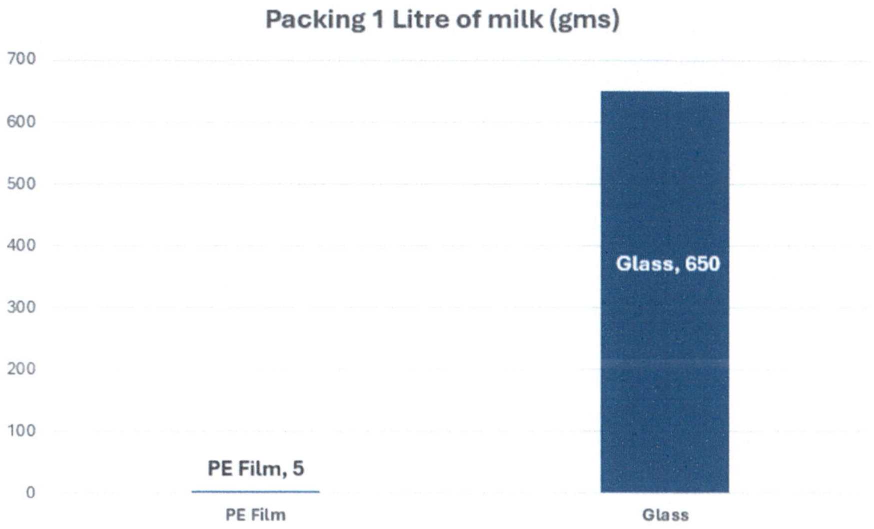

Fig. 4 Comparative Material Footprint for Milk Packaging. *Source* Indian Center for Plastics in the Environment (ICPE)

Milk delivered in the past in a 10-ton capacity delivery vehicle was carrying 6 tons of milk and 4 tons was the weight of packaging material (glass). Now, dairies are able to deliver 9.5 tons of milk in the delivery vehicle of same capacity since the packaging material (plastics) weighs only 0.5 tons. This saves nearly 40% of trips for delivering the same quantity of milk and a corresponding reduction in the emission. An additional benefit is the total elimination of liquid effluent generated while washing glass bottles in past before filling and, not to mention the safety hazards often encountered with broken glass bottles.

The use of plastics in most of the applications eventually results in lower energy consumption and hence lowers GHG emissions as compared to alternatives.

Life Cycle Assessment (LCA) carried out for many plastic products has arrived at similar conclusions. This includes an LCA study of few Single Use Plastic Products (SUPs) carried out by the United Nations Environment Program (UNEP). One such meta study on plastic carry bags concluded that the environmental footprint of plastic carry bags is the lowest as compared to the alternatives like paper, cloth, or bags made from biodegradable plastics [3].

Climate change caused by global warming is a threat to humanity and biodiversity on the planet. An effective means to counter climate change is to reduce our carbon footprints. This can be achieved by reducing the use of fossil fuels and expanding green coverage. Unfortunately, in last seventy years, nearly 20% of forest cover has been lost to meet our needs for paper and timber. These are often offered as alternatives to plastics. The rate of deforestation has not yet abated making mitigation of climate change a daunting task. All these do not justify more use of plastics. It is imperative that we reduce our use of all materials, including plastics, to attain a more sustainable future.

4 Plastic Pollution: Challenges and Solutions

Despite all the benefits listed above for plastics, there are challenges. These are in the generation and accumulation of plastic waste in different environmental compartments. Numerous studies have identified and assessed the sources, pathways, and fate of plastic waste. Being light in weight and highly visible in waste piles, plastic has drawn global attention. Tackling "Plastic Pollution" has become an urgent imeprative. UNEP is spearheading a global effort to forge a legally binding docuemnt with member countries to contain plastic pollution.

However, plastic waste is a part of all solid wastes generated by people. The quantity of waste generated is also influenced by socioeconomic status of households in a particular community, the level of affluence, and the country. An Indian study has found plastic waste, on an average, accounts for 7% of all solid waste generated. In affluent households, the share of plastic waste generated is relatively higher and is close to 17%. Thus, a holistic approach to contain and manage all solid waste would be necessary to address the issue of plastic pollution. This would also need substantial investment in upgrading solid waste management infrastructure and behavioral changes in consumers to address the issue of littering, particularly in the underdeveloped and developing countries.

Plastic waste, when well segregated, is highly amenable to recycling. The waste can be recycled through thermomechanical processes or through chemical processes. Chemical recycling can produce virgin quality material from plastic waste. Greater focus on recycling technology, improving waste management infrastructure, and making behavioral changes in people can lift up recycling rates and contribute to a more sustainable growth. Finally, at the end of its life, the energy content embedded in plastic waste can be recovered through incineration.

4.1 Biodegradable Plastics

An argument often offered in favor of a drastic reduction in demand for traditional plastic is the possible use of alternative material, in particular bio-based material. This is on the premises that bio-based materials are more planet-friendly. This is a fallacious argument. Products made from many of these alternative materials fall short on functionality, durability, and cost.

There is also a general lack of clarity and consensus surrounding bioplastics, biodegradable plastics, and compostable plastics. All bioplastics are not necessarily biodegradable. And all compostable plastics need not be bio-based. Many conventional plastics can be made from bio-based feedstocks.

The argument in favor of exploring the use of bio-based plastic materials is their impact on the carbon cycle of the planet. Since these are made from novel carbon atoms, instead of fossilized carbon of fossil fuel source, their contribution to GHG build-up is minimal. These novel carbons remain in circulation in the environment throughout the life cycle of these plastics over a much shorter time cycle compared to fossil source carbon in the carbon cycle of the planet.

Any discussion on plastic pollution would be incomplete without looking at the emerging areas of micro and nano plastics.

4.2 Micro and Nano-plastics

Narrative on microplastics has gained extraordinary momentum after a researcher reported their presence in drinking water. Since then, numerous research papers reported the presence of microplastics in various environmental compartments including in various parts of the human biota [4]. These have logically raised alarm of potentially serious health impact of plastics on human health.

Microplastics are fragments of synthetic materials measuring below 5 mm and nano-plastics are particles in nano-scale size. Primary microplastics are micro-sized plastic particles intentionally created for use as exfoliants in cosmetic products or as microbeads in blast cleaning of ocean-going vessels. In later case it is often paints containing polymeric materials that chemically the same molecule but are not plastics as commonly understood. The secondary sources of microplastics are presumed to be fragmented pieces of plastics generated due to forces of abrasion or UV degradation of plastic products. This process is yet to be fully understood given the complex nature of the interaction of varied climatic forces on different plastic products. However, there had been a spurt of research papers on microplastics ingression into marine and terrestrial biota. Moving through the food chain, it's impact on human health, is also being extensively researched. This has heightened the concern over commonly used and understood plastic products which may not be the primary sources of microplastics.

Scientific evidence suggests that microplastic could be a small portion of the body of microparticles widely distributed in different environmental compartments. Besides dust, shoots, and pollen (naturally occuring microparticles), other microparticles are largely microfibers emanating from synthetic, semi-synthetic, and natural fibers. There may also be nonsynthetic microfibers categorized as microplastics due to measurement errors. Major sources of synthetic microparticles of human origin are the release of microfibers from washing activities and vehicular tire abrasion. Thus, an incorrect terminology (microplastics) has the potential to derail our effort to manage various microparticles in the environment that have potential risk factor.

Little is known about the human health risks of NMPs (nano and microplastics), and what is known is surrounded by considerable uncertainty. There is no evidence of widespread risk to human health from NMPs (plastics of sizes below 5 mm) at present [5]. NMPs are evolving fields of enquiry and it is necessary to approach this with an open mind.

4.3 Circular Economy

Carbon economy is foundational to the plastic industry. The industry depends predominantly on fossilized carbon. Over 90%, of plastics are derived from crude oil, natural gases, or coal based sources. During production as well as conversion of plastics, energy is expended contributing to greenhouse gas release into the environment. During use, plastic products offer substantial benefits due to their superior functionality, substantially lower weight coupled with lower energy and water footprints during conversion, use and thermomechanical recycling. This substantially negates the initial handicap of starting as a highly energy-intensive material.

Technically all plastics can be recycled. This brings us to the paradigm of circularity in the plastic economy. Keeping plastics in circulation through reuse and recycling (mechanical, chemical, and feedstock recycling) reduces the demand for new feedstocks. Recycling of plastics also has the least energy and water footprints as compared to alternatives. The challenges are in keeping plastic waste well segregated by putting in place a robust waste management infrastructure and changes in consumer behavior to make the process more efficient and effective.

Recycling is critical to circularity in plastics. Shifting from a linear path of "Procure—Produce—Use and Throw," recycling retains material in the loop thus minimizing waste (Fig. 5).

To combat plastic pollution, it is imperative to shine more light on creating circularity in the plastic value chain than exploring upstream solutions of limiting production that could have other unintended negative impacts on the environment. Mindful use of products and responsible management of waste can significantly reduce plastic waste in the environment.

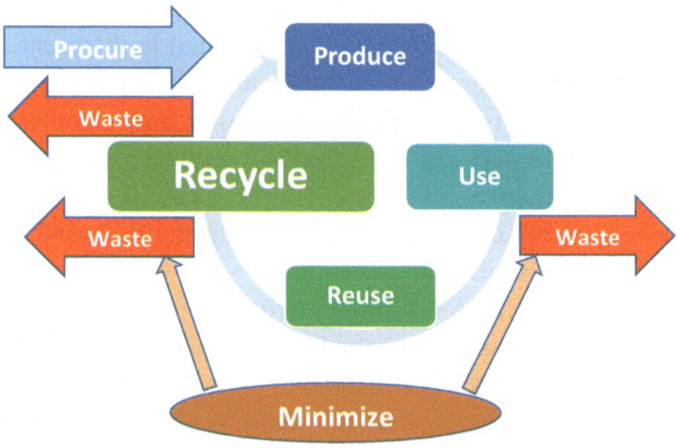

Fig. 5 Circular Economy model in Plastics. *Source* Indian Center for Plastics in the Environment (ICPE)

5 Conclusion

Plastics bring numerous economic, social, and environmental benefits—savings in material, water, and energy. These reduce food losses and protect people from diseases. These also help reach food, water, medicines, and temporary shelters to people facing any unfortunate natural calamities. Plastic products are convenient to use and are relatively more affordable as compared to alternatives.

All human activities leave some environmental footprints. Production of steel or aluminum involves leaving mining tails mostly in eco-sensitive zones. Processing of ores uses large volume of water and generates waste which often degrades soil and nearby water bodies besides releasing particulates in the atmosphere. It is always prudent to carry out an informed cost benefit both in economic and ecological terms to ensure that the benefits outweigh the costs. In majority of cases of their current uses, plastic products would be found as better alternatives.

The world has made significant progress in reducing hunger and malnutrition, containing infectious diseases, and extending the average human lifespan. While advances in food and medical sciences had been the primary contributors, material platforms like plastics played a critically supporting role in delivering these benefits.

The first tenet of sustainability is reduction of demand for resources—material, water, and energy. Demand for plastics is largely linked to demand for goods and services availed by consumers and the society. Contrary to general perception, plastic products with their ability to use less resources like material, water, and energy, offer a more sustainable pathway to development. This is the sustainability paradox of plastics.

References

1. John, B. (1999). *Plastics materials* (7th, edn., pp. 1–5). Elsevier. ISBN 81-8147-581-X.
2. United Nations General Assembly Resolution # 70/1 adopted on 25th, September 2015.
3. Ekvall, T., Liptow, C., & Miliutenko, S., (IVL Swedish Environmental Research Institute AB); United Nations Environment Programme. (2020). Single-use plastic bags and their alternatives. *Recommendations from Life Cycle Assessments*, 57–58.
4. Bank Michael, S., & Hansson Sophia, V. (2021). *Microplastic in the environment: Pattern and process* (pp. 4–9). ISBN 978-3-030-78626-7-4 (eBook).
5. Science Advice for Policy by European Academies; A Scientific Perspective on Microplastics in Nature and Society (p. 12). ISBN 978-3-9820301-0-4.

Plastics in the Indian Ocean: Survey, Transport, and Removal

R. Venkatesan, R. Keerthivasan, Yazhini Sivaram, and S. Ramasundaram

1 Introduction

Over the last thirty years, a significant increase in plastic production, driven by population growth and economic development, has led to a substantial increase in plastic in the ocean [1, 2]. The proliferation of low-cost single-use plastic, particularly in packaging, has emerged as a prominent contributor to this crisis. Mismanagement of plastic waste on land stands out as the primary cause of oceanic plastic pollution, exacerbated by the practices of both developing nations and those with significant plastic footprints [3–5]. Moreover, rivers serve as conduits, transporting plastic waste from land to sea [6–8]. Around 8 million metric tons of land-based plastic are estimated to find their way into the ocean annually, constituting approximately 80% of total marine pollution [9]. Projections suggest that unless current consumption patterns change, the quantity of plastic in marine ecosystems could nearly triple over the next twenty years, reaching alarming levels [10, 11].

The "Global Commitment 2021 Progress Report" is a comprehensive document that outlines the progress of over 1,000 organizations that have pledged to create a circular economy for plastic, ensuring that it never becomes waste. This report, compiled by the Ellen MacArthur Foundation with contributions from the United Nations Environment Programme (UNEP), presents three key findings. First, the use of virgin plastic by global commitment brands and retailers appears to have peaked and is projected to decline more rapidly by 2025. Second, while recycling has been the primary driver of progress, it is insufficient for addressing plastic pollution. There

R. Venkatesan (✉)
Institute of Ocean Management, Anna University, Chennai 600025, India
e-mail: dr.r.venkatesan@gmail.com

R. Keerthivasan · S. Ramasundaram
National Institute of Ocean Technology, Chennai 600100, India

Y. Sivaram
Former Intern, University of Liverpool, Liverpool L69 3BX, UK

M. Goel and N. G. Tripathi (eds.), *Plastic Pollution*,
https://doi.org/10.1007/978-981-97-5528-8_13

is an urgent need to focus more on eliminating single-use packaging. Finally, many businesses and countries support a global agreement on plastic pollution, acknowledging that voluntary initiatives alone are insufficient. This report provides valuable data and insights into the performance of plastic signatories in various aspects of the circular economy, such as elimination, reuse, recyclability, recycling, decoupling, and transparency. It also includes sector-based insight papers and individual reports for each signatory, offering a detailed view of their targets and progress on various metrics. This report serves as a critical resource for understanding the current state of plastic use and management and the steps being taken toward a more sustainable future [12].

Plastics have become ubiquitous in modern society, finding their way into nearly every aspect of our daily lives [13]. Unfortunately, this widespread usage has also led to a significant environmental issue: plastic pollution in our oceans. The pathways of plastics to the ocean are numerous and varied, including direct littering, inadequate waste management systems, and runoff from landfills and urban areas [14–16, 1]. Once in the ocean, plastics can persist for hundreds of years, breaking down into smaller pieces known as microplastics [17]. The contributions of plastics to ocean pollution have spurred global concern and efforts to reduce plastic usage, improve waste management practices, and develop innovative solutions for cleaning up our oceans [18].

2 Influence of Seasonal Reversal of Circulation Features on Microplastic Concentrations in the Northern Indian Ocean

The Indian Ocean experiences the impact of the monsoon system, which plays a crucial role in shaping surface currents. During the southwest monsoon (June to September), strong winds blow from the southwest to the northeast, resulting in the development of the Southwest Monsoon Current. In contrast, during the northeast monsoon (November to March), the winds reverse, forming the Northeast Monsoon Current. These wind-driven currents influence the movement of water masses and contribute to circulation patterns in the Indian Ocean. Monsoonal winds, characterized by seasonal reversals in direction, exert a profound influence on the Indian Ocean's dynamics, including the transport of plastic debris. During the southwest monsoon from June to September, winds blow from the southwest, pushing surface waters toward the Indian subcontinent. Conversely, the northeast monsoon, prevailing from October to April, brings winds from the opposite direction, dispersing surface waters away from land [19]. This alternating pattern facilitates the movement of plastic debris, often originating from coastal regions, across vast expanses of the Indian Ocean. Plastic waste carried by these winds can travel long distances, posing threats to marine ecosystems and wildlife [20–22]. Understanding the interplay between monsoonal winds and plastic transport is crucial for effective management and mitigation strategies in combating marine pollution in the Indian Ocean.

The surface currents in the North Indian Ocean exhibit pronounced seasonal reversals driven by monsoon wind forcings (Fig. 1a, b). The reversals in coastal currents could cause the transportation of relatively high concentrations of microplastics observed in coastal waters into the open ocean. The anticyclonic currents could converge the microplastics to the center of the circulations. The concentration of microplastics along the three transects could exhibit seasonal patterns driven by changes in circulation features. Further analysis based on microplastic datasets from the other two seasons could provide more insights into the impact of North Indian Ocean circulations on the distribution of microplastics. The seasonal reversal of wind over the Indian Ocean creates a semipermanent upwelling zone in the western Indian Ocean and seasonal upwelling zones in the Arabian Sea and Bay of Bengal.

Under the influence of the southwestern monsoon, microplastics travel between the East Indian Ocean and the South China Sea via the Strait of Malacca [24]. The open ocean currents seasonally reverse during the summer monsoon current, the current flows toward the east of the Bay of Bengal from the west of the Arabian Sea, and during the winter monsoon current, it flows west of the Arabian Sea from the eastern boundary of the Bay of Bengal [19]. The East Indian coastal current and West Indian coastal currents also seasonally reverse [19]. This reversal of current systems helps transport microplastics from the Bay of Bengal to the Arabian Sea [25] and shows spatial variation in microplastics during various seasons.

Differences in polymer type, diversity, and aspect ratio among the stations were also observed. The transport of MPs to the deep sea is suggested to occur through different processes depending on oceanographic parameters [21]. MPs from coastal areas are also transported to hadal depths near the coast by turbidity currents along submarine canyons [26]. The need to understand the current status and behavior of MPs in the deep sea is highlighted, as they may threaten deep-sea ecosystems and biodiversity. The importance of reducing plastic production and consumption and controlling plastic discharge into the ocean is also emphasized.

3 Estimation of Marine Plastic Through Riverine Discharge (Sampling of the Muttukadu Estuary)

3.1 Pathways of Plastic Debris in Marine Environments—Drifter Study

Drifters are an essential source of information relevant to the transport processes and distribution patterns of floating marine litter (FML) on regional to global scales. The low-cost and robust design of a satellite-tracked drifter is applicable for studying complex pathways and submesoscale dynamics of floating litter in tidally influenced coastal and estuarine systems [27, 28]. The tracker study was conducted at

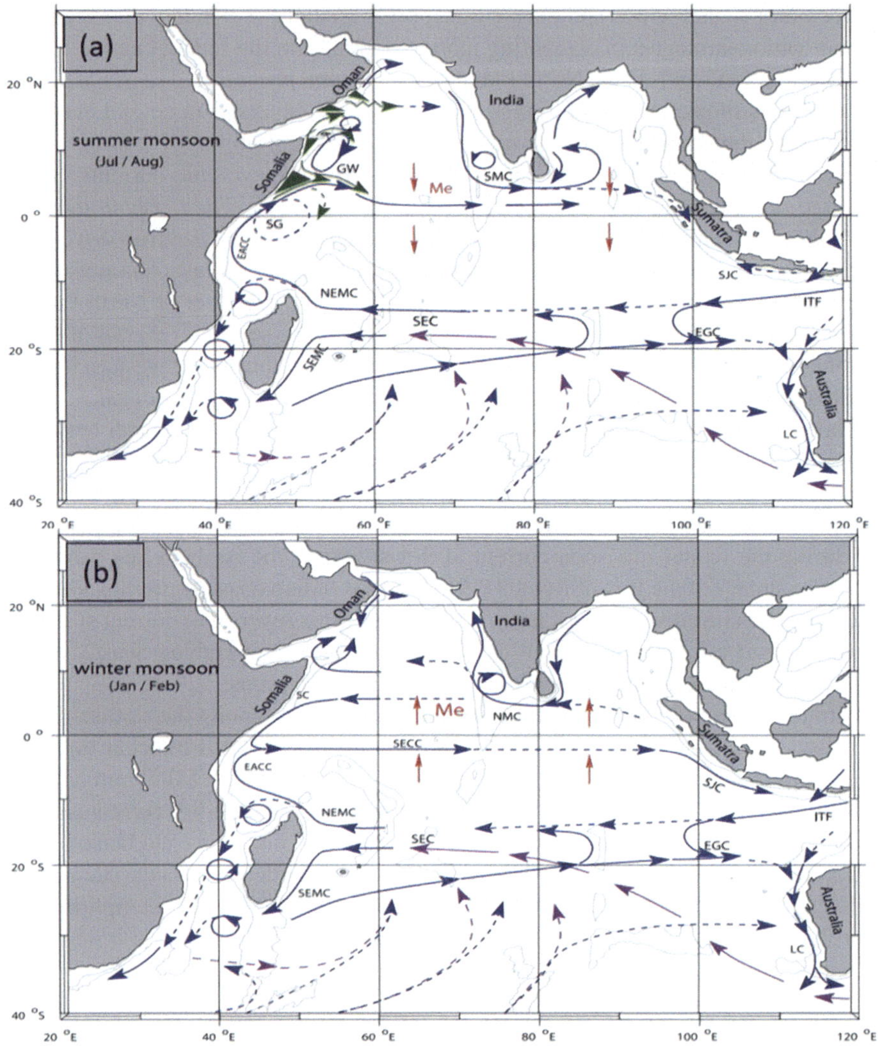

Fig. 1 **a** Schematic representation of the identified current branches during the summer (southwest) monsoon. Current branches indicate the Southwest and Northeast Monsoon Currents (SMC and NMC) and **b** Winter monsoon (Image courtesy: [23])

the Muttukadu estuary with a buoyant PVC tube equipped with onboard GPS data loggers and a position recorder. The PVC tube drifts along the surf zone while its position is recorded, and it is reminiscent of the popular technique of measuring the flow patterns in nearshore currents using a GPS-tracked surf zone drifter. Surface drift tests have significantly improved our understanding of complex surface current systems, such as floating marine litter (FML) pathways. A similar study can aid in marine litter build-up prediction [29], oil spill dispersion modeling and prediction [30].

This study describes the use of satellite-tracked drifters to study the transport processes and distribution patterns of floating marine litter (FML), including plastic pollution, in coastal and estuarine systems. Understanding these patterns is crucial for developing effective strategies to prevent plastic pollution from entering the sea (Fig. 2).

Pilot studies were carried out at the Muttukadu estuary site to estimate the pathways through which plastics reach the ocean through riverine discharge. The samples were collected regularly during the neap and spring tides every month for one year (May 2019–March 2020). As a standard practice, surface and sediment samples were collected using a Manta Net (300 microns) and a grab sampler, respectively. Also, to assess the flow of water through the net, a current meter sensor was installed on the boat. Similarly, to compute the salinity and temperature, a CT sensor was installed on

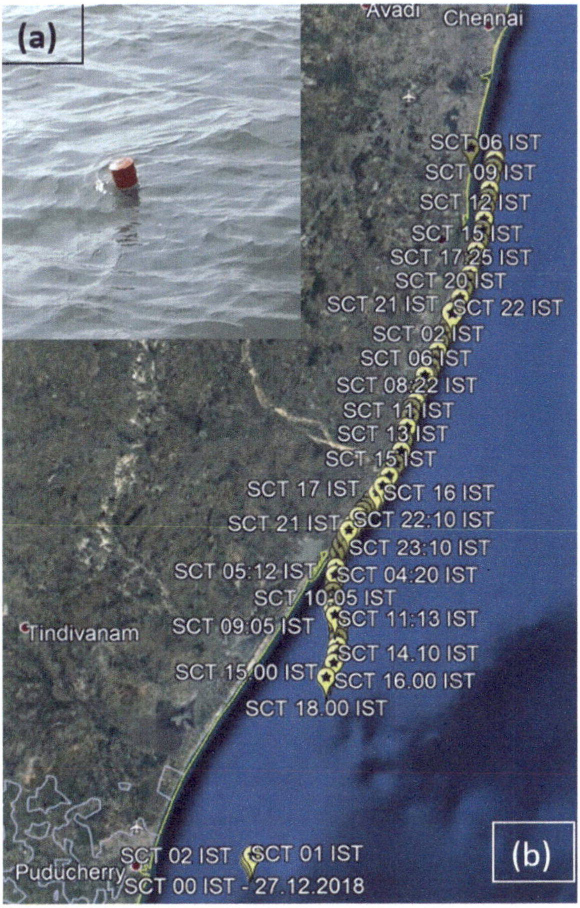

Fig. 2 **a** Drifter and **b** pathways of the drifter

the boat. The experiments were carried out for one year, and the collected samples were first subjected to laboratory analysis, microscopic examination, SEM–EDS, and FT-IR analysis.

This study helps monitor the process of understanding and assessing the presence of plastic pollution in a specific estuarine environment, addressing plastic pollution at its source, and preventing it from entering the sea (Fig. 3).

Surface samples and sediment samples were collected from the river mouth and off the Surf Zone-Muttukadu Estuary during the High Tide and Low Tide periods. Surface and sediment samples were collected every month beginning in May, and the collected samples were oven-dried. Macroplastics are segregated and quantified based on weight (Figs. 4, 5 and 6).

Surface water samples were collected by a 300-micron mesh Manta net with a rectangular opening of 60 cm × 40 cm. This net is towed for 20 min while the ship

Fig. 3 Sample collection at the river mouth and in the Off-Surf zone

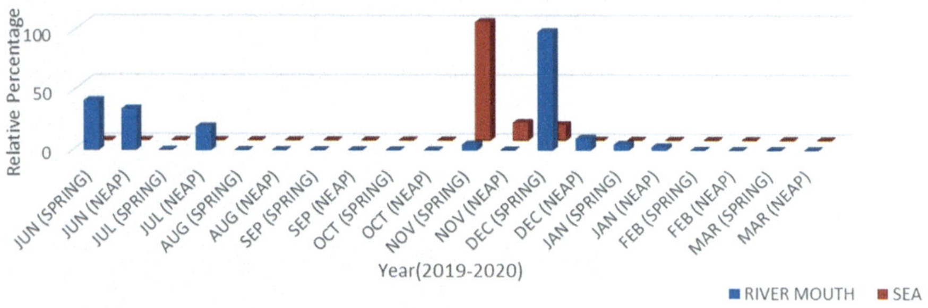

Fig. 4 Relative percentage of macro plastics in the river mouth and sea

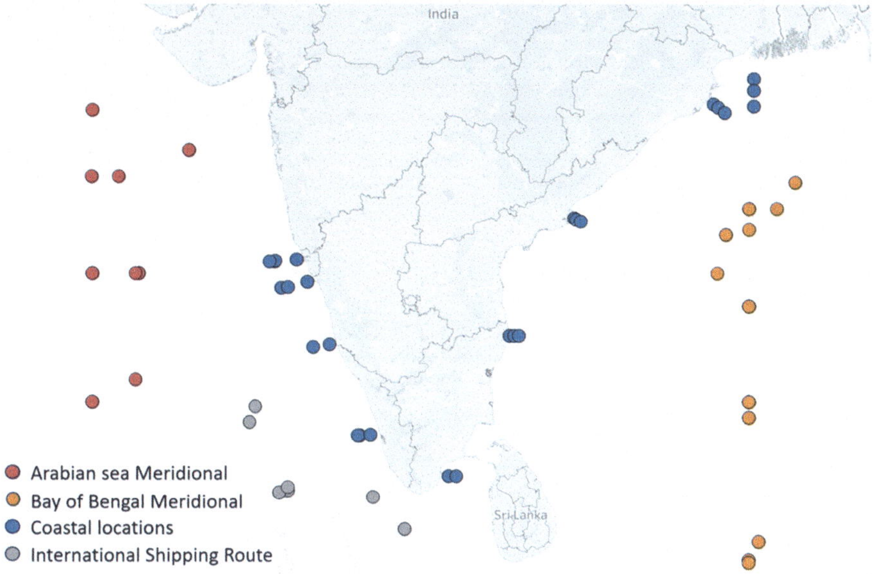

Fig. 5 Microplastic sampling transects in the open ocean waters of the northern Indian Ocean—Bay of Bengal meridional, Arabian Sea meridional, and international shipping routes

is sailing at 2–3 knots. A mechanical flow meter was attached to the center of the net to calculate the volume of the filtered water. A spinnaker boom was used to tow the Manta net along the side of the ship to avoid turbulence during sample collection. The cod end of the sample was transferred to a glass bottle for further analysis.

This study [31] involved seasonal sampling (pre-monsoon, monsoon, and post-monsoon) to assess the concentration of microplastics in the northern Indian Ocean, both in the open ocean and at coastal stations. The recorded microplastic concentrations were lower throughout all seasons than those observed in other seas and open ocean regions. Notably, the concentration in anticyclonic eddies was found to be less than that in the North Atlantic subtropical gyre and the South Pacific subtropical gyre. The reversal of the monsoon climate significantly influenced the concentration differences in the northern and southern locations of the Bay of Bengal (BoB) and the eastern central and southeastern locations of the Arabian Sea (AS).

The dominance of discarded fishing lines, nets, and ropes was identified as a significant factor contributing to the prevalence of fiber/line microplastics. This study represents the first investigation into temporal variations in the open ocean and coastal stations of the Arabian Sea (AS), Bay of Bengal (BoB), and Indian Ocean (IS), specifically at contour depths of 40, 200, and 800 m. Additionally, this study includes a comparative analysis with other open oceanic regions. This research also investigated the hydrodynamic influences arising from the seasonal reversal of monsoon currents [31].

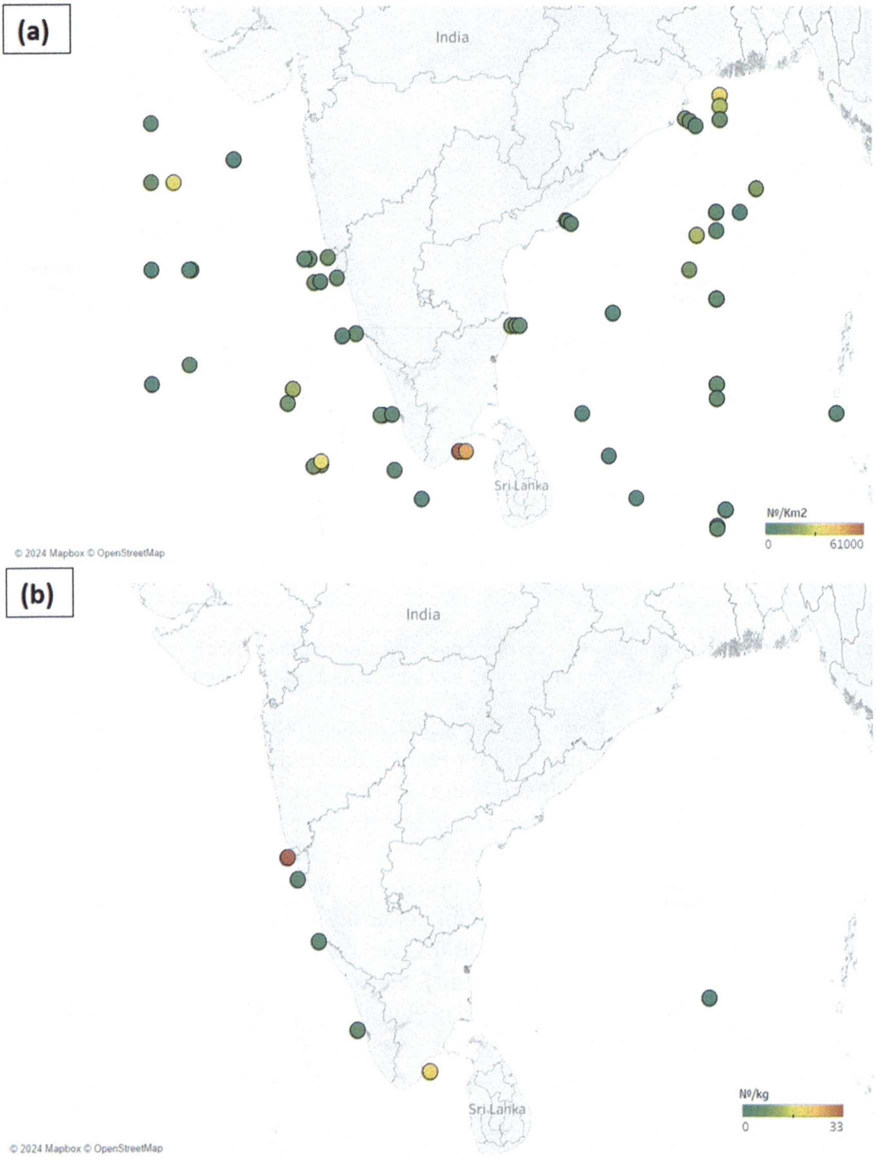

Fig. 6 **a** Sea surface sample collected from 2018 to 2021, **b** sea bed sediment sample

The deep-sea sediment samples from the BoB (11°N/88°E) exhibited a notable absence of plastic pollutants, indicating a relatively pristine environment at this depth (3030 m). In contrast, coastal locations off Chapora and Tuticorin had higher concentrations of particulate matter, suggesting a more significant influence of anthropogenic activities in these nearshore areas.

4 Classification of Techniques for Plastic Pollution Mitigation

The growing concern over pollution in water bodies has spurred the development of numerous technologies to mitigate this environmental challenge. These technologies range from simple mechanical devices to sophisticated automated systems, each designed to address specific types of waste in various aquatic environments. This paper provides an overview of the current state of water body cleanup technologies, categorizing them based on their operational environment and waste target.

Efforts to mitigate plastic pollution have led to the development of diverse technological solutions, which can be primarily classified into two distinct categories: prevention and collection. According to the comprehensive review by Schmaltz et al. [32], addressing plastic pollution requires diverse collection technologies. These technologies, classified into various categories, are designed to mitigate the impact of plastic waste in different environmental settings. This section provides an in-depth analysis of these categories, highlighting the technological innovations and their implications in the context of environmental management.

Prevention Technologies: This category encompasses systems and devices designed to prevent plastics from entering the environment. Key innovations in this group include various stormwater and wastewater filters, such as the In-line Litter Separator (ILLS) and the StormTrap TrashTrap. These technologies are pivotal for intercepting macroplastic particles in urban drainage systems. Additionally, unique solutions such as CLEVER-Volume and Cora Ball exemplify the breadth of this category. While CLEVER-Volume focuses on maritime waste management, Cora Ball is a consumer-focused product designed to capture microfibers in laundry machines. These prevention technologies collectively address a wide range of plastic pollution sources, from industrial and urban runoff to everyday household activities.

Collection Technologies: The second major category comprises technologies developed to remove plastics that have already contaminated various ecosystems. This group includes large-scale mechanical systems such as the Ocean Cleanup System, which uses natural oceanic forces to gather floating plastics. At the more localized level, technologies such as the Seabin and WasteShark demonstrate targeted approaches to collect debris in specific environments such as marinas and coastal waters. Additionally, specialized tools such as the Marine Microplastic Removal Tool are designed to address the less visible but equally important issue of microplastic pollution. This category signifies the adaptability and scalability of technologies to address plastic pollution in different environmental contexts, from open oceans to urban waterways (Table 1).

Table 1 Classification of plastic pollution cleanup technologies

Category	Technologies	Description
Stormwater management and urban waterway cleanup	In-line litter separator (ILLS), StormTrap TrashTrap, PumpGuard, Netting TrashTrap System, StormX Netting Trash Trap, Watergoat Trash Trap, Bandalong Litter Trap, Clear River Litter Trap, SCG Litter Trap, Clean River Project River Boom, Bandalong Boom, The Litterboom Project	Systems are installed in urban drainage networks and waterways to trap debris and prevent pollution from entering natural water bodies
Floating and surface debris collection	Holy Turtle, FRED, WasteShark, Seabin, BluePhin, Marina Trash Skimmer, Plastic Fischer Trash, AlphaMERS Floating Barrier	Technologies designed to collect debris floating on water bodies' surfaces include automated and AI-driven systems
Large-scale water body cleanup	Ocean Cleanup System, SeaVax, Inner Harbor Water Wheel, Versi-Cat Trash Skimmer Boat, ERVIS, One Earth-One Ocean SeeHamster, One Earth-One Ocean, Manta, The Interceptor, MariClean	Advanced systems for large-scale cleanup efforts in rivers, lakes, and oceans, tailored to specific environments and extensive pollution areas
Microplastic and fiber collection	GoJelly Project, Cora Ball, Fibre Free, Lint LUV-R, Showerloop, Marine Microplastic Removal Tool, Unnamed Invention by Anna Du, Unnamed Invention by Fionn Ferreira	Solutions targeting the capture of microplastics and fibers, with applications ranging from household to industrial scales
Beach and coastal cleanup	Barber Surf Rake, Barber Sand Man, Hoola One	Mechanical and vacuum-based solutions designed for removing waste from beaches and coastal areas
Innovative and experimental solutions	Stow it Don't Throw It, CLEVER-Volume, Unnamed Invention by Gering High School, Jellyfishbot, Malolo I, Unnamed GPS Device, NetTag, Wikilimo, OC-Tech, The Great Bubble	Emerging technologies featuring grassroots initiatives, advanced robotics, AI, and novel approaches to pollution control

4.1 Stormwater Management and Urban Waterway Cleanup

Stormwater runoff is a significant source of pollution in urban waterways. Technologies such as the In-line Litter Separator (ILLS), StormTrap TrashTrap, and PumpGuard exemplify solutions targeting this issue. These systems are primarily installed in urban drainage networks, trapping debris before reaching natural water bodies. The Netting TrashTrap System and StormX Netting Trash Trap use mesh nets to capture smaller particles, demonstrating versatility in dealing with varying sizes of waste.

4.2 Floating and Surface Debris Collection

Floating debris collection systems such as the Watergoat Trash Trap and the Bandalong Litter Trap are also crucial in urban settings. These devices are strategically placed in waterways to intercept floating debris, preventing it from reaching larger water bodies. Integrating these technologies into urban infrastructure reflects a proactive approach to pollution management, emphasizing prevention at the source.

For more extensive water bodies, floating debris collection systems are indispensable. Technologies such as Holy Turtle, FRED (Floating Robot for Eliminating Debris), and WasteShark focus on collecting debris floating on the water surface. Seabin and BluePhin are more advanced iterations that often involve the incorporation of automated or AI-driven mechanisms for enhanced efficiency. For example, the Marina Trash Skimmer shows an adaptive approach that is suitable for marina environments where floating debris is a common issue.

4.3 Large-Scale Water Body Cleanup

The Ocean Cleanup System, SeaVax, and Interceptor represent the forefront of large-scale cleanup efforts. These technologies are designed for deployment in rivers, lakes, and oceans, targeting extensive areas affected by pollution. The Inner Harbor Water Wheel and the Versi-Cat Trash Skimmer Boat are examples of how these systems can be tailored to specific environments, whether closed water bodies or open seas.

4.4 Microplastic and Fiber Collection

The advent of microplastics has presented a new challenge in water pollution. Solutions such as the GoJelly Project, Cora Ball, and Fiber Free focus on capturing these minuscule particles. Technologies such as the Lint LUV-R and the Showerloop extend this functionality to household applications, indicating a shift toward integrating pollution prevention into everyday activities. The innovative approaches of the unnamed invention by Anna Du and Fionn Ferreira highlight the evolving nature of technology in addressing microplastic pollution.

4.5 Beach and Coastal Cleanup

Specialized cleanup technologies are required in beach and coastal areas due to their unique environments. The Barber Surf Rake and the Barber Sand Man are mechanical solutions designed to efficiently remove waste from sandy beaches. Hoola One represents a more advanced approach that uses vacuum technology to separate debris from sand, demonstrating the adaptability of cleanup technologies to different terrains.

4.6 Innovative and Experimental Solutions

Emerging technologies in this field are marked by innovation and experimentation. The Stow it Don't Throw It project, CLEVER-Volume, and the Unnamed Invention by Students at Gering High School illustrate grassroots and educational initiatives in pollution control. Advanced systems such as the Jellyfish bot and Great Bubble Barrier represent the cutting edge in pollution control technology, employing robotics, AI, and novel physical principles to address pollution.

4.7 Initiatives in India

India recognized the threat plastic poses to its biodiversity, and people and various initiatives have been taken at various levels to address this issue. The central government, state governments, organizations, NGOs, and the general public all play vital roles, but there has been a lack of collective approaches. Several initiatives are briefly described in the following sections:

4.7.1 Marine Plastic Survey in India

Despite efforts to mitigate plastic pollution, a significant knowledge gap exists in implementing monitoring systems to make mitigation policies more effective. This significant knowledge gap has led to an increase in plastic waste generation. Sixty major Indian cities collectively produce 15,343 tons of solid waste that is disposed into the South Asian Seas every day [33, 34]. The average per capita plastic consumption in India is approximately 11 kg [33].

Despite efforts to address plastic pollution, we lack the monitoring systems needed to make policies truly effective. This knowledge gap severely hinders our fight against the growing plastic waste crisis. India is a major contributor to this problem, with 60 major cities generating over 15,000 tons of daily solid waste that pollutes the South Asian Seas [33, 34]. Additionally, plastic consumption per person in India stands at roughly 11 kg annually, and increasing plastic production means millions of tonnes of waste with poor recycling rates [33].

A long-term ocean monitoring program (COMAPS) revealed plastic waste buildup along India's Andaman coastline, likely caused by ships and fishing vessels dumping trash into ocean currents [35]. This groundbreaking study sparked further research into India's plastic pollution problem. Sadly, plastic makes up 8% of India's waste, with Delhi, Kolkata, and Ahmedabad being major pollution sources. Plastic production in India has skyrocketed, but recycling rates are tragically low. This means most plastic ends up in landfills, incinerators, or polluting rivers and the ocean. India alone contributes a shocking amount of ocean plastic pollution through its waterways. Rivers worldwide are major plastic pollution pathways, and the Ganges River carries a massive amount of plastic into the Bay of Bengal annually [1, 36].

4.7.2 Swachh Sagar, Surakshit Sagar Campaign

This campaign was a national initiative to clean 75 beaches in 75 days in India [37]. It was conducted as part of the Azadi ka Amrit Mahotsav celebrations, and volunteers removed 1500 tons of litter from beaches [37]. The campaign also provided a way to assess how much plastic debris Indian coastlines accumulate, estimated to be approximately 0.98 metric tons of debris per km [34].

Commemorating the seventy-fifth year of India's independence, this unique first-ever national campaign culminated in the "International Coastal Cleanup Day" on 17th September 2022, celebrated as the "Sewa Divas" in the country. The target of this drive is to remove 1,500 tonnes of garbage from the seacoast, which will be a massive relief to marine life and the people staying in coastal areas.

This event was initiated by the Union Ministry of Earth Sciences and supported by the Ministry of Science and Technology, Ministry of Environment Forest and Climate Change (MoEFCC), Indian Coast Guard, National Disaster Management Authority (NDMA) and many other ministries, along with other social organizations and educational institutions. The mobile app "EcoMitram" was launched to spread awareness about the campaign and for the common people to volunteer registration for beach cleaning activities on September 17, 2022.

Development of Cleaning Marine Plastic Pollution

The deployment of large-scale passive collection systems represents a significant stride in the fight against marine plastic pollution. Organizations **such as Ocean Cleanup** have pioneered these systems, which employ advanced engineering to harness the power of ocean currents. By strategically positioning these systems in high-density plastic areas, they can effectively concentrate and capture plastic debris.

This passive approach reduces the need for more energy-intensive and environmentally harmful cleanup efforts, ensuring that plastic waste is intercepted before it breaks down into harmful microplastics.

Autonomous robots equipped with cutting-edge sensors and artificial intelligence have also become instrumental in addressing marine plastic pollution. These robots are designed to patrol water bodies and autonomously identify and collect plastic waste. Envirobo is a robotic device with artificial intelligence that can collect and dispose of plastic and debris from sewers and stormwater drains before they enter and pollute water bodies. The device has a self-unloading receptacle and a cognitive ability for periodic cleaning. Envirobo saves 5600 tons/year of GHG or CO_2 equivalent and tons of plastic debris otherwise pollute the aquatic environment, hinder ocean carbon fixation capacity, and exacerbate plastic-related climate issues (indiaai.gov.in).

The development of **biodegradable plastics** represents another vital avenue in the quest to combat marine plastic pollution. Researchers and companies are actively working to create plastics that breakdown more rapidly in marine environments. By designing plastics with reduced environmental persistence, the hope is that they will have a significantly reduced impact on marine ecosystems. As these innovative materials become more prevalent in various industries, the transition to biodegradable plastics could help mitigate the long-term effects of plastic pollution [38].

Moreover, recycling technologies are being developed to transform recovered marine plastics into new, usable products. This approach removes plastic waste from the ocean and decreases the demand for new plastic production. Recycling and repurposing plastics can reduce the strain on natural resources and energy while mitigating the pollution crisis. Innovations in recycling processes ensure that more plastic waste can be given a second life, reducing the overall environmental impact of plastic materials [39].

In conclusion, the battle against marine plastic pollution is to be fought on multiple fronts, with innovative technologies taking center stage. These groundbreaking solutions offer a comprehensive and proactive approach to tackling this issue, from passive collection systems, debris traps in riverine discharge, and autonomous robots to biodegradable plastics and recycling technologies. By intercepting plastic waste before it further harms our oceans, reducing the demand for new plastic production, and developing more environmentally friendly alternatives, these technologies drive the fight against marine plastic pollution toward a more sustainable and hopeful future.

References

1. Lebreton, L. C., Van Der Zwet, J., Damsteeg, J. W., Slat, B., Andrady, A., & Reisser, J. (2017). River plastic emissions to the world's oceans. *Nature Communications, 8*(1), 15611.
2. Geyer, R., Jambeck, J. R., & Law, K. L. (2017). Production use and fate of all plastics ever made. *Science Advances, 3*(7). https://doi.org/10.1126/sciadv.1700782
3. Law, K. L., Morét-Ferguson, S., Maximenko, N. A., Proskurowski, G., Peacock, E. E., Hafner, J., & Reddy, C. M. (2010). Plastic accumulation in the North Atlantic subtropical gyre. *Science, 329*(5996), 1185–1188.
4. Nøklebye, E., Adam, H. N., Roy-Basu, A., Bharat, G. K., & Steindal, E. H. (2023). Plastic bans in India–Addressing the socio-economic and environmental complexities. *Environmental Science & Policy, 139*, 219–227.
5. Mihai, F.-C., Gündoğdu, S., Markley, L.A., Olivelli, A., Khan, F.R., Gwinnett, C., Gutberlet, J., Reyna-Bensusan, N., Llanquileo-Melgarejo, P., Meidiana, C. et al. (2022). Plastic pollution, waste management issues, and circular economy opportunities in rural communities. *Sustainability, 14*, 20. https://doi.org/10.3390/su1401002
6. Honorato-Zimmer, D., Kiessling, T., Gatta-Rosemary, M., Campodónico, C. K., Núñez-Farías, P., Rech, S., & Thiel, M. (2021). Mountain streams flushing litter to the sea–Andean rivers as conduits for plastic pollution. *Environmental Pollution, 291*, 118166.
7. Windsor, F. M., Durance, I., Horton, A. A., Thompson, R. C., Tyler, C. R., & Ormerod, S. J. (2019). A catchment-scale perspective of plastic pollution. *Global Change Biology, 25*(4), 1207–1221.
8. van Emmerik, T. H., Frings, R. M., Schreyers, L. J., Hauk, R., de Lange, S. I., & Mellink, Y. A. (2023). River plastic transport and deposition amplified by extreme flood. *Nature Water, 1*(6), 514–522.
9. Abreu, A., & Pedrotti, M. L. (2019). Microplastics in the oceans: The solutions lie on land. *Field Actions Science Reports. The Journal of Field Actions*, (Special Issue 19), 62–67.
10. Stephanie B. Borrelle et al. (2020). Predicted growth in plastic waste exceeds efforts to mitigate plastic pollution. *Science, 369*,1515–1518.https://doi.org/10.1126/science.aba3656
11. La Daana, K. K., Asmath, H., & Gobin, J. F. (2022). The status of marine debris/litter and plastic pollution in the Caribbean Large Marine Ecosystem (CLME): 1980–2020. *Environmental Pollution, 300*, 118919.
12. Ellen MacArthur Foundation. (2023). The Global Commitment 2022 Progress Report312. Retrieved from: https://www.ellenmacarthurfoundation.org/assets/downloads/Global-Commitment-2022-Progress-Report.pdf
13. Jambeck, J. R., et al. (2015). Plastic waste inputs from land into the ocean. *Science, 347*(6223), 768–771.
14. Pattiaratchi, C., van der Mheen, M., Schlundt, C., Narayanaswamy, B. E., Sura, A., Hajbane, S., Wijeratne, S., et al. (2022). Plastics in the Indian Ocean–sources, transport, distribution, and impacts. *Ocean Science, 18*(1), 1–28.
15. Ritchie, H., & Roser, M. (2023). Where does the plastic in our oceans come from?. *Our World in Data.*
16. Russell, R. (2017). UN resolves to end ocean plastic waste. DW. Oceans.
17. Geyer, R., et al. (2017). Production, use, and fate of all plastics ever made. *Science Advances, 3*(7), e1700782.
18. Lebreton, L., & Andrady, A. (2019). Future scenarios of global plastic waste generation and disposal. *Palgrave Communications, 5*(1), 1–11.
19. Shankar, D., Vinayachandran, P. N., & Unnikrishnan, A. S. (2002). The monsoon currents in the north Indian Ocean. *Progress in Oceanography, 52*(1), 63–120.
20. Kim, I. S., Chae, D. H., Kim, S. K., Choi, S., & Woo, S. B. (2015). Factors influencing the spatial variation of microplastics on high-tidal coastal beaches in Korea. *Archives of Environmental Contamination and Toxicology, 69*, 299–309.

21. Venkatesan, R., Ramanamurthy, M. V., Latha, G., Ramasundaram, S., Mathew, M. V., Keerthivasan, R., Janani, R., et al. (2022, February). Marine plastic studies in the mid-water column of the North Indian Ocean. In *OCEANS 2022*, Chennai (pp. 1–6). IEEE.
22. Cózar, A., Echevarría, F., González-Gordillo, J. I., Irigoien, X., Úbeda, B., Hernández-León, S., ... & Duarte, C. M. (2014). Plastic debris in the open ocean. *Proceedings of the National Academy of Sciences, 111*(28), 10239–10244.
23. Schott, F. A., Xie, S. P., & McCreary Jr, J. P. (2009). Indian Ocean circulation and climate variability. *Reviews of Geophysics, 47*(1).
24. Li, C., Wang, X., Liu, K., Zhu, L., Wei, N., Zong, C., & Li, D. (2021). Pelagic microplastics in surface water of the Eastern Indian Ocean during monsoon transition period: Abundance, distribution, and characteristics. *Science of the Total Environment, 755*, 142629.
25. Veerasingam, S., Saha, M., Suneel, V., & Vethamony, D. P. (2017). Microplastic pollution: A serious threat to the marine ecosystem. *Blue Waters, 18*(1), 2–5.
26. Tsuchiya, M., Kitahashi, T., Nakajima, R., Oguri, K., Kawamura, K., Nakamura, A., Fujikura, K., et al. (2023). Distribution of microplastics in bathyal-to hadal-depth sediments and transport process along the deep-sea canyon and the Kuroshio Extension in the Northwest Pacific. *Marine Pollution Bulletin*, 115466.
27. Van Der Mheen, M., Van Sebille, E., & Pattiaratchi, C. (2020). Beaching patterns of plastic debris along the Indian Ocean rim. *Ocean Science, 16*(5), 1317–1336.
28. Meyerjürgens, J., Badewien, T. H., Garaba, S. P., Wolff, J.-O., & Zielinski, O. (2019). A state-of-the-art compact surface drifter reveals pathways of floating marine litter in the German Bight. *Frontiers in Marine Science, 6*, 58. https://doi.org/10.3389/fmars.2019.00058
29. Van Sebille, E., Wilcox, C., Lebreton, L., Maximenko, N., Hardesty, B. D., Van Franeker, J. A., Eriksen, M., Siegel, D., Galgani, F., & Law, K. L. (2015). A global inventory of small floating plastic debris. *Environmental Research Letters, 10*(12), 124006.
30. Reed, B., Brown, J., VanderZee, D., Loveland, T., Merchant, J., & Ohlen, D. (1994). Measuring phenological variability from satellite imagery. *Journal of Vegetation Science, 5*, 703–714. https://doi.org/10.2307/3235884
31. Janakiram, R., Keerthivasan, R., Janani, R., Ramasundaram, S., Martin, M. V., Venkatesan, R., Sudhakar, T., et al. (2023). Seasonal distribution of microplastics in surface waters of the Northern Indian Ocean. *Marine Pollution Bulletin, 190*, 114838.
32. Schmaltz, E., Melvin, E. C., Diana, Z., Gunady, E. F., Rittschof, D., Somarelli, J. A., Virdin, J., & Dunphy-Daly, M. M. (2020). Plastic pollution solutions: Emerging technologies to prevent and collect marine plastic pollution. *Environment International, 144*, 106067. https://doi.org/10.1016/j.envint.2020.106067
33. Bhattacharya, R. R. N., Chandrasekhar, K., Roy, P., & Khan, A. (2018). Challenges and opportunities: Plastic waste management in India.
34. CS, S. G. (2023). Intelligent debris mass estimation model for autonomous underwater vehicle. arXiv:2309.10617
35. Dharani, D. G., Abdul Nazar, A. K., Venkatesan, R., & Ravindran, M. (2003). Marine debris in Great Nicobar. *Current Science, 85*(5), 574–575.
36. Lechthaler, S., Waldschläger, K., Sandhani, C. G., Sannasiraj, S. A., Sundar, V., Schwarzbauer, J., & Schüttrumpf, H. (2021). Baseline study on microplastics in Indian rivers under different anthropogenic influences. *Water, 13*(12), 1648.
37. Robin, R. S., Karthik, R., Nithin, A., & Purvaja, R. (2023). Removal of marine litter and its impact along the coast of India. *Records of the Zoological Survey of India, 123*(1S), 67–86.
38. Moshood, T. D., Nawanir, G., Mahmud, F., Mohamad, F., Ahmad, M. H., & AbdulGhani, A. (2022). Biodegradable plastic applications towards sustainability: A recent innovations in the green product. *Cleaner Engineering and Technology, 6*, 100404.
39. Kassab, A., Al Nabhani, D., Mohanty, P., Pannier, C., & Ayoub, G. Y. (2023). Advancing plastic recycling: Challenges and opportunities in the integration of 3D printing and distributed recycling for a circular economy. *Polymers, 15*(19), 3881.
40. Wang, X., Li, C., Liu, K., Zhu, L., Song, Z., & Li, D. (2020). Atmospheric microplastic over the South China Sea and East Indian Ocean: Abundance, distribution and source. *Journal of Hazardous Materials, 389*, 121846.

Microbial Biodegradation of Plastics and Microplastics: Recent Development

Ayesha Nazeer, Atiya Ahmad, Faisal Ahmad, and Shamim Ahmad

Abbreviations

BPA	Bisphenol A
DES	Diethylstilbestrol
EG	Ethylene Glycol
HDPE	High-Density Polyethylene
LCC	Leaf-branching Compost Cutinase
LDPE	Low-Density Polyethylene
MPs	Microplastics
MW	Molecular Weight
OTUs	Operational Taxonomic Units
PA	Phthalic Acid
PAEs	Phthalates
PBS	Poly Butylene Succinate
PBSA	Poly Butylene Succinate-co-Adipate
PCL	Poly-Caprolactone
PE	Polyethylene
PES	Polyethersulfone
PHB	Polyhydroxybutyrate
PLA	Poly Lactic Acid
PET	Polyethylene Terephthalate
POPs	Persistent Organic Pollutants
PP	Polypropylene
PS	Polystyrene
PVC	Polyvinyl Chloride
TPA	Terephthalic Acid

A. Nazeer · A. Ahmad · F. Ahmad · S. Ahmad (✉)
Brentwood Tower, Eros Garden, Charmwood Village, Faridabad, Haryana, India
e-mail: drsahmad@email.com

M. Goel and N. G. Tripathi (eds.), *Plastic Pollution*,
https://doi.org/10.1007/978-981-97-5528-8_14

1 Introduction

Plastics have emerged as a significant global concern, contributing to widespread ecological problems. While these materials bring convenience to our lives, their persistent nature allows them to linger in the natural environment for extended periods, causing environmental contamination. The global production of plastics has reached alarming levels, resulting in substantial environmental challenges. The disposal of plastic waste, with only a small percentage being recycled or incinerated, exacerbates these issues. Plastics, including polyethylene (PE), polypropylene (PP), polyethylene terephthalate (PET), polystyrene (PS), and polyvinyl chloride (PVC), are major contributors to environmental pollution. These high molecular weight polymers contribute to the formation of "white pollution," causing ecological damage globally. The slow degradation of high polymers further complicates the environmental impact, necessitating a thorough investigation into plastic degradation pathways and efficiencies.

Plastics comprise various compounds, including monomers, oligomers, polymers, and additives like plasticizers, antioxidants, heat stabilizers, and pigments. Microbial degradation of plastics, particularly those with additives and adsorbed chemicals, requires special attention. Microplastics (MPs), containing a mix of chemical components, pose additional concerns, potentially bio-accumulating in the food chain. The release of plastics and MPs into the environment initiates degradation processes such as hydrolysis, thermal-oxidative degradation, biodegradation, electrochemistry, and photodegradation. Among these processes, microbial biodegradation stands out as a promising solution. Microbial enzymatic degradation, especially bacterial degradation, holds substantial potential for addressing plastic pollution. Microorganisms metabolize carbon sources in the form of organic matter, producing non-toxic by-products, and providing energy.

Understanding the impact of plastics on biodiversity loss and their life cycle is crucial. Microbial degradation emerges as a significant strategy to mitigate negative consequences, given the diverse composition of microbial communities. This review emphasizes the advantages of biodegradation, particularly through bacterial degradation. Despite limited knowledge of environmental plastic degradation, exploring the ability of bacteria and the interaction between bacterial enzymes and plastics is essential. The article delves into the viability of biodegradable plastics, highlighting current research on various microorganisms and their enzymes in synthetic plastic biodegradation. The challenges associated with microbial degradation and corresponding solutions are addressed, emphasizing the significance of pre-treatment in enhancing biodegradation efficiency. Microplastics (MPs) and plastics pose significant challenges to biodegradation due to lack of rich functional groups and hydrolyzable bonds. Their intrinsic composition, characterized by durability and resistance to degradation, makes them unsuitable substrates for microbial attachment and enzymatic reactions.

Recalcitrant polymers, with their less flexibility and long chains, undergo biodegradation into small monomers. However, this process requires prolonged enzymatic activities to target specific sensitive bonds within the polymer side chain or polymer chain, ultimately enhancing their fracture. In the biodegradation process, conventional petroleum-based plastics face challenges in being fully decomposed and assimilated by microorganisms, especially under abiotic factors like UV radiation, temperature variations, and physical stress over an extended period.

To address these challenges, there is a pressing need to enhance microbial degradation abilities and efficiency, potentially through the discovery of more suitable and efficient microbial strains. However, the current landscape faces obstacles such as undeveloped technologies and cost concerns, with technological development and capital consumption not aligning with social development. The quest for efficient plastic degradation remains a complex task, requiring advancements in microbial capabilities and the exploration of innovative technologies to match the scale of the issue at hand.

Furthermore, plastic degradation has become a focal point in environmental research due to the challenge of releasing toxicity when breaking plastics into smaller particles. A significant concern is the widespread use of phthalates (PAEs) as plasticizers to impart unique characteristics such as plasticity, flexibility, and durability to plastic products. Human exposure to PAEs, primarily through food, can lead to serious issues such as endocrine disruption, metabolic disorders, and reproductive toxicity.

Another notable compound contributing to endocrine disruption is Bisphenol A (BPA), a major component of synthetic polymers and additive plasticizers. Produced globally and employed in various polymer materials and polycarbonate synthesis, BPA poses health risks on a large scale. While plastics, as macromolecular substances, initially exhibit limited toxicity and are not easily absorbed by the human body, the peak of plastic toxicity arises from the addition of toxic additives during processing or incomplete reactions during synthesis. Notably, small particulate compounds, such as monomers, exhibit high toxicity.

Microplastics (MPs) and persistent organic pollutants (POPs) share a complex relationship influenced by their properties and environmental conditions, leading to the adsorption of POPs on MPs. The toxicity escalates during the monomer stage when a significant number of POPs is adsorbed on the MP surface. The efficiency of transferring these ingested POPs in the food chain remains uncertain but encompasses a wide range. Human exposure to monomers and POPs can disrupt the natural balance of gut microbiota through inhalation, ingestion, and skin contact. The wide-reaching toxicity and pollution caused by plastics and MPs have profound implications for ecology and human health in various aspects. It is imperative to urgently investigate the reasons for microbial biodegradation to address these environmental concerns comprehensively as discussed elsewhere as well [1–3].

2 Viability of Plastics and Microplastics (MPs) Biodegradations

A crucial prerequisite for successful biodegradation lies in the diversity of microorganisms. These microorganisms encompass bacteria, fungi, and a select few algae, boasting inherent traits like small size, rapid absorption and metabolism, robust adaptability, easy variability, and widespread distribution. Their innate or acquired survival abilities enable them to thrive even in exceptionally harsh environments. Microbial communities exhibit adaptability to new surroundings, adhering to them and releasing specific enzymes that enable them to utilize persistent plastic pollution as their sole carbon source.

Biodegradation is a complex interplay of biological and abiotic factors. Biological factors involve bacteria, fungi, biofilms, and microbial communities serving as the carbon source for microorganisms. Abiotic factors encompass aspects of natural photooxidation and human-induced physical and chemical degradation. These factors catalyze intricate metabolic reactions, shaping the ecosystem structure and function of biodegradable plastics.

The microbial degradation of plastics and MPs is influenced by the growth of microorganisms and external environmental factors. Environmental factors and the habitat under investigation play pivotal roles in shaping microbial diversity and richness during the biodegradation process. In addition to the microorganisms at play, the circumstances influencing plastic biodegradation depend on the surface properties and structure of the plastic materials involved. These properties encompass characteristics like roughness, electrostatic interaction, topography, hydrophobicity, and free energy. A comprehensive understanding and strategic manipulation of these factors are crucial for delving into the feasibility and optimization of biodegradation processes for plastics and microplastics (MPs) [1, 4–9].

In marine environments, the formation of biofilms by microbial populations presents promising avenues for the effective degradation of plastics and MPs. This process involves the initial attachment of microorganisms, culminating in the creation of a microbial biofilm on the plastic surface, commonly referred to as a plastisphere. These biofilms harbor diverse microbial communities, including potential plastic-degrading microorganisms. For example, the hydrophobicity of polyethylene (PE) can impede biofilm formation, yet micro-bacterial communities with high cell surface hydrophobicity demonstrate superior adhesion to PE surfaces, facilitating biofilm development.

In nature, collaborative studies reveal that bacteria often work in communities, enhancing efficiency compared to individual microorganisms. Microbial communities or consortia exhibit the capability to break down complex compounds into single monomers, showcasing the synergy of multiple microorganisms coexisting in an environment. Isolated pure cultures of bacteria from various environments demonstrate substantial diversity and functionality. Extensive data identification indicates that bacterial species belonging to diverse phyla, including Proteobacteria, Firmicutes, and Actinobacteria, possess the capacity to degrade plastics. Notably, bacteria with plastic-biodegrading potential are frequently isolated from contaminated sites such as landfills.

Leveraging the inherent ability of bacteria to break down long-chain fatty acids enables plastic degradation. Bacterial genera like Pseudomonas, Escherichia, and Bacillus exhibit significant potential, particularly in degrading recalcitrant polymers like PE, PET, and PS. Strains such as Pseudomonas aeruginosa, Bacillus megaterium, and Rhodococcus ruber, among others, have been identified for their ability to break down thermoplastics like PE and PET. Furthermore, fungi equipped with potent enzymatic systems contribute to plastic degradation. Filamentous fungi play a pivotal role in the degradation and mineralization of plastic pollutants, showcasing proficiency in degrading PE and PET. Ongoing research has uncovered diverse microorganisms with plastic-degrading capabilities, highlighting variations in diversity, abundance, and activity, thus providing valuable evidence for the plastic degradation potential within microbial ecosystems as discussed in enclosed references as well [1, 4, 10–13].

3 Mechanism of Plastics/MPs Biodegradation

Microorganisms play a crucial role in the depolymerization of microplastics (MPs), generating extracellular enzymes and free radicals that catalytically cleave biodegraded plastics into smaller elements. The microbiological degradation of plastics typically involves breaking down polymers into shorter chains or smaller molecules, such as oligomers, dimers, and monomers. Once polymers reach a small enough size, they undergo depolymerization into monomers essential for microbial uptake, growth, and eventual mineralization within cells. Monomers within the cells are further mineralized into CO_2 and H_2O under aerobic conditions or CO_2, H_2O, and CH_4 under anaerobic conditions, producing biomass for energy.

The initial step in microbial degradation is the colonization of the plastic surface through adhesion or exposure, serving as the foundation for enzymatic degradation. The subsequent step involves hydrolysis, where enzymes and the polymer matrix combine, leading to catalytic hydrolysis and cracking. The bound enzyme, known as hydrolase, catalyses the hydrolysis of organic matter. Extracellular enzymes, including depolymerases and hydrolases, induce hydrolytic cleavage of the polymer chain, resulting in small oligomers or monomers that bacteria can integrate into the cell and absorb after metabolism by intracellular enzymes.

Enzymatic degradation, in contrast to oxidative degradation, can be measured by weight loss and the addition of functional groups. Significant changes during microbial plastic biodegradation, such as weight loss, are indicative of the process's effectiveness. The by-products produced during microbial degradation and metabolism contribute positively to the environment.

Polymer morphology, particularly crystallinity, is a critical factor influencing biodegradability, affecting the rate of biodegradation. Microbial enzymes, primarily targeting the amorphous domain of the polymer, play a key role in environmental remediation and ecological health. Their involvement in biotransformation and biodegradation processes combines efficiency and functionality, making biocatalytic processes more competitive, clean, safe, and environmentally friendly [1, 9, 12–14].

4 Microbial Degradation of Commonly Used Plastics/MPs

The degradation of commonly used plastics and microplastics (MPs) can be confirmed primarily by observing three key features: (1) Alterations in the structure of the plastics/MPs, (2) measurable physical loss of mass, and (3) the production of metabolites derived from the plastics/MPs.

4.1 Microbial Degradation of Recalcitrant Plastics/MPs

Nonhydrolyzable plastics, such as PE, PP, PVC, and PS, are primarily composed of C–C bonds in their molecular chains, offering a wide range of applications but exhibiting relatively low levels of biodegradation. Microorganisms typically convert nonhydrolyzable microplastics (MPs) into small organic molecules through redox reactions before utilizing them. In contrast, hydrolyzable plastics like PA and PET are more easily decomposed, with growing evidence indicating microorganisms' ability to degrade plastic fragments.

PET, the most common thermoplastic plastic produced globally at nearly 70 million tons annually, presents challenges for microbial degradation. Only a few bacterial phyla, notably Actinobacteria, particularly the genera Thermobifida and Thermomonospora, host most bacterial isolates capable of degrading PET. Enzymes associated with PET metabolism, including PETase, MHETase, TPA dioxygenase, and PCA dioxygenase, have been identified. Ideonella sakaiensis 201-F, a bacterium capable of utilizing PET as its primary energy and carbon source, is a ground-breaking discovery. This unique bacterium secretes PETase, efficiently hydrolyzing PET into environmentally benign monomers. The strain produces two specialized enzymes, PETase and MHETase, playing a pivotal role in PET hydrolysis and mineralization. Although the binding mechanism of PETase remains unknown, bacteria like Pseudomonas nitroreducens S8 and Pseudomonas monteilii S17 have demonstrated the ability to colonize PET surfaces, forming transparent biofilm layers. Together, these bacteria exhibit a strong synergistic effect, disrupting PET surfaces and making them highly accessible for PET hydrolases [15–18, 43].

Polystyrene (PS), a high molecular weight synthetic hydrophobic polymer, poses challenges for microbial attack and exhibits low biodegradation potential. However, research has revealed the unique ability of mealworms' intestinal tracts to degrade PS, with the mealworm gut microbiome featuring operational taxonomic units (OTUs), such as Citrobacter sp. and Kosakonia sp., showing strong connections with plastics like PE and PS. Further exploration of mealworm gut microbiomes could provide insights for enhancing PS biodegradation.

Similarly, polyethylene (PE), with its carbon–carbon backbone and non-hydrolysable covalent bonds, is highly resistant to degradation. However, microbial consortia, including Pseudomonas sp., Bacillus sp., Mycobacterium sp., and Nocardia sp., have demonstrated the ability to degrade PE. Fungi have proven more effective in degrading PE compared to bacteria, with fungal strains like Penicillium oxalicum NS4 and Penicillium chrysogenum NS10 showing promising plastic-degrading abilities. Aspergillus niger isolated from a plastic dump wasteland displays the potential in breaking down HDPE.

Understanding the enzyme systems of these fungi, especially A. niger, is crucial for a deeper understanding of their role in HDPE biodegradation. Ongoing research in this field holds promise for developing efficient and environmentally friendly methods for plastic biodegradation [4, 10, 12, 19–23].

4.2 Microbial Degradation of Toxic Components of Plastics/ MPs

Since the 1950s, with the advent of mass plastic production, the manufacturing of diverse polymers incorporating various additives and fillers has witnessed exponential growth. Notably, phthalates (PAEs) and Bisphenol A (BPA) are significant plastics that can have detrimental effects, particularly in marine environments where they pose toxicity to a wide range of animals and vertebrates. BPA, recognized as potentially carcinogenic due to its structural resemblance to diethylstilbestrol (DES), exhibits heightened toxicity when plastic particles or smaller stages are exposed.

In the biodegradation of plastics, toxic substances like PAEs and BPA play a role. Studies on the biodegradation of PAEs commonly involve bacteria such as Pseudomonas, Arthrobacter, Rhodococcus, Bacillus, Mycobacterium, Delfia, and Gordonia. The microbial degradation of PAEs occurs through ester bond hydrolysis, producing monoester, which further decomposes into phthalic acid (PA) and alcohol. Ultimately, short-chain acids are formed, engaging the Krebs cycle. Key enzymes in the biodegradation process include hydrolase and oxygenase, with esterase demonstrating optimal activity in phthalate degradation at a pH range of 7–10 and a temperature range of 30–70 °C.

Reduction in weight is found to be particularly effective in pre-treatment, complementing the role of hydrolase. In specific environments, such as estuary waste electronics dismantling areas, Bacillus sp. GZB, an isolated facultative anaerobic BPA-degrading bacterium, has shown promise in accelerating BPA biodegradation in water sediment. The addition of Bacillus sp. GZB has been closely linked to improved biodegradation efficiency.

Enhanced BPA biodegradation has also been observed through the addition of substances such as electron donors, co-substrates, yeast extract, NaCl, humic acid, and glucose, showcasing the potential for bioaugmentation and bio-stimulation in water–sediment microcosms [14, 24, 25].

4.3 Microbial Degradation of Common Plastics/MPs

Biodegradable plastics, often referred to as BPs, represent a promising solution to environmental pollution and challenges posed by synthetic plastics. Notably, polymers like polycaprolactone (PCL), polybutylene succinate-co-adipate (PBSA), polybutylene succinate (PBS), and polylactic acid (PLA) exhibit a propensity for degradation, distinguishing them from high polymers. This heightened degradability is particularly beneficial for addressing environmental concerns associated with conventional plastics. Certain biodegradable polymers, including PCL, polyhydroxybutyrate (PHB), PLA, and PBS, actively contribute to composting processes by undergoing significant degradation.

The chemical structure of these polymers, featuring ester linkages, renders them susceptible to breakdown by microbial enzyme systems. This susceptibility is a key factor in their ability to overcome environmental challenges and underscores their role in fostering sustainability in the realm of plastic materials.

Thermophilic actinomycetes exhibiting activity toward polyhydroxybutyrate (PHB), polycaprolactone (PCL), or polyethersulfone (PES) were isolated from diverse environments, including compost, in Taiwan. Two thermophilic bacteria, isolated from compost, demonstrated effective synergy when exposed to PCL wastes at 50 °C. This synergistic action accelerated the breakdown of PCL, significantly increasing the disintegration of the polymer [26].

Given that PCL is a thermoplastic crystalline polyester, it undergoes hydrolysis into small molecules by lipase and is subsequently assimilated by microorganisms. For instance, Aspergillus sp. strain ST-01, isolated from soil, completely degraded PCL within six days of incubation at 50 °C, showcasing the efficacy of thermotolerant PCL-degrading microorganisms. As enzymes, Aspergillus sp. strain ST-01 secretes catalase and protease.

Moreover, marine organisms, including Pseudomonas, Alcanivorax, and Tenacibaculum, isolated from deep-sea sediments, have proven to be effective in degrading PCL. This diverse range of microorganisms highlights the potential for various ecological niches to contribute to the biodegradation of PCL, emphasizing its versatility and environmental adaptability [27].

4.4 Enzymatic Degradation of Plastics/MPs

A diverse array of enzymes, including cutinases, esterases, lipases, laccases, peroxidases, proteases, and ureases derived from both bacterial and fungal sources, have demonstrated the ability to break down polyethylene (PE), polyethylene terephthalate (PET), and polypropylene (PP). Fungal cellulase systems, observed in cellulose depolymerization, allow free enzymes to act directly on solid polymeric substrates, ultimately converting monomeric constituents (e.g., cellobiose hydrolysis to glucose). The properties of fungal enzymes, particularly depolymerases, enable them to degrade a variety of polymers, including PET and PE.

The distribution and penetrative ability of fungal hyphae play a significant role in their initial colonization before depolymerization, enhancing their enzyme capacities for attaching to hydrophobic substrates. Laccases derived from actinomycetes, such as Rhodococcus ruber, and fungi like Aspergillus flavus and Pleurotus ostreatus, have displayed notable degradation capabilities, potentially arising from the oxidation of the PE hydrocarbon backbone.

Additionally, in the biodegradation of conventional low-density plastic LDPE, various fungal species, including Fusarium solani, Alternaria solani, Aspergillus fumigatus, Spicaria spp., Geomyces pannorum, Phoma sp., Penicillium spp., and more, exhibit substantial plastic-degrading properties. Microorganisms involved in the degradation of polyethylene materials predominantly include bacteria and fungi, with F. oxysporum and Aspergillus fumigatus being noteworthy fungi.

F. oxysporum strains exhibit enhanced tolerance to high concentrations of polyethylene, leading to potent oxidation phenomena and observable changes in the morphology of polyethylene films. Aspergillus fumigatus forms biofilms to efficiently degrade LDPE, achieving notable experimental success in the fungal degradation of this plastic [4, 28, 29].

4.5 Enhanced PET Degradation by Bacterial Enzymes

Certain hydrolyzable plastics, including PET, undergo efficient degradation facilitated by specialized bacteria such as Ideonella sakaiensis 201-F. Bacterial enzymes, such as hydrolase and keratinase, play a pivotal role in the hydrolysis of these plastics, with specific enzymes like PETase and MHETase combining synergistically for enhanced degradation.

Traditionally, esterases, lipases, keratinases, and other hydrolytic enzymes have been utilized in previous studies to degrade PET. When an enzyme specifically designed for PET hydrolysis is identified, it is termed PETase. Recent research has revealed a step-by-step breakdown of PET polymer by bacteria-secreted enzymes, where the synthesis and synergistic action of PETase and MHETase result in a sixfold increase in degradation efficiency.

During PET depolymerization by I. sakaiensis strain enzymes, the liberation of Terephthalic Acid (TPA) and Ethylene Glycol (EG) is crucial. This process requires the concerted action of PETase and MHETase. Both enzymes have garnered significant attention for their potential in degrading PET and reducing plastic waste.

PETase, with its esterase catalyzing ester bonds, effectively impacts the stability of PET molecules. The hydrolysis of polymeric PET molecules by PETase releases the dimer BHET and monomer MHET. Subsequently, MHETase converts MHET into TPA and EG, completing the PET depolymerization process [28, 30, 31, 43].

5 Enzymatic Biodegradation: Advances in Multienzyme Systems

At the forefront of research in pollutant biodegradation, both internally and externally, lies the exploration of synergistic enzymes secreted by various microorganisms and microbes. These enzymes, with their potent degrading ability and unique metabolic types, offer promising avenues for addressing environmental challenges. An emerging area of interest is the development of multienzyme systems for the depolymerization of plastic waste. Illustrating this, the I. sakaiensis strain boasts a dual enzyme system comprising PETase and MHETase, evolving the capability to utilize crystalline polyester substrates.

Recent research has demonstrated that scientists can design enzymes from the I. sakaiensis strain to enhance degradation efficiency. Engineered enzymes, optimized through genetic modifications, can outperform natural evolution. Notably, a PET hydrolase from the bacterium I. sakaiensis strain 201-F6 has shown remarkable efficiency, achieving over 90% PET depolymerization within a 10-h period. This breakthrough enzyme combination, featuring PETase and MHETase, proved superior to all previously reported PET hydrolases, leading to the near-complete destruction of PET films within a few weeks at mesothermic temperatures (around 30 °C).

MHETase's presence is crucial for PET degradation, and even at low concentrations, PETase enhances the efficiency of degradation. Commercial practices like enzyme immobilization, entrapping, or encapsulation hinder optimal enzyme–substrate interaction, resulting in a relatively low turnover rate for PET hydrolases. However, heat-resistant PET hydrolase allows effective enzymatic PET degradation at temperatures between 65 and 75 °C.

Research on PET hydrolase can extend to explore the substrate scope of PETase, screening PETase variants to enhance thermal stability and PET degradation ability. Further investigations into the molecular mechanisms of PET hydrolase and the development of new PET-degrading enzymes are crucial for advancing our understanding of PET biodegradation. Multiple cutinase enzymes, prevalent esterase enzymes in nature, have been identified, showcasing the diverse enzymatic arsenal available for PET depolymerization [5, 15, 28, 32, 44].

6 Advancements in Fungal-Mediated Plastic Degradation

Fungi, spanning 10 genera of the Ascomycota and Mucor, exhibit superior ability in degrading polyethylene (PE) compared to bacteria. This distinction arises from fungi's capability to adhere to polymer surfaces, especially those with hydrophobic properties. Aspergillus, Fusarium oxysporum, and Penicillium, when cultured with PP and PE as the sole carbon source, demonstrate promising degradation potential over three months, accompanied by alterations to the surfaces of PP and PE and the presence of biofilms.

Fusarium oxysporum, among the ascomycetes, secretes keratinase capable of depolymerizing polyethylene terephthalate (PET). LC-cutinase, derived from a leaf's fosmid library and produced in Escherichia coli, proves effective in degrading PET and polycaprolactone (PCL), showcasing the potential of cutinases in modifying PET surfaces. Further exploration into the mechanisms of cutinase enzymes for degrading specific plastics is essential. Transformation and cloning of bacteria and enzymes hold promise for enhancing the degradation of various plastics.

Recent studies highlight LC-cutinase and Thermobifida fusca cutinase as high-performing enzymes, with T. fusca cutinases demonstrating superior benefits in the presence of surfactants and organic solvents. Two cutinases, TfU 0882 and TfU 0883, isolated from T. fusca, show 93% similarity, and possess the ability to metabolize synthetic polyesters. Notably, TfU 0883 can break down PET at 60 °C. LC-cutinase, compared to BC lip, exhibits specific PCL-degrading activity, emphasizing its potential for efficient plastic degradation.

In PET degradation experiments, LC-cutinase outperforms other cutinases, achieving significantly higher degradation rates. LC-cutinase's hydrolytic activity plays a pivotal role in PET deterioration, breaking it down into terephthalic acid (TPA) and ethylene glycol (EG). Understanding the protein structures of these enzymes opens avenues for improving their catalytic efficacy in degrading various polymers, including PET.

Microorganisms, through evolution, have developed enzymes like mutated keratinase capable of rapid PET decomposition. Comprehensive studies on cutinase structure and function are crucial for developing more efficient enzymes. Analyzing these enzymes at a molecular level holds the key to advancing our understanding and application of fungal-mediated plastic degradation [33, 44].

7 Biodegradation Associated Enzymes Produced by Microorganisms in the Extreme Environments

Extreme environmental microorganisms, such as halophiles and psychrophiles, demonstrate the potential for plastic degradation. Bacteria thriving in thermophilic, alkaliphilic, halophilic, and psychrophilic conditions across various extreme environments possess the capability to degrade synthetic plastics. Screening for more effective degradation bacteria is imperative in such specialized environments. Plastic-contaminated sites often exhibit extreme environmental conditions, including low or elevated temperatures, acidic or alkaline pH levels, high salt concentrations, or high pressure. Thermophilic and halophilic enzymes are known for their extended lifespan, allowing for room temperature storage without significant loss of enzymatic activity. Consequently, the exploration of thermophile and extremophile microbiomes holds significant promise as a source of plastic-degrading enzymes and microorganisms.

The most thermally stable leaf-branching compost cutinase (LCC) exhibits the highest PET depolymerization rates at 65 °C. Several thermophiles have shown considerable potential for polymer degradation at elevated temperatures, akin to high-temperature plastic degradation agents. These bacteria can produce numerous enzymes with heightened activity, enhancing substrate bioavailability and solubility. For instance, Chelatococcus sp. E1, isolated from a compost sample, demonstrated the capability to degrade PE for the first time. When pre-treated PE samples were exposed to 60 °C, Chelatococcus sp. E1, as a thermophilic strain, effectively shifted the molecular weight distribution toward the lower molecular weight side, thereby increasing the biodegradability of HDPE and LDPE.

The development of extremozymes and the proliferation of extremophiles in severe environments hold promise for addressing one of society's primary challenges [16, 21, 34, 35].

8 Improvement of Plastics/MP Biodegradation

Pre-treating plastic waste involves recycling or collecting microplastics from sewage to mitigate the toxicity of additives, prevent the adhesion of POPs, and break down depolymerized polymer chains. Enhancing the hydrophilicity of the polymer is essential, achieved through chemical or biological oxidation processes, to facilitate bacterial attachment and degradation. Different pre-treatment methods yield varying efficiencies in biodegradation. Recycling and biological processing significantly reduce plastic pollution toxicity and enhance biodegradation efficiency. However, to further degrade polymers, exploring different plastic characteristics for degradation is necessary.

Biodegradable plastics are more amenable to pre-treatment than traditional plastics due to their favorable physicochemical properties, including relatively low molecular weight, high flexibility, and increased functional groups. Various physical and chemical agents can pre-treat plastics, altering their structural and morphological characteristics by reducing molecular weight, breaking chemical bonds, creating surface cracks, and enhancing functional groups. These modifications facilitate subsequent biodegradation processes.

Assessing the impact of different pre-treatments on the biodegradation of plastics and MPs is crucial to increase degradation percentages. Pre-treating plastic polymers with high temperatures, photooxidation catalysis, and microbial enzyme catalysis can promote efficient biodegradation [12, 36].

8.1 Polymer Pre-treatment—Enhances Microbial Biodegradation

The efficacy of microbial biodegradation is heavily influenced by polymer pre-treatment, with tailored techniques for different plastics. Pyrolysis emerges as a crucial pre-treatment method for polyethylene (PE), making it more conducive to microbial metabolism. Conversely, temperature-based pre-treatment stands out as optimal for enhancing the biodegradation of polyethylene terephthalate (PET). Temperature optimization, particularly through pyrolysis, has proven highly effective in improving the efficiency of PET-hydrolyzing enzymes like PETase. Ambient temperature significantly impacts the biodegradation activity of these enzymes, with reduced PET molecule flexibility at lower temperatures enhancing catalytic performance. Hence, selecting appropriate temperature conditions is vital for maximizing the activity of microbial enzymes involved in PET degradation.

Furthermore, pre-treatment with elevated temperatures has been shown to enhance plastics' biodegradability by certain bacteria. Thermal treatment not only boosts enzyme efficacy but also intensifies its impact on long-chain polymer molecules, leading to notable breakage and damage. For example, the efficient biodegradation of thermally pre-treated high-density polyethylene (HDPE) by Klebsiella pneumoniae CH001 demonstrates the positive outcomes of temperature-based pre-treatment. Thermal pre-treatment not only facilitates biodegradation but also results in a substantial reduction in HDPE tensile strength, indicating significant alterations in the polymer's structural integrity [9, 12, 37].

8.2 Enhancing PE Degradation: A Green Approach

The reduction of polyethylene (PE) weight via degradation can be significantly enhanced by incorporating biodegradable additives, photo-initiators, or engaging in copolymerization. An environmentally friendly method that has garnered rapid development in recent years is photocatalysis. In this domain, various iron salts have been utilized for their catalytic prowess to facilitate the oxidative bond-breaking process. Notably, when subjected to 400 nm LED irradiation in the presence of pure oxygen or air, ferrous chloride has emerged as the optimal photocatalyst for achieving efficient PE degradation. Aromatic hydrocarbons like isopropylbenzene and cyclohexylbenzene have also showcased catalytic potential, leading to effective C–C bond breaking and conversion into benzoic acid within this catalytic system. Importantly, experiments have confirmed the necessity of light, ferrous chloride ($FeCl3$), and air for the efficient degradation process.

This method has demonstrated high effectiveness, even in breaking down commercial samples of high molecular weight polystyrene (PS). Such an innovative approach offers a promising avenue for sustainable PE degradation through environmentally friendly means [38–41].

8.3 Advancing Plastic Surface Properties Through Enzymatic Modification

Enzymatic modification of polymer fibers, involving alterations to the surface functional groups of plastic fibers, brings about notable enhancements in plastic characteristics. This process has been shown to improve wettability, fastness, dyeing, and pilling resistance, ultimately elevating the hydrophilicity of plastics. Noteworthy enzymes such as lipase from Candida antarctica and keratinase from Aspergillus oryzae are employed to enhance the hydrophilicity of fabrics.

For plastics characterized by C–C backbones, whole-cell enzyme biodegradation emerges as a potentially superior method. Adopting a multidisciplinary approach, PET monomer biocatalysis, through methods like whole-cell catalysis, transforms into a biopolymer, contributing to a sustainable biological cycle. Whole-cell catalysis proves effective in reducing the hydrophobicity of plastics, thereby enhancing the binding strength between enzymes and high-crystalline substrates like PET. This process involves two fundamental steps: adsorption and hydrolysis. The anaerobic thermophile Clostridium thermocellum, acting as a whole-cell biocatalyst, seamlessly integrates enzyme synthesis with PET hydrolysis in a single process. Remarkably, a PET film experiences a weight loss of almost 60% over a 14-day incubation at 60 °C under the influence of this approach.

Consequently, whole-cell catalysis emerges as an ideal additive for pre-treatment catalysts. The pre-treatment of diverse substrates using various parameters, including photo-treatment, temperature, enzymatic pre-treatment, and various additives, has demonstrated significant enhancements in the biodegradation of plastics and microplastics [21, 31, 42].

9 Conclusions

Enzymatic plastic biodegradation demands thorough investigation. Plastic and microplastic (MP) pollution present enduring challenges, prompting serious concerns across multiple domains due to their adverse effects on ecological systems and human health. Despite these concerns, a significant knowledge gap persists regarding the capacity and potential of enzymes, including hydrolases, cutinase, and bacteriophilic enzymes, to catalyze the breakdown of plastics through enzymatic pathways. Therefore, further research is essential to optimize effective enzymatic conditions for plastic degradation. This necessitates focused exploration into the structural analysis of relevant enzymes and understanding their reaction mechanisms to achieve desired outcomes.

Recent years have witnessed a surge in studies focusing on enzymes derived from bacteria, exemplified by PETase and cutinases. The selection of appropriate microorganisms for plastic treatment, tailored to different types of plastics, is of paramount significance. Delving into the intricacies of how microorganisms operate in plastic biodegradation not only enhances our understanding but also reveals their potential to mitigate plastic accumulation in the environment. Moreover, these studies lay the groundwork for the development of superior enzymes to address challenges associated with plastic waste.

This chapter underscores the role of microorganisms in breaking down commonly used plastics and explores their enzymes as potent tools for degradation, laying the foundation for the creation of sustainable bioremediation methods. Furthermore, investigations into the impact of various pre-treatment techniques and additives on the microbiological breakdown of synthetic plastics are expected to yield more effective and comprehensive outcomes.

Acknowledgements The authors express their gratitude to Dr. Malti Goel for organizing a panel discussion on plastic pollution, during which several experts shared insights on various aspects of the issue. Inspired by this initiative to raise public awareness, the authors have endeavored to scrutinize the current approach to tackling plastic pollution on a broader scale.

References

1. Cai, Z., Li, M., Zhu, Z., Wang, X., Huang, Y., Li, T., Gong, H., & Yan, M. (2023). Biological degradation of plastics and microplastics: A recent perspective on associated mechanisms and influencing factors. *Microorganisms., 11*(7), 1661.
2. Fu, L., Li, J., Wang, G., Luan, Y., & Dai, W. (2021). Adsorption behaviour of organic pollutants on microplastics. *Ecotoxicology and Environmental Safety, 217*, 112207.
3. Zhou, B., Zhao, L., Wang, Y., Sun, Y., Li, X., Xu, H., Weng, L., Pan, Z., Yang, S., Chang, X., et al. (2020). Spatial distribution of phthalate esters and the associated response of enzyme activities and microbial community composition in typical plastic-shed vegetable soils in China. *Ecotoxicology and Environmental Safety, 195*, 110495.
4. Amobonye, A., Bhagwat, P., Singh, S., & Pillai, S. (2021). Plastic biodegradation: Frontline microbes and their enzymes. *Science of the Total Environment, 759*, 143536.
5. Papadopoulou, A., Hecht, K., & Buller, R. (2019). Enzymatic PET degradation. *Chimia, 73*, 743–749.
6. Wang, G. X., Huang, D., Ji, J. H., Völker, C., & Wurm, F. R. (2020). Seawater-degradable polymers-fighting the marine plastic pollution. *Advancement of Science, 8*, 2001121.
7. Yang, H., Chen, G., & Wang, J. (2021). Microplastics in the marine environment: Sources, fates, impacts and microbial degradation. *Toxics, 9*, 41.

8. Yang, Y., Liu, W., Zhang, Z., Grossart, H. P., & Gadd, G. M. (2020). Microplastics provide new microbial niches in aquatic environments. *Applied Microbiology and Biotechnology, 104*, 6501–6511.
9. Yuan, J., Ma, J., Sun, Y., Zhou, T., Zhao, Y., & Yu, F. (2020). Microbial degradation and other environmental aspects of microplastics/plastics. *Science of the Total Environment, 715*, 136968.
10. Bardají, D. K. R., Moretto, J. A. S., Furlan, J. P. R., & Stehling, E. G. (2020). A mini-review: Current advances in polyethylene biodegradation. *World Journal of Microbiology & Biotechnology, 36*, 32.
11. Ganesh Kumar, A., Anjana, K., Hinduja, M., Sujitha, K., & Dharani, G. (2020). Review on plastic wastes in marine environment—Biodegradation and biotechnological solutions. *Marine Pollution Bulletin, 150*, 110733.
12. Matjašič, T., Simčič, T., Medvešček, N., Bajt, O., Dreo, T., & Mori, N. (2021). Critical evaluation of biodegradation studies on synthetic plastics through a systematic literature review. *Science of the Total Environment, 752*, 141959.
13. Sánchez, C. (2020). Fungal potential for the degradation of petroleum-based polymers: An overview of macro- and microplastics biodegradation. *Biotechnology Advances, 40*, 107501.
14. Carmen, S. (2021). Microbial capability for the degradation of chemical additives present in petroleum-based plastic products: A review on current status and perspectives. *Journal of Hazardous Materials, 402*, 123534.
15. Kawai, F., Kawabata, T., & Oda, M. (2019). Current knowledge on enzymatic PET degradation and its possible application to waste stream management and other fields. *Applied Microbiology and Biotechnology, 103*, 4253–4268.
16. Lear, G., Kingsbury, J. M., Franchini, S., Gambarini, V., Maday, S. D. M., Wallbank, J. A., Weaver, L., & Pantos, O. (2021). Plastics and the microbiome: Impacts and solutions. *Environmental Microbiome, 16*, 2.
17. Yan, Z. F., Xu, K. W., & Wu, J. (2022). Synergism between multi-pseudomonas and cutinase for biodegradation of crude oil-based derivatives. *Current Microbiology, 80*, 30.
18. Zhang, B., Chen, L., Chao, J., Yang, X., & Wang, Q. (2020). Research progress of microplastics in freshwater sediments in China. *Environmental Science and Pollution Research International, 27*, 31046–31060.
19. Brandon, A. M., Gao, S. H., Tian, R., Ning, D., Yang, S. S., Zhou, J., Wu, W. M., & Criddle, C. S. (2018). Biodegradation of polyethylene and plastic mixtures in mealworms (larvae of tenebrio molitor) and effects on the gut microbiome. *Environmental Science and Technology, 52*, 6526–6533.
20. Gao, R., & Sun, C. (2021). A marine bacterial community capable of degrading poly(ethylene terephthalate) and polyethylene. *Journal of Hazardous Materials, 416*, 125928.
21. Khandare, S. D., Chaudhary, D. R., & Jha, B. (2021). Marine bacterial biodegradation of lowdensity polyethylene (LDPE) plastic. *Biodegradation, 32*, 127–143.
22. Roager, L., & Sonnenschein, E. C. (2019). Bacterial candidates for colonization and degradation of marine plastic debris. *Environmental Science and Technology, 53*, 11636–11643.
23. Sridharan, R., Krishnaswamy, V. G., & Kumar, P. S. (2021). Analysis and microbial degradation of low-density polyethylene (LDPE) in Winogradsky column. *Environmental Research, 201*, 111646.
24. Das, R., Li, G., Mai, B., & An, T. (2018). Spore cells from BPA degrading bacteria Bacillus sp. GZB displaying high laccase activity and stability for BPA degradation. *Science of the Total Environment*, 640–641, 798–806. 379.
25. González-Márquez, A., Loera-Corral, O., Santacruz-Juárez, E., Tlécuitl-Beristain, S., García-Dávila, J., Viniegra-González, G., & Sánchez, C. (2019). Biodegradation patterns of the endocrine disrupting pollutant di(2-ethyl hexyl) phthalate by Fusarium culmorum. *Ecotoxicology and Environmental Safety, 170*, 293–299.
26. Al Hosni, A. S., Pittman, J. K., & Robson, G. D. (2019). Microbial degradation of four biodegradable polymers in soil and compost demonstrating polycaprolactone as an ideal compostable plastic. *Waste Management, 97*, 105–114.

27. Feng, S., Yue, Y., Chen, J., Zhou, J., Li, Y., & Zhang, Q. (2020). Biodegradation mechanism of polycaprolactone by a novel esterase MGS0156: A QM/MM approach. *Environmental Science Processes & Impacts, 22*, 2332–2344.
28. Knott, B. C., Erickson, E., Allen, M. D., Gado, J. E., Graham, R., Kearns, F. L., Pardo, I., Topuzlu, E., Anderson, J. J., Austin, H. P., et al. (2020). Characterization and engineering of a two-enzyme system for plastics depolymerization. *Proceedings of the National academy of Sciences of the United States of America, 117*, 25476–25485.
29. Spina, F., Tummino, M. L., Poli, A., Prigione, V., Ilieva, V., Cocconcelli, P., Puglisi, E., Bracco, P., Zanetti, M., & Varese, G. C. (2021). Low density polyethylene degradation by filamentous fungi. *Environmental Pollution, 274*, 116548.
30. Prata, J. C., Silva, A. L. P., da Costa, J. P., Mouneyrac, C., Walker, T. R., Duarte, A. C., & Rocha-Santos, T. (2019). Solutions and integrated strategies for the control and mitigation of plastic and microplastic pollution. *International Journal of Environmental Research and Public Health, 16*, 2411.
31. Tiso, T., Narancic, T., Wei, R., Pollet, E., Beagan, N., Schröder, K., Honak, A., Jiang, M., Kenny, S. T., Wierckx, N., et al. (2021). Towards bio-upcycling of polyethylene terephthalate. *Metabolic Engineering, 66*, 167–178.
32. Tournier, V., Topham, C. M., Gilles, A., David, B., Folgoas, C., Moya-Leclair, E., Kamionka, E., Desrousseaux, M. L., Texier, H., Gavalda, S., et al. (2020). An engineered PET depolymerase to break down and recycle plastic bottles. *Nature, 580*, 216–219.
33. Wróbel, M., Szymańska, S., Kowalkowski, T., & Hrynkiewicz, K. (2023). Selection of microorganisms capable of polyethylene (PE) and polypropylene (PP) degradation. *Microbiological Research, 267*, 127251.
34. Atanasova, N., Stoitsova, S., Paunova-Krasteva, T., & Kambourova, M. (2021). Plastic degradation by extremophilic bacteria. *International Journal of Molecular Sciences, 22*, 5610.
35. Orellana, R., Macaya, C., Bravo, G., Dorochesi, F., Cumsille, A., Valencia, R., Rojas, C., & Seeger, M. (2018). Living at the frontiers of life: Extremophiles in Chile and their potential for bioremediation. *Frontiers in Microbiology, 9*, 2309.
36. Li, X., Chen, L., Ji, Y., Li, M., Dong, B., Qian, G., Zhou, J., & Dai, X. (2020). Effects of chemical pretreatments on microplastic extraction in sewage sludge and their physicochemical characteristics. *Water Research, 171*, 115379.
37. Narancic, T., & O'Connor, K. E. (2019). Plastic waste as a global challenge: Are biodegradable plastics the answer to the plastic waste problem? *Microbiology, 165*, 129–137.
38. Huang, Z., Shanmugam, M., Liu, Z., Brookfield, A., Bennett, E. L., Guan, R., Vega Herrera, D. E., Lopez-Sanchez, J. A., Slater, A. G., McInnes, E. J. L., et al. (2022). Chemical recycling of polystyrene to valuable chemicals via selective acid-catalysed aerobic oxidation under visible light. *Journal of the American Chemical Society, 144*, 6532–6542.
39. Oh, S., & Stache, E. E. (2022). Chemical upcycling of commercial polystyrene via catalyst-controlled photooxidation. *Journal of the American Chemical Society, 144*, 5745–5749.
40. Song, Y. K., Hong, S. H., Eo, S., & Shim, W. J. (2022). The fragmentation of nano- and microplastic particles from thermoplastics accelerated by simulated-sunlight-mediated photooxidation. *Environmental Pollution, 311*, 119847.
41. Zhang, L., Sintim, H. Y., Bary, A. I., Hayes, D. G., Wadsworth, L. C., Anunciado, M. B., & Flury, M. (2018). Interaction of Lumbricus terrestris with macroscopic polyethylene and biodegradable plastic mulch. *Science of the Total Environment, 635*, 1600–1608.
42. Yan, F., Wei, R., Cui, Q., Bornscheuer, U. T., & Liu, Y. J. (2021). Thermophilic whole-cell degradation of polyethylene terephthalate using engineered clostridium thermocellum. *Microbial Biotechnology, 14*, 374–385.
43. Maity, S., Banerjee, S., Biswas, C., Guchhait, R., Chatterjee, A., & Pramanick, K. (2021). Functional interplay between plastic polymers and microbes: A comprehensive review. *Biodegradation, 32*, 487–510.
44. Maity, W., Maity, S., Bera, S., & Roy, A. (2021). Emerging roles of PETase and MHETase in the biodegradation of plastic wastes. *Applied Biochemistry and Biotechnology, 193*, 2699–2716.

45. Urbanek, A. K., Rymowicz, W., & Mirończuk, A. M. (2018). Degradation of plastics and plastic-degrading bacteria in cold marine habitats. *Applied Microbiology and Biotechnology, 102*, 7669–7678.

Plastic Waste to Value: Desirable Technology Interventions

Pranshu Chhabra, Ruchi Gupta, K. Nidhi, Meenu Talwar, and Rakesh K. Khandal

Abbreviations

AI	Artificial Intelligence
BR	Butadiene Resin
EEE	Electrical and Electronic Equipment
EP	Epoxy Resins
EVA	Ethylene Vinyl Acetate
GHG	Greenhouse Gases
HDPE	High-Density Polyethylene
ICs	Integrated Circuits
LDPE	Low-Density Polyethylene
LED	Light-Emitting Diode
MPs	Micro-Plastics
NIR	Near Infrared
OECD	Organisation for Economic Co-operation and Development
PA	Polyamide
PC	Polycarbonate

P. Chhabra
Self Employed Scientist, Ahmedabad, India

R. Gupta
Chemcred, Delhi, India

K. Nidhi
CFRA, NIFTEM-K, Sonipat 131028, Haryana, India

M. Talwar
Material Science Division, Shriram Institute for Industrial Research, Delhi 110007, India

R. K. Khandal (✉)
SDGI Global University, Ghaziabad, U.P., India
e-mail: dr.r.k.khandal@gmail.com

M. Goel and N. G. Tripathi (eds.), *Plastic Pollution*,
https://doi.org/10.1007/978-981-97-5528-8_15

PCBs	Printed Circuit Boards
PE	Polyethylene
PEFs	Process-Engineered Fuels
PET	Polyethylene Terephthalate
PF	Phenolic Resins
PMMA	Polymethylmethacrylate
POM	Poly-Oxy-Methylene
PP	Polypropylene
PS	Polystyrene
PUR	Polyurethane Rigid
PVC	Polyvinyl Chloride
RDF	Refuse-Derived Fuel
SBR	Styrene Butadiene Rubber
TG	Glass Transition Temperature
UNEP	United Nations Environment Programme
UP	Unsaturated Polyester Resins

1 Introduction

The growth rate of products made of plastics is much higher than any other material because plastics are replacing conventional materials used in almost every application. It is due to the unique virtues of plastics that industry and domestic users prefer them as one of the options for designing new products. One has to understand the reasons as to why there is an increasing demand for plastics. It is due to their durability, lightweight, neutrality, insulating behavior, etc., that consumers prefer to opt for plastics in place of conventional materials [1]. Plastics have become the most used commodity in the world. It all started about seven decades back when the first plastics were a resource for fibers and textiles [2]. The consumption of plastics crossed the volume of 380 MMT in 2015 and 367 MMT in 2020 from less than 2 MMT in 1950, which signifies the impact of plastics on human life [3]. During COVID-19, the production and consumption of plastics registered an overall decline, but in certain applications, plastics proved to be the most vital material [4]. Imagine how plastics played a lifesaving role during COVID-19 when PPE was the most essential safeguard from this epidemic. Until the other day, it was the developed world where the use of plastics per capita was on the higher side, but today, due to globalization, it has been catching up almost everywhere, even in the developing countries [5].

As per a recent report by the United Nations Environment Programme [6], the generation of plastic waste is more than 400 MT each year, globally. This report also highlights that out of the total waste generated (6.3 billion tonnes) all these years since 1950, recycling was only 9% of it. Almost 80% of this waste remains unrecycled and hence, mostly disposed of in landfills or left in the environment. A report from the Organisation for Economic Co-operation and Development (OECD) showed that the quantity of plastic waste in the world almost doubled during the period from the year 2000 to 2020 [7]. The latest data on plastic waste published suggest that global plastic waste generated now is ~585 Mt a year. Looking at the scenario emerging for the future, several UN sustainable Development Goals (Good Health and Well Being; Clean Water and Sanitation; Sustainable Cities and Communities; Responsible Consumption and Production; Climate Change; Life below water; Life on land) have a direct link with plastic waste [8]. That is why dealing with plastic waste is a challenge as well as the most important priority action plan for all countries.

For devising strategies and techniques for managing plastic wastes, it is essential to envisage that they have a complex structure; on the one hand and on the other hand, there are constituents that may result in harmful emissions as well as toxic ash on burning. The aim of processing plastic waste should therefore be to extract the valuable substances as well as to minimize the hazards of reprocessing on the environment and ecology [9]. In order to address this fundamental criterion of plastic waste management, The American Society of Plastics Industry created a standard coding system, categorizing 50 different types of plastic waste into seven groups. Based on their chemistry and criticality of their behavior on reprocessing, seven categories are as follows: (a) polyethylene terephthalate (PET); (b) high-density polyethylene (HDPE); (c) low-density polyethylene (LDPE); (d) vinyl/polyvinyl chloride (PVC); (d) polypropylene (PP); (e) polystyrene (PS); and (f) several more types of plastic products. This document is exhaustive as it not only describes the chemical structure of the polymeric resin, but also highlights the toxic elements present so that recycling and reprocessing remain safe. In addition, this chapter also covers aspects such as applications as a virgin material, advantages, challenges, and utilization of recycled plastics [10]. This type of standard guidelines would be of great help to developing countries such as India, where there is a large work force engaged in the collection, segregation, supply, and storage of plastic waste from the informal sector. Kemal et al. [10] present a good guide and reference point to solve issues related to the identification and traceability of different types of plastic waste. Everyone is talking today about the utilization of waste, to achieve the goal of a circular economy which will ensure growth without putting too much pressure on resources, in a way that there is no adverse impact on the environment [11].

2 Why Plastics?

The use of plastics in all verticals of various industries, especially in sectors pertaining to agriculture, packaging, automobile, building construction, infrastructure, information technology, food, etc., are driving the demand for plastics. According to Plastindia Foundation [12] other than the engineering plastics and thermoplastics, demand for different plastics during 2016–18 had the following order: PP (32%) > Polyethylene (PE) (33%) > PVC (21%). Waste plastics will have the following order: PP > HDPE > Linear Low-Density Polyethylene (LLDPE) > Polyethylene terephthalate (PET) [13]. Global per capita consumption of plastics, a decade ago, was

28 kg, the USA leading the list with 109 kg and India standing far behind at 11 kg [14]. Even though the per capita consumption of plastics in India was much lower, India stood in second position in the volume of plastic waste produced (14.17 MMT) in the world. Each year, if there is an increase in per capita consumption by 1 kg, India would require at least 1.25 MMT of additional quantity of polymers [15]. The following sectors are leading ones, as users of plastics as also, producers of waste:

Packaging Industry: The packaging industry accounts for ~60% of packaging materials of plastics. Packaging is to act as a protective barrier, shielding products from potential damage or contamination during transportation and storage. Plastics act as a barrier against moisture and oxygen, essential for packaging materials. Because plastics are of high strength and lightweight, they are the preferred packaging materials for all products, including those made up of plastics. Due to this reason, the carbon footprint ascribed to the distribution of products remains as low as it is possible since a greater number of products are packed in lightweight packaging for transport. Out of the packaging waste generated, the share of different types of plastics are 33% PE, 29% PP, 17% PET, 7% PVC, and 14% others [16]. Single-use plastics for packaging food, beverages, etc., are the ones that create maximum disposal-related problems. They are large in volume but lightweight.

Agricultural Sector: For various applications related to agriculture and horticulture, plastics used are PE, LLDPE, LDPE, HDPE, PVC, and PP [12].

Construction and Infrastructure: Plastic waste from these two industries is majorly LDPE, followed by a significant share of HDPE and PP [16].

Automobile Manufacturing Sector: This industry sector uses high-performance plastics such as PP, ABS, PBT, PA, and PE based on their features such as functionality and enhanced fuel economy [17]. All these plastics turn into waste only after usage for a long period.

Biomedical Sector: For applications like gloves, syringes, IV tubes, catheters, inflatable splits, etc. Heart valves, hearing aids, spectacles, prosthetics, etc., use plastics such as PE, PP, PS, polyester, PC, PVC, polyethersulfone, polyacrylate, polysulfone, and polyetheretherketone. Poly acetals, PC, PETG, HDPE, and PP are the plastics used for medical equipment [18, 19]. A large amount of plastic waste from this sector is single-use plastics and is mostly incinerated.

Electrical and Electronic Equipment (EEE) Industry: For devices like washing machines, refrigerators, computers, mobile phones, smart TVs, notebooks, digital cameras, investors and UPS, USB, STBs, and LCD and several electronic goods, PS and ABS plastics are the key drivers [12, 20]. Plastics in this sector are long-lasting and hence, disposed off after a long time.

Solar PV and LED Manufacturing Sector: Transparent plastics like Ethylene vinyl acetate (EVA) are the key plastics for solar cells. Polycarbonates are the other transparent plastics used for light-emitting diodes (LED) [12, 20].

3 What Are Plastics?

Derived from the Greek word "Plastikos" (meaning "to mold"), plastics are moldable materials. It is easy to design products of different shapes and sizes using plastics as the base material for the purpose. That is the reason that almost all types and forms of products are available based on plastics. The most popular plastics are organic polymers [21] with different chemistries. One must note that all plastics are polymers, but not all polymers are plastics. This understanding will help while processing plastics.

Polymers, having a long chain structure with repeated units of monomer, are the base materials for plastics [22]. When formulated with special additives to render desired properties, polymers turn into plastics. For example, cellulose, the most abundant natural polymer [23], when converted into cellulose nitrate, became the first synthetic plastic molded for commercial utilization [24]. Moldable materials that flow during heating or under stress but retain the given shape during cooling and removing stress are plastics.

Historically, coal was the first raw material explored for the production of synthetic polymers. In present times, petroleum crude is the main resource for all key starting materials for polymers and thus, for plastics [25–27]. In order to understand the basics of plastics, it is essential to know about the chemical composition, manufacturing process, types of applications, morphology, property profile, etc., of polymers. One gets a fair idea from Figs. 1 and 2 about the chemical composition, manufacturing process, morphology, and life cycle of plastics.

Plastics are either thermoplastics or thermosets. Thermoplastics melt on heating but solidify on cooling because the association of polymer chains is through weak Vander Waal forces. For this reason, it is easy to recycle them several times. Thermoplastics can be either amorphous or crystalline in nature, but molded plastics can never be crystalline due to the complex nature of the long molecular chains in their structure; rather, they have a low degree of crystallinity or are amorphous [28]. Polyvinyl chloride, polystyrene, polycarbonate, poly acrylonitrile–butadiene–styrene copolymer, etc., are some of the common amorphous polymers, and polyethylene, polypropylene, polyamide, fluoropolymers, etc., are some of the crystalline plastics. Due to their low density, resistance to corrosion, and ease of permeability, thermoplastics can easily replace metals in several engineering applications, hence, the term engineering plastics is used for them [29]. Nylon, poly-acetal, polycarbonate, polysulphones, thermoplastics polyesters, poly-phenylene oxide, etc., are some of the engineering plastics.

During polymerization [30], repeated units of monomers combine to produce polymers of the long chain, but if the long chains start reacting with each other, it results in a cross-linked structure. This crosslinking or the interlinking of chains gives rise to a very rigid structure, which does not soften on heating, rather, it degrades when subjected to high heat; such polymers are thermosets. A few examples of thermoset plastics are phenol formaldehyde, melamine formaldehyde, epoxies, etc.

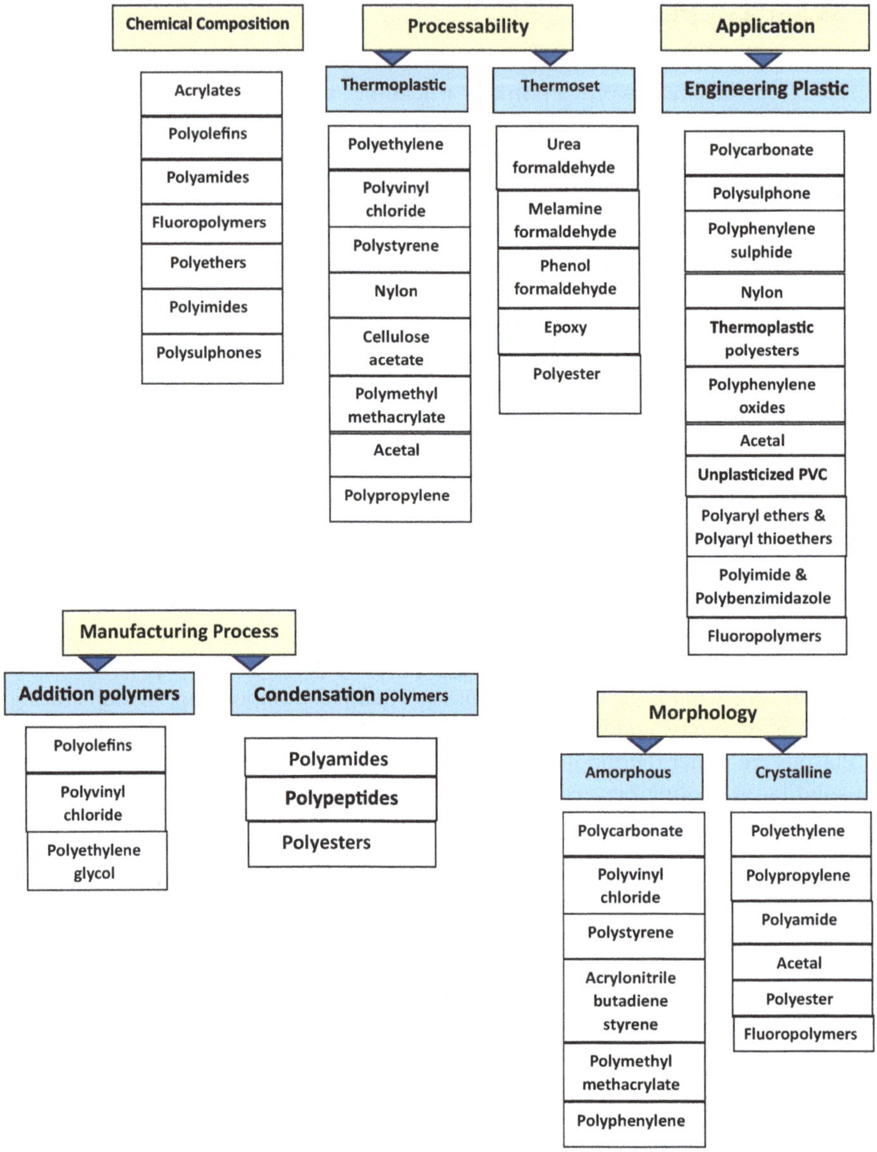

Fig. 1 Classification of polymers

Thermoset plastics have numerous applications due to their high thermal and chemical resistance, dimensional stability, low-flammability, non-melting behavior, etc. Examples of high-performance engineering plastics are polyacrylates, polyimides, fluoropolymers, polybenzimidazoles, etc.

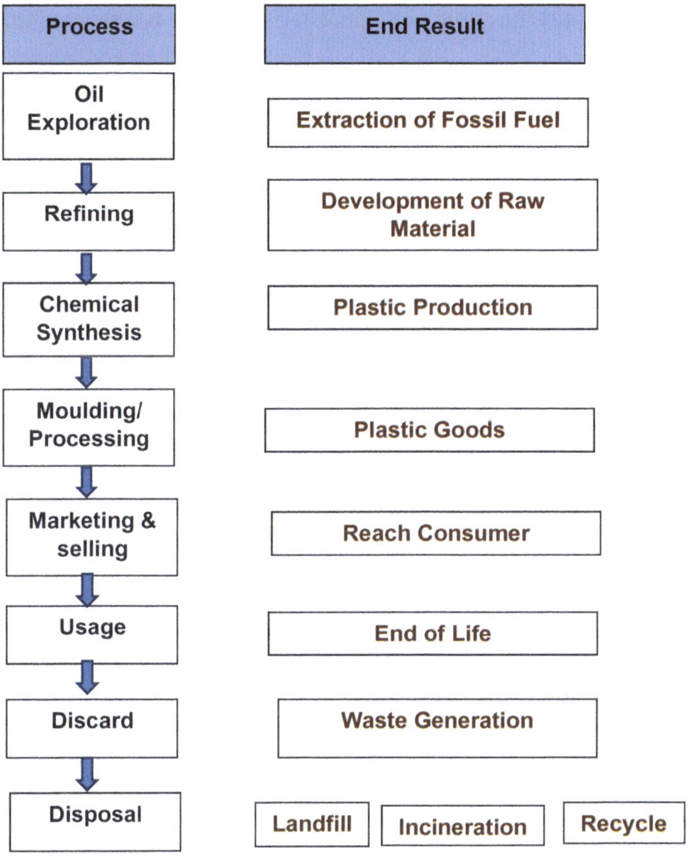

Fig. 2 Life cycle of plastic

Plastic Composites: Plastic composites, made through fiber reinforcement, are novel materials designed for specialized engineering domains such as aerospace and automobiles [31, 32]. For reinforcing, both natural fiber such as jute, mesta, etc., and man-made fiber such as glass, carbon, polyesters, etc., are employed. Composites of both thermoplastics and thermosets are possible; the former with long fibers and the latter with short fibers.

Plastic Alloys or Blends: A synergistic combination of different types of plastics, known as plastic alloys and blends [33], provides materials with unique properties. When two different plastics join, they complement each other and render extraordinary properties. It is easier, both technically and economically, to blend polymers to achieve desired properties than to develop new monomers.

Applications of Plastics: Plastic is a highly versatile material and finds usage in every industrial sector starting from packaging [34, 35], building and construction [36, 37], textiles [38, 39], electronics [40, 41], aerospace [31, 32], equipment and machinery, marine [42], furniture and upholstery [43, 44], healthcare [45, 46], and energy [47].

The packaging sector is the largest sector for applications of plastics [48], whether it is film or bottle/jar or drums or any other shape and whether it is single-use or multiple-use plastic, the consumption of plastic as a packaging material is the highest. Plastic packaging has the capability of acting as a protective material not only for goods packed in plastics, but also for the environment.

4 Difference Between Plastics and Polymers

It is essential to highlight here the differences between plastics and polymers so that there is no confusion while dealing with plastic waste. Polymers, as the term suggests, are nothing but compounds made up of numerous repeated units of monomers. Monomer molecules must have essential features to be able to bond with others to form a long-chain polymer of high molecular weight. Polymers will have their physical and chemical properties completely different from their monomers. Since there exist many monomers, it is possible to design polymers of a wide variety and characteristics. One can design polymers of both types, i.e., Homo (having monomers of the same chemistry) and Hetero (having monomers of more than one chemistry).

Conformations of Polymer Chains: Depending upon the way the monomers combine with each other, polymers of various conformations are possible [49, 50], as shown in Fig. 3.

The cis-link of carbon atoms has the highest bonding energy while the trans-link has the lowest (Fig. 3). Depending on the type of bonding patterns several chain conformations are possible, such as zigzag conformation (e.g., PE or PVC) or helix conformation (e.g., PP, POM, or PTFE).

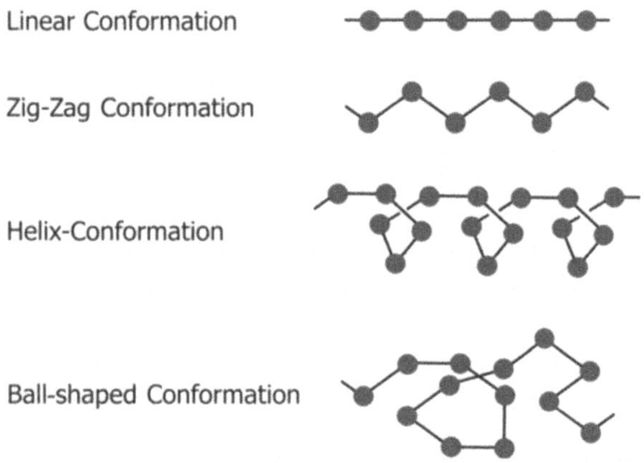

Fig. 3 Conformations of polymers (*Source* [49])

Besides the synthetic polymers, there exists a variety of naturally occurring polymers as well, but the latter are not moldable and hence, they are not plastics. Differences between plastics and polymers are as given in Table 1.

The key point to note here is that plastics are a specific type of polymer that is moldable. In order to have plastics for a particular application, it would be essential to design polymers of the desired characteristics.

Types of Plastics

The basic structure and the method used to synthesize polymers would determine the nature, characteristics, and type of plastics. The following description may help in understanding this aspect of plastics [49, 50].

Categorization of Plastics

In order to understand the various application areas of plastics, first we need to categorize polymers based on different criteria [52].

Origin: Natural and Synthetic: Natural polymers are the ones obtained from plants and animals whereas synthetic ones are those synthesized by complex chemical processes. Cellulose is a natural polymer, which is not plastic, but Nylon, a synthetic polymer, is a plastic.

Behavior on Heating: Thermoplastics, Elastomers, and Thermosets: Thermoplastics are easy to mold polymers, once they are in their molten form. It is possible to remold such polymers many times. They can be either amorphous or semi-crystalline in nature.

Amorphous thermoplastics are, in general, optically transparent and mostly brittle. Typical amorphous thermoplastics are polycarbonate (PC), polymethylmethacrylate (PMMA), polystyrene (PS), or polyvinylchloride (PVC). Semi-crystalline thermoplastics consist of statistically oriented macromolecule chains as amorphous phases with embedded crystalline phases, built by near-order forces. They are usually opaque and tough. Typical semi-crystalline thermoplastics are polyamide (PA), polypropylene (PP), poly-oxy-methylene (POM), etc.

Table 1 Difference between polymers and plastics

Polymers	Plastics
Long-chain compounds are made up of a large number of repeating units of monomers	Only mouldable polymers are plastics and not all polymers
Both natural and synthetic are possible	Only synthetic ones are plastics
Polymers of long as well as short chains are possible	Plastics essentially have very long polymer chains
Polymers can be liquid as well as solids ranging in strength and thickness	Only solid polymers are plastics

Source Rolf Klein [51]

Elastomers are polymers having a wide net like a cross-linked structure. They do not melt on heating without degradation of their molecular structure and hence, they are not remoldable. Examples of elastomers are butadiene rubber (BR), styrene-butadiene rubber (SBR), or polyurethane rigid (PUR). Above their glass transition temperature (Tg), elastomers are soft and elastic, but below Tg, they behave like hard elastic to brittle. The value of the glass transition temperature increases with an increasing number of crosslinks.

Thermosets are polymers that solidify into a permanent design and shape and are not remoldable. Thermosets have narrow cross-linked molecular chains. Examples of thermosets are epoxy resins (EP), phenolic resins (PF), or polyester resins (UP). Due to the strong resistance of molecular movement caused by crosslinking, mechanical strength, and elasticity are not temperature dependent, unlike thermoplastics or elastomers.

Recyclability: Recyclable or non-recyclable: Remoldable polymers are recyclables whereas the rest are non-recyclable. The above discussion is useful in determining which of the wastes is recyclable and which one is not.

Homogeneous structures are examples of copolymers or thermoplastic elastomers, built by the chemical composition of two or more different monomer units in macromolecules.

Heterogeneous structures are, for example, polymer blends or thermoplastic elastomers, built on the physical composition of separate phases of different polymers. The heterogeneous plastics can be further divided into various types, i.e., polymer blends with different forms in itself [52, 53].

Plastic composites consist of a polymeric matrix with organic or inorganic fillers such as mineral pigments, short fibers, long fibers, continuous fibers, paper, or fabrics. Composites made from thermoplastics are recyclable as they melt on heating. Composites from thermosets do not melt on heating and hence, they are not recyclable.

5 Properties of Plastics

One of the important properties of plastics is their temperature of phase change [54]. Unlike materials like metals, phase changes of plastics occur in certain temperature ranges. The width of such temperature ranges is dependent on the homogeneity of the material's structure. During the processing of plastics, the orientation of molecular segments can result in the orientation of segments of the macromolecular chain. The addition of fillers to the polymer matrix leads to altering various properties of the native polymer or plastic. Particles like mineral powder, wood flour, or carbon black, when added, may increase the stiffness of the matrix. The following key characteristics are important to keep in mind while devising strategies for the management of plastic waste.

I. **Thermal properties**

- Glass transition
- Flow temperature
- Crystalline melting temperature
- Thermal decomposition

II. **Mechanical properties**

- Stiffness
- Tensile strength
- Shear strength
- Elongation
- Compressive strength

III. **Rheological properties**

- Melt flow Index

IV. **Specific gravity**

6 Why is Plastic Waste a Cause of Concern for Policymakers?

One of the challenges posed by plastic waste pertains to the pollution of water. Through rivers, it reaches the oceans. With a coastline of >7500 km, a huge quantity (8 MT) of plastics is polluting the ocean [55]. Six plastics, i.e., PP, PVC, PE, PET, polyurethane (PU), and PS are the leading pollutants in the ocean. Waste from single-use plastics makes up most of the waste going into the oceans [56]. Based on their size, plastic wastes are either macro plastics (of a size of 25 mm and above) or microplastics (of size 5 mm and below). Micro and nano-plastics are toxic and detrimental to the species and organisms of the oceans. In fact, through organisms, they end up getting into the food chain. The biggest problem is that, due to their high surface-to-volume ratio, these plastic wastes can adsorb various other pollutants, including heavy metals and persistent organic pollutants, and carry them into the food chain [55, 57].

Assessment on Plastic Waste (PW) Management and Generation in India: Before talking about plastic waste, it is important to understand that, in terms of value, the plastic industry in India is 110,000 crores, involving ~30,000 units. The country is almost self-reliant on plastics [58]. Recycling of plastics is also at a significantly notable level. However, there are many areas of concern related to the hazards associated with this [59]. Plastic scrap collected manually in numerous ways remains not appropriately segregated and contaminated with undesirable impurities. The infrastructure of municipalities is neither adequate nor of proper quality. As a result, waste, which could be useful as a resource material, remains a waste and burden on the environment. The availability of technologies for recycling plastic waste is also an issue in this regard. What is more striking here is that imported plastic waste, properly segregated, was also flooding the market, affecting the local recycling industry adversely. All this is of significance, considering the fact that policymakers here would want plastic waste to be the broken supply chain of plastics.

(https://www.undp.org/content/dam/india/docs/evaluation-reports/HUL%20Brochure.pdf).

PW Generation in India: The volume of plastic waste generated in India is of the order of ~15 MMT [60]. Recycling of the waste is at a much lower level than desired. A major part of plastic waste remains a burden on the environment, posing a risk of contaminating groundwater [61]. This is in spite of the fact that the recycling of plastics in India is the highest (as high as ~60%), much higher than China (10%), Japan (12%), Europe (7%), the USA (10%), and South Africa (16%) [62]. The main cause of concern is the fact that many of the states are not even reporting the amount of waste generated, as per the Pollution Control Board of India. States such as Uttar Pradesh, Maharashtra, and Tamil Nadu are the ones leading in the generation of plastics [63]. The national average per capita plastic waste generated is ~3 kg per year. The highest per capita plastic waste generated per year is ~22 kg per year, in the state of Goa. Another cause of concern is the import of plastic waste [64]. In India, like many other countries (>120 of them), decisions like the closure of illegal recycling units besides restricting imports are useful to deal with this issue. There is a complete ban on the import of plastic waste as per the latest notification [65].

There is a ban on single-use plastics. However, this is not good enough to control pollution because the quantity of plastics used for single-use plastic products is very small. Such a ban is of significance only in controlling the littering of plastic bags. As per one report, the quantity of recycled plastics was ~7 MMT in 2000 which increased to ~30 MMT in 2020. A growth of more than four times within this period may look good, but the fact is that it is insignificant as it is only 6% of the total waste generated. Moreover, the consumption of plastics doubled (to ~355 MMT) in the same period. What is disturbing is the fact that ~70% of these have a life of less than 5 years [64].

Out of 380 million tons produced globally, at least 8 million tons of it ends up in the ocean causing pollution. Further, there is a global shift from reusable to single-use containers resulting in a growth of plastic waste in municipal solid waste from less than 1% in 1960 to 10% by 2005 in middle and high-income countries [52]. The breakdown of larger plastic items into microplastics presents unique challenges, as these tiny particles spread widely, potentially causing ecological and health implications. Microplastics (MPs) present an ongoing and nuanced environmental challenge, acknowledged for their significant impact on aquatic habitats and human health. While MP pollution spans land, sea, and air, these environments, often viewed in isolation, are interconnected. UV radiation and low temperatures contribute to the breakdown of conventional plastics into microscopic particles, entering marine ecosystems through runoff. Recent findings reveal heightened MP levels in aquatic ecosystems near urban centers, with organisms in these areas accumulating significant MP levels. Additionally, MPs readily attract water pollutants like dyes and heavy metals, acting as carriers of diverse contaminants in aquatic species. This exploration delves into the distribution of MPs and their adverse effects on ecosystems and marine biodiversity, concluding with insights into future research opportunities. This research provides a foundation for scientists and policymakers, inspiring further efforts to mitigate the negative impacts of MPs [66, 67]. Micro-plastics today exist everywhere in the environment including in water bodies like rivers, oceans, etc. The reason of formation of micro-plastics is due to fragmentation of plastic waste into particles, of small size of ≤ 5 mm, which are of high molecular weight and do not biodegrade. How much micro-plastics exist in different countries, is difficult to quantify but their existence is a matter of concern for policy makers. One of the reasons for their formation is inappropriate disposal of waste. All types of processing of plastic waste, i.e., recycling, pyrolysis, and incineration also generate micro-plastics [68].

Because of their micro-size, at times, even on the Nano-scale, it is not possible to eliminate them from the environment; they get into the food chain and organisms ingest them easily, causing damage to organs. Accumulation of microplastics in marine and aquatic species is the cause of their mortality [69]. There are several reports that highlight the cases of complex toxicity in fish and even in human bodies [70]. Regulators of all countries in the world are focussing on addressing issues of eco-toxicity due to microplastics [71]. Developing standard methods for monitoring, occurrence, distribution, movement, and ways to control microplastics in the environment are the key areas.

Here, it will be worthwhile to state the notable measures taken by certain countries as follows (Table 2).

1. The Korean Ministry of Environment prepared a plan for recycling (in 2018) and for managing microplastics (in 2019).
2. The Korean Ministry of Oceans and Fisheries prepared a plan for reducing marine plastics in 2019.
3. The United States of America and France regulated banning the intentional use of primary microplastics in cleaning products.
4. The EU adopted the "European Plastics Strategy" in 2018, to increase recycling of and a reduction of plastic waste.
5. The United Kingdom and France banned the use of single-use disposable plastics.

Microplastics are of two categories; primary (intentionally added as microbeads in consumer products like detergents, cosmetics, etc.) and secondary (produced from plastic waste including fibers) [72]. It is easy to assess the toxicity of primary microplastics because they are mostly of the same size and shape. Since secondary microplastics vary a lot in their shape and size, it is difficult to conduct toxicology studies for them. Microplastics adsorb toxic heavy metals and organic pollutants which can stimulate the release of endocrine disruptors in humans [73].

Exposure of humans to microplastics happens at various stages during the use of plastics in the form of packaging materials, paints, coatings, textiles, and while processing waste. Through the food chain, humans are ingesting and inhaling microplastics. Ingestion of microplastics may cause inflammation and several

Table 2 Domestic and international statuses of regulation on banning the use of microbeads and microplastics

Country	Year	Regulations
USA	2014	Banned use of microbeads in personal care products (Illinois)
	2015	Banned use of microbeads in cosmetic products (California, New Jersey, New York, etc.)
Canada	2016	Banned use of microbeads in exfoliation and cleansing products
New Zealand	2018	Prohibition of the manufacture and sales of personal care products containing microbeads
Northern Ireland	2019	Prohibition of the manufacture and sales of personal care products containing microbeads
Italy	2020	Prohibition of the manufacture and sales of personal care products containing microbeads
UK	2021	Prohibition of the manufacture and sales of personal care products containing microbeads
Sweden	2019	Amended acts to ban the use of microbeads in cosmetic products
Taiwan	2020	Amended acts to ban the use of microbeads in cosmetic products
Korea	2017	Designation of microbeads as raw materials that cannot be used in cosmetic products
	2019	Preparation of a test method for detecting microbeads in cosmetic products by the Ministry of Food and Drug Safety in Korea
	2021	Specific regulations for microplastic contents in some household chemical products: cleaning solutions, deodorant agents, laundry detergent, bleach, and fabric softener
India	2017	Banned the use of microbeads

Source Yongjin et al. [68]

gastrointestinal complications; change in the intestinal microbiome, leading to an imbalance between beneficial and harmful bacteria, causing abdominal pain, bloating, etc. Due to the ingestion of heavy metals and organic pollutants through microplastics, humans may suffer from chemical toxicity effects. Workers in the synthetic fiber industry regularly inhale microplastics and as a result they suffer from respiratory and lung diseases. As per certain other studies, inhalation of Nano-sized plastics caused mitochondrial damage in human respiratory cells. Today, exposure of humans to micro-plastics is to the extent of several milligrams of micro-plastics per day, through inhalation of indoor air and drinking water in plastic bottles; especially the exposure of newborns through feeding bottles [68].

Unfortunately, a significant portion of this plastic waste often finds its way to landfill sites or marine environments [74]. The extensive proliferation of plastic waste poses a significant threat to ecosystems, manifesting in diverse forms of pollution throughout its life cycle [75, 76]. Open burning of plastic waste is also a serious concern for policymakers as it emits substantial amounts of greenhouse gases (GHG), COx, NOx, and various toxic gases, exacerbating environmental concerns. Open dumping further compounds issues by disseminating odors and fostering the spread of infectious diseases. Land dumping renders soil infertile and contaminates groundwater, while sea dumping disrupts marine life, with plastic persisting for hundreds of years [76, 77].

7 How to Convert Plastic Waste into Valuable Products?

It is evident from the discussion above that meeting the rising demand for plastics by producing more polymers and plastics will not be sustainable. Utilizing waste to make products will only be the solution to avoid a burden on the environment due to plastic waste on the one hand and to reduce the demand for virgin plastics on the other hand. It would be worthwhile discussing the life cycle of plastic waste. Since plastics are not biodegradable, ideally the devices made of plastic remain in use for as long as possible. Once plastic devices turn into waste, they will remain either on the land or in the water bodies without degradation. It is thus, essential to convert the waste into valuable products. Recycling of waste would be the best option. If not, incineration of waste for energy is another option. Before talking about options for this purpose, it would be better to ponder over the advantages of recycling. How long the plastic materials remain in use and how fast they get into the dustbin are the two factors policymakers envisage while planning strategies for managing waste.

The product-to-waste life cycle of plastic encompasses the journey of plastic materials from their initial production as consumer goods through their usage and eventual disposal as waste [78]. This life cycle involves several stages, including extraction of raw materials, manufacturing, packaging, distribution, consumer use, and the end-of-life disposal phase. Understanding this life cycle is crucial for assessing the environmental impact of plastics, considering factors such as resource extraction, energy consumption, and the potential for pollution at each stage. Exploring the product-to-waste life cycle provides insights into sustainable practices, recycling opportunities, and the challenges associated with managing plastic waste in an environmentally responsible manner [79]. Recently, global recycling technology trends have given valuable insights into the recycling industry's latest advancements. Notably, chemical and advanced mechanical recycling technologies are attracting significant investments.

Recycling of plastic waste will mean that there will be a break in the requirement for raw materials. It means that there would be a reduction in the extraction of raw materials coming from fossil resources (crude oil). Since fossil reserves will not last forever, recycling of plastic waste will help conserve resource materials. The demand for plastics will not recede as per global trends and forecasts for per capita consumption of plastics. The only alternative, therefore, is to reduce the requirement for raw materials by creating plastics from waste. Further, avoiding the extraction of raw materials by converting waste into products again would be of much lower carbon footprint. Treating plastic waste as a resource is the most environment-friendly alternative. Similarly, there will be no need to produce more virgin plastic, thereby saving energy and eliminating GHG emissions. The time taken for plastics to turn into waste varies widely; daily used packaging materials turn into waste within the shortest time whereas plastics used in making devices would remain in use for a long duration and hence, would take the longest to end up as waste. Single-use plastics, which are more in volume but least in weight, are the fastest to reach in the dustbin, making waste management a tough task. Single-use plastics are the biggest burden on the environment. Consumers must reuse the waste before the waste goes for recycling or recovery and finally for disposal.

In India, culturally, discarding packaging materials is not as common as elsewhere in the world. Reuse of plastic packaging materials is prevalent and this keeps the carbon footprint of plastics low. Collection, sorting, and processing the waste into pellets or flakes are the steps involved in making the re-entry of waste into products again for use. For non-recyclable plastics, incineration is a better option than landfilling. Incineration of plastics is of advantage, especially in the case of treatment of clinical wastes and hazardous wastes to destroy toxins and pathogens. This, however, would need technological intervention. It is important to note that burning of plastic waste in the open is dangerous as it releases toxic pollutants such as dioxins, sulfur dioxide, mercury, polychlorinated biphenyls, and heavy metals as detailed later in the section on incineration. Landfill and dumpsite disposal of plastics contribute to environmental pollution through leachate contamination, soil, and air pollution. This may not be immediate but on a long-term basis, it will result in soil and groundwater contamination due to additives and breakdown of by-products in plastics, potentially transforming into persistent organic pollutants [80, 81].

The World Economic Forum, five years ago, emphasized the need to go for recycling in order to deal with plastic waste, which will be twice the size that exists presently. It expressed concern about such low levels of recycling of waste in developed countries like Europe and the USA. With the advancement of science, technology, and innovation, the development of new devices is the need of the hour. This creates the demand for resource materials without which it is not possible to introduce new devices. Since plastics are an integral part of new and novel materials, dependence on resource materials, which are non-renewable, is increasing. In fact, the rate of new discoveries is falling short due to the high rate of consumption. This massive consumption of plastics is causing an adverse impact on the environment because of energy consumption, emissions of greenhouse gases, and a high carbon footprint. All this puts pressure on land use as well as the sustainable availability of water. Thus, on one hand, renewable resources are depleting in meeting the increasing demand for plastics. On the other hand, there is continuous generation of huge amounts of waste. It will no longer be possible to dedicate land for landfilling of plastic waste. It is important to note that the key starting materials for manufacturing of plastics are from non-renewable (fossil) resources. Therefore, recycling of plastic waste means conserving resources which are available in limited quantities. Further, the whole world is now focusing on sustainability, which demands preserving resources and using them judicially. In the future, therefore, the emphasis will be on the circular economy rather than a linear economy, in order to meet the rising demand for plastics as well as their key starting resources. Recycling serves that purpose very well. Large multinational corporates, especially those dealing in petrochemicals and polymers, created alliances to boost efforts to recycle plastic waste [82].

Different Types of Recycling of Plastic Waste: One of the most desirable processes is recycling when dealing with waste, which means converting the waste into a useful material as well as in its new form. Plastics are recyclables and hence, recycling of their waste should be the first and foremost option for waste management [83]. It is like producing materials not from base materials but from waste, making the process of recycling the most environment-friendly [84]. Retrieving the material back from its waste without losing the inherent characteristics of its virgin form should be the goal. There is an urgent need to re-evaluate the present outlook on the usage of resources and to reassess ways to minimize the damages to the environment. For this, the following key actions would be necessary: increasing the lifetime of products, lesser use of resources; reuse of products and recycling of waste. The extension of the lifetime of plastic products by their reuse will always have its limitations and ultimately the product will become a waste material. The lesser use of resource material will be possible only by retrieving plastic material from waste. This is again to show that recycling of plastic waste is the best option as it enables getting the product back from its waste at the end of the product lifecycle. This means that we consider waste as a resource rather than a burden on the environment. Certain virtues like mouldability, resistance to decomposition and corrosion, and retaining inherent characteristics even after recycling plastic many times are the advantages. The key strategy to manage plastic waste is recycling as emphasized above. Recycling of plastic waste will not only help in fighting waste management problems, but will also reduce the dependence on fossil fuels, depleting at a fast rate.

The presence of additives used in plastic formulations and the use of plastic composites and blends make the recycling of plastics a complex process. Technological interventions would be necessary to address issues arising out of these complexities. Table 3 [85] describes the types of plastics, their applications, and recyclability.

The recycling process involves three steps; collection and separation of recyclable materials from waste; processing the recyclable material into a raw material and then converting the latter into a finished product. There are three types of recycling as described in the following sections.

Mechanical Recycling: This is mainly used for waste from industrial scrap or domestic or commercial disposals and especially for PE and PP. Mechanical recycling includes processes such as grinding, washing, separating, drying, re-granulating, and compounding. There is no change in the chemical structure of the polymers, which remain reusable multiple times for making products of a similar type [86]. Popular products obtained by this process are garbage bags, floors, hoses, car parts, packages, etc.

Energy Recycling: Plastic waste can be a good resource for the generation of both thermal and electric energy. This is nothing but incineration of plastics, which are a storehouse of fuel. How much energy recycling is environmentally friendly remains

Table 3 Types of plastics and their recyclability

Polymer	Abbreviation	PLASTIC TYPE	Example usage	Recyclable
Teraftal polyethylene	PET	1	Beverage packaging	Yes
High density polyethylene	HDPE	2	Toys, shampoo bottles, pipes	Yes
Polyvinyl chloride	PVC	3	Window frame, floor covering, pipes, cable insulation	Difficult
Low density polyethylene	LDPE	4	Reusable bags, food packaging film	Difficult
Polypropylene	PP	5	Food packaging, automotive components	Not
Polystyrene	PS	6	Food packaging, insulator, electrical equipment	Not
Polyurethane	PURE	7	Insulator, mattresses	Not most of the time
ABS, elastomer, plastic bio, PBT, PC, PMMA, PTFE	Others		Tyres, packaging, electronics, transport	
Fibres made from different polymers	Fiber		Textile applications and other sectors	

Source [85]

a tough question to answer. Non-recyclable plastics are the best resource for manufacturing process-engineered fuels (PEFs), which are energy-combustible materials used in place of fossil fuels. Cement plants or any other process using kilns would use PEFs. Likewise, non-recyclable plastics in combination with municipal solid waste can form refuse-derived fuel (RDF). The Ministry of Housing and Urban Affairs prepared guidelines for the use of RDF in industries in 2018. As per this, non-recyclable plastic waste of calorific value of 1500 kcal/kg or more must be a resource for PEF or RDF; should not go to landfills. The section concerning incineration in this chapter covers all aspects related to this in detail [87].

Chemical Recycling: Chemical recycling involves modification of chemical structures of plastics. It is the most complex process and needs technological intervention; still in its early stages of evolution.

Let us discuss here a case of recyclable plastic, i.e., polyethylene terephthalate (PET), used for making bottles for beverages, including packaged drinking water and mineral water. One mineral water bottle comprises three different plastic components; a bottle made from PET, a cap made of polypropylene (PP), and a label made of polyethylene (PE). For recycling of a bottle, therefore, it is essential to separate each of these components. There is an established industry set up for recycling PET. There are two major options for PET recycling [88, 89], one is the conversion of PET bottles into fibers, which finds their use in the textile industry for the replacement of cotton and wool and the other option is the breakdown of PET into its monomeric components and then remanufacture of PET called recycled PET. This recycled PET has a huge market internationally due to compliances imposed by authorities for the use of recycled PET along with fresh PET. The industry is importing waste PET bottles and exporting recycled PET granules.

Printed circuit boards (PCBs) are an example of multi-component plastic composites, for which recycling is quite complex. To understand the intricacies of its recycling, it is first important to appreciate the type of material present in PCBs. PCBs comprise a base, which has laminated sandwich structures comprising one or more conductive metals (e.g., copper) and insulating layers (which are glass–epoxy composites). The base, joined with electronic components or integrated circuits (ICs) mechanically and also through soldering, makes recycling a tough task. Further, ICs may comprise of precious metals depending on the design criteria of the electronic gadget made up of the former. Another important aspect here is the difference in PCBs of different electronic components. The type of material present in PCBs also decides their cost and, accordingly, the recyclers plan and prioritize recycling of those waste plastics. Presently, there is a large sector of the industry dedicated to electronic waste recycling [90] involving multiple steps due to different materials present in PCBs as listed in Table 4. As evident from this data, the value of waste will depend upon the content of metals. That is why recycling is more for recovering metals than for plastics.

To begin with, the practice of recycling of electronic waste was mostly in the unorganized sector, on a very small scale with many units operating at individual levels concentrated in the district of Moradabad, Uttar Pradesh. Recycling by unorganized sector involved burning of electronic waste, separation of precious metals, and dumping of residual ash on riverbanks, which contributed in a big way to the contamination of surface and groundwater bodies with heavy metals. Due to the imposition of strict norms by regulators, such practices are no longer prevalent. Today, recycling of electronic waste consisting of plastics is mostly in the organized sector [93–95]. The process begins with the first step, which is the segregation of components. PCB recycling is a combination of manual and mechanized operations including the following steps: Collection of wires cut out from PCBs manually. Separation of wires mechanically, into two portions, those containing metal from those having plastics. After this, break down both these portions into small pieces, which recyclers procure for further processing.

Table 4 Composition of electronic waste

S. no	Electronic items	Plastic (%)	Iron (%)	Aluminium (%)	Copper (%)	Silver (ppm)	Gold (ppm)	Pd (ppm)
1	TV board	28	28	10	10	280	20	10
2	PC board	23	7	5	20	1000	250	110
3	Mobile phone	56	5	2	13	3500	340	130
4	Portable audio	47	23	1	21	150	10	4
5	DVD-Player	24	62	2	5	115	15	4
6	Calculator	61	4	5	3	260	50	5

Source Cui and Zhang [91], Sahil [92]

- The next step in dismantling the machine dislodges the circuit components on PCBs (ICs, condensers, etc.) from the PCB plate, at a temperature of ~120 °C to soften the PCB plate. Attrition of the material under these conditions separates circuit components from bare PCB board. After this, it is easy to separate the two to get components for further processing.
- Separate the circuit components manually into two categories: one having precious metals (gold, silver, and palladium) and the other having metals like aluminium, copper, iron, brass, and solder (tin and lead). For extraction of precious metals, use the leaching process.
- For items containing plastic parts (e.g., ABS) and metal parts use magnetic separation to separate iron from the bulk. Crush the left-over portion and heat it to 200 °C to separate out the solder. After this, crush the leftover that consists of aluminium, copper, and brass, which makes it easy to pick up aluminium particles, which are bigger. The leftovers contain a mixture of copper and brass, which is a saleable product.
- The bare PCB board crushed to powder itself is a saleable product from recycling.

Unlike other plastics, PET (polyethylene terephthalate) presents challenges during recycling due to its susceptibility to degradation. Additionally, PET contains antimony trioxide, a carcinogen, and thus, its release into the environment is undesirable. HDPE waste is the most recycled plastic, deemed safe with a low risk of leeching into foods and water by FPA. PVC, on the other hand, is not safe to recycle due to the presence of alkylene chloride, which is harmful, linked with bone and liver diseases. Specialized programs to recycle PVC into flooring, paneling, roadside gutters, etc. need technology intervention. Commercially named as Styrofoam used in beverage cups, packaging materials, egg cartons, etc., is not safe to recycle. Plastics under this category of miscellaneous plastics often contain bisphenol, a toxic chemical. In addition, the recyclability of these plastics is low; it is difficult to recycle.

Regulators categorized plastic waste into seven groups given with codes written in the manner as shown below in Fig. 4, for recycling:

1. Polyethylene Terephthalate (PET)
2. High Density Polyethylene (HDPE)
3. Polyvinyl Chloride (PVC)

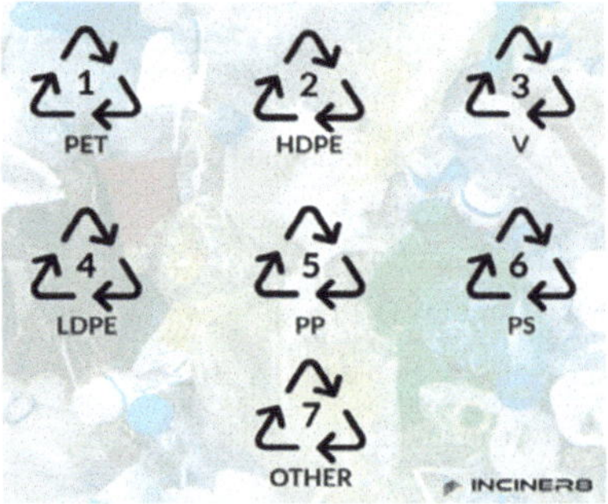

Fig. 4 Plastic polymer codes. *Source* Jeffrey [96]

4. Low Density Polyethylene (LDPE)
5. Polypropylene (PP)
6. Polystyrene (PS)
7. Other such as polycarbonate, acrylic, and nylon

Here, the number given to each category, in the center of a triangle formed by three arrows, defines the ease of recycling; the lower the number more recyclable that product is (RR), (SS). For example, PP, PS, PC, etc., are not easy to recycle. Further, it is important to mention here that recycling of plastic waste becomes complex and tough when the waste consists of mixed plastics. Therefore, incineration of mixed-plastic waste would be a better option than landfilling.

Recycling of plastic waste by any process or technology would involve the following steps: Primary path: Plastic waste re-extruded into new goods is the process of primary recycling. Secondary path: mechanical recycling which involves the following steps: Cutting/Shredding; Separation of contaminants; Milling; Washing and Drying and Extrusion. The great challenge here pertains to the sorting and segregation of different types of waste [97].

Sorting and Separation Techniques in Mechanical Recycling of Plastic Waste:

- Air Classifier (Zig Zag Separator): Utilized to [98] segregate light films, contaminants like paper or foil, or fine dust from recovered materials such as plastic flake, the air classifier proves highly effective, typically post-granulation, or after dry cleaning operations.
- Air Tabling: a density concentration technique separates particles of mixed sizes, shapes, and densities due to differential settling in an upward airflow.
- Ballistic Separator: Designed for the separation of solid waste at the inlet, the ballistic separator relies on factors like size, density, and shape for efficient sorting.
- Dry and Wet Gravity Separation (Sink Float Tank): This method segregates different plastics based on their density. Heavier fractions sink to the tank's bottom, while lighter fractions float. Widely used in recycling plastic bottles and caps, it facilitates effective material separation.
- Froth Flotation: Involving an intensive rinsing process with the addition of fluid or steam, froth flotation extracts significant fluid volumes. It is particularly effective for cleaning mixed plastics and highly contaminated films, making it an ideal choice for treating soiled plastics during recycling.
- Electrostatic Separation (Triboelectric Separation): charged particles pass through a tribocyclone, causing negatively charged particles to gravitate toward a positively charged plate and vice versa. This process classifies three or more resins simultaneously.

Incineration

Results of a study, [99], shows that 46% of plastic waste from households in UK, goes for incineration. Only 12% of the plastic waste gets recycled, while the rest of it goes to landfill (25%) and exports (17%). Incineration is obviously the preferred option for managing the waste here. In comparison to this, in India, incineration of plastic waste is at an insignificant level of 4%. Crude oil is the base material for the manufacture of plastics and hence, burning the latter is just like burning of petroleum fuels (diesel, petrol, or natural gas). In fact, plastics like PP, LLDPE, HDPE, and PS are hydrocarbons like petroleum fuels and there should be no harmful emissions except CO_2 and CO from their burning. The problem is that when the whole world is trying to minimize GHG emissions, the release of CO_2 is also not acceptable. Each tonne of plastic waste would release ~2 tonnes of CO_2, which will add to global warming [77]. At present, incineration of plastic waste in the UK releases 750,000 MT of CO_2 and in the future, with the increase in volume of the waste, this will rise high. Incineration of waste of halogenated plastics emits gases like dioxins and furans which are toxic and carcinogenic. In spite of the measures taken to capture toxic gases and avoiding them getting into the atmosphere, incineration is not an acceptable option due to the release of GHGs (Table 5). All countries will have to stop the incineration of plastic waste [100, 101].

Besides getting rid of the waste, the main purpose of the incinerator is to generate energy from heat produced during burning. Plastic waste, mixed with municipal solid waste, burns in incinerators designed accordingly. Since incinerators will produce emissions of CO_2, etc., gaseous pollutants, fly ash, and solid waste residue, the design of incinerators should comply with regulatory norms. Often, the content of plastics is not sufficient to burn municipal solid waste, which has high moisture and low calorific value. In such cases, the addition of fossil fuels, e.g., coal, to the waste, becomes necessary. This is not an environment-friendly option and hence, incineration is not a wise solution to solve the problem of plastic waste [99].

In China, for example, incinerator operators routinely add coal to the municipal waste to make it combustible. Incinerator companies, therefore, have an incentive to keep plastic in the municipal waste stream, rather than implement strategies to reduce, reuse or recycle plastic waste [103].

Table 5 Plastic waste management

Country	Mismanaged and uncollected (%)	Land filled (%)	Incinerated (%)	Recycling (%)
India	46	36	4	13
Europe	5	37	44	14
USA	4	73	19	4
Middle East and North Africa	40	54	1	5
Other Africa	64	30	1	6
China	27	36	24	13
Latin America	42	47	1	10
Whole World	22	49	19	9

Source [102]

The World Energy Council endorses the view that burning waste to produce energy is a good option for dealing with the menace of plastic waste, providing enough measures to scrub toxic gases and other pollutants. Further, there are reports that suggest that instead of burning, pyrolysis of plastic waste would be a better option. Neither of these options can put a break to the rising quantity of plastics produced nor do they help in curbing climate change. The fact is that the best option is recycling as this can only help reduce the need to produce more virgin plastics and, at the same time, it will reduce GHG emissions [104].

If one thinks that incineration can eliminate plastic waste forever, a recently published study [105] should serve as a caution. Specimens of ash from the bottom of several incinerators of different types showed that they contained particles of microplastics. Each metric ton of ash contains hundreds of thousands of plastics. Out of the nine plastics studied, PP and PS were prominent in particles of microplastics. It is therefore evident that incineration may reduce the quantity of waste, but it leaves a huge number of particles of microplastics [105]. Considering the fact that microplastics are toxic to organisms, it will not be a good idea to opt for incineration [106]. This could perhaps be the reason as to why, in the USA, the number of incinerators is on the decline [106] as obvious from Fig. 5:

Globally, the distribution of management of total plastic waste is as follows: recycling: 9%, incineration: 19%, landfilling: 49%, and mismanaged and uncollected: 22% (Table 5). The most disturbing facts, as of 2019, about plastic waste are: ~6.1 MMT of it leaked into aquatic environments (of this ~1.7 MMT ended up in the ocean); ~30 MMT was already in the oceans and ~109 MMT the remained accumulated in rivers and would ultimately flow into oceans in the future. All this suggests that principles of circular economy need to be adopted for plastics in order to address issues related to the management of plastic waste [107].

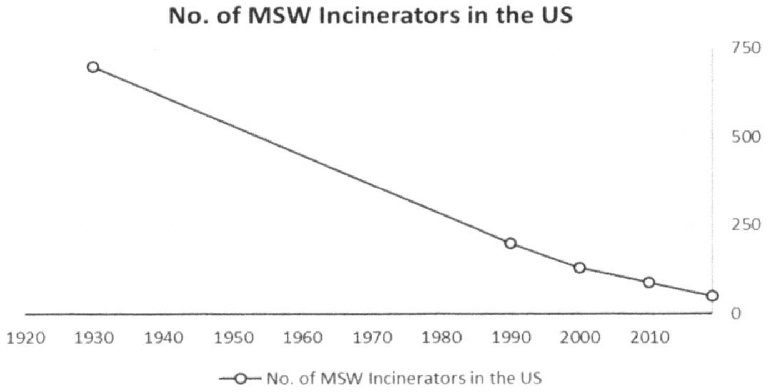

Fig. 5 Declining number of Incinerators in the USA with time. *Source* [106]

8 Technology Interventions

From the discussion above, following technology interventions are necessary for meeting the challenges:

Sensor-Based Separation Technologies:

- Color Sorting: Utilizing ultraviolet light and advanced digital camera technology, this technique identifies materials as small as 0.04 mm. Plastic color sorting machines cover a range of material throughputs and demonstrate high sorting accuracy, often up to 99.99%.
- Near Infrared (NIR): NIR sorting involves illuminating materials, with the NIR sensor distinguishing between different materials based on their reflection of light in the near-infrared spectrum. Commonly used for whole bottles and jugs, it requires a minimum particle size of 50 mm.

Pyrolysis:

Tertiary path: chemical and thermal recycling: Chemical recycling is a procedure in which a polymer undergoes decomposition to extract its key starting materials, known as monomers. It has been successful in recovering monomers from materials like PET, nylons, and polyurethanes. The process, conducted in an oxygen-free reactor with catalytic additives, reaches a maximum temperature of 350 °C. The outcome includes liquid Refuse-Derived Fuel (RDF), gases, and solids. Gases and residual solids are usable as fuel in the cement industry. This method offers the potential to convert "challenging-to-recycle" products into valuable feedstock, restoring them to a quality similar to virgin resin [53].

Pyrolysis involves thermal breakdown, in the absence of oxygen, yielding valuable products such as elemental components, liquid fuels, and gases [108].

Plasma Arc Gasification: It utilizes high-temperature plasma to disintegrate plastics into their elemental components. It also offers a means to obtain useful syngas, a mixture of hydrogen and carbon monoxide, for subsequent applications.

Conventional Gasification: Conventional gasification entails transforming plastic waste into gas through a controlled reaction with oxygen and steam. The resulting syngas serves as a versatile feedstock for synthesizing various chemical products, including new plastics.

Typically, these methods yield:

Gases with Elevated Calorific Content- The chemical recycling techniques frequently produce gases distinguished by a heightened calorific content used for diverse applications, enhancing resource recovery [109]. Efficient plastic categorization for recycling and real-time annotated data collection with a Tensor Flow object detection model. https://iopscience.iop.org/article/10.1088/2515-7620/acfecf/pdf.

Oils with Elevated Calorific Content- Similarly, the procedures generate oils that possess an increased calorific content. These oils can function as valuable energy sources, amplifying the overall efficiency of the chemical recycling pathway.

Quaternary path: Quaternary recycling involves energy recovery, which encompasses processes such as energy from waste or valorization [53]. This is the last option. Additionally, biodegradable plastics offer another facet of tertiary recycling, particularly in composting. This form of recycling is often referred to as organic or biological recycling, as highlighted in the work of Song et al. [87].

Dissolution Recycling

Dissolution recycling stands out as a distinctive process in the recycling realm, with a specific focus on separating polymers from plastic waste and restoring them to their original chemical structural properties. This unique method involves dissolving the polymers found in plastic waste using solvents, thereby facilitating their separation and reverting them to their purest form. Polymers like Polyvinyl chloride (PVC), Polystyrene (PS), Nylon (PA), and Polypropylene (PP) undergo a meticulous breakdown into individual high-quality monomers [109], good enough to produce polymers again.

Recycling Robots

The use of recycling robots can address challenges of pollution and workforce shortages faced during the mechanical recycling process. Recycling robots can overcome human limitations, leading to accelerated trash-picking speeds, minimized errors, especially in waste sorting and improved picking efficiency. This technological advancement brings significant benefits by reducing operational expenses, enhancing quality control over waste streams, and gaining better visibility into waste flows [110].

Artificial Intelligence (AI)

Incorporation of AI not only automates procedures but also safeguards the well-being of workers by minimizing direct exposure to perilous waste streams [109]. Through the amalgamation of AI into recycling workflows, emerging enterprises can leverage machine learning and computer vision techniques. This empowers them to identify patterns, comprehend pivotal information, make precise predictions and execute intricate tasks with minimal human intervention. The adoption of automated workflows has yielded substantial improvements across diverse operations within the recycling business. Startups harnessing AI have witnessed heightened productivity, reduced expenditures, and accelerated scalability in their undertakings. This paradigm shift underscores the potential of AI to revolutionize recycling practices, rendering them more sustainable and efficient [79, 111].

The innovative approaches, such as waste heat vaporization and waste-to-energy solutions, underscore a growing commitment to extracting value from waste streams while minimizing the reliance on virgin materials [78, 112–120].

9 Path Forward

For addressing the complex issues associated with the management of plastic waste, the policymakers will have to devise strategies for the following:

(a) Banning single-use plastics is necessary, but emphasis is also necessary on the systems procedures, regulations, and infrastructure for appropriate disposal of plastic waste at the municipality level.
(b) Awareness of consumers, not only about reuse but also about segregation of waste, at the point of disposal so that sorting before recycling becomes less challenging.
(c) Adoption of modern technologies for sorting and segregation of waste as well as for optimization of the process of recycling.
(d) Use of process technology that scrubs emit hazardous gases using the most advanced instrumentation control systems and IT tools.
(e) Adopt the safe chemical recycling process and recover monomers.
(f) Design composites using waste use the latest concepts of creating novel materials.

The abovementioned action points as the path forward will serve a great deal in managing plastic waste. In addition to this, there are several success stories that demonstrate the use of plastic waste, especially of polyolefinic type, in place of bitumen for building roads. Process-engineered fuel and certain composite materials using plastic waste are other viable options as outlets for plastic waste. The reuse, recycling, reprocessing, and depolymerization would then be the targeted focus areas for researchers finding solutions to the problems envisaged due to plastic waste. Once there are systems in place for managing plastic waste and technological interventions for this purpose, many of the problems would disappear. Then, it would be possible to create value from the waste.

References

1. Andrady, A. L., & Neal, M. A. (2009). Applications and societal benefits of plastics. *Philosophical Transactions of the Royal Society B, 364*, 1977–1984. https://doi.org/10.1098/rstb.2008.0304
2. Brydson, J. A. (1999). *Book on plastic materials* (7th edn.). Elsevier, ISBN: 0080514081, 9780080514086.
3. Geyer, R., Jambeck, J. R., & Kara Lavender Law. (2017). Production, use, and fate of all plastics ever made. *Science Advances, 3*(7), e1700782. https://doi.org/10.1126/sciadv.1700782
4. Parashar, N., & Hait, S. (2021). Plastics in the time of COVID-19 pandemic: Protector or polluter? *Science of the Total Environment, 759*, 144274. https://doi.org/10.1016/j.scitotenv.2020.144274
5. Rakesh, K., Anurag, V., et al. (2021). Impacts of plastic pollution on ecosystem services, sustainable development goals, and need to focus on circular economy and policy interventions. *Sustainability, 13*(17), 9963. https://doi.org/10.3390/su13179963

6. UNEP. (2023). Chemicals in plastics: A technical report. Available on the UNEP website at https://www.unep.org/resources/report/chemicals-plastics-technical-report
7. Dussaux Damien, Agrawala Shardul, OECD Environment Working Papers No. 199, Sep 21, 2022. Quantifying environmentally relevant and circular plastic innovation: Historical trends, current landscape and the role of policy. https://doi.org/10.1787/1f6dbd07-en
8. Walker Tony, R. (2021). (Micro)plastics and the UN sustainable development goals. *Current Opinion in Green and Sustainable Chemistry, 30*, 100497. ISSN 2452-2236. https://doi.org/10.1016/j.cogsc.2021.100497
9. Vanapalli, K. R., Samal, B., Dubey, B. K., & Bhattacharya, J. (2019). 12 - Emissions and environmental burdens associated with plastic solid waste management. In S.M. Al-Salem (Ed.), *In plastics design library, plastics to energy* (pp. 313–342). William Andrew Publishing. ISBN 9780128131404. https://doi.org/10.1016/B978-0-12-813140-4.00012-1
10. Özkan, K., Ergin, S., Işık, Ş., Işıklı, İ. (2015). A new classification scheme of plastic wastes based upon recycling labels. *Waste Management, 35*, 29–35. ISSN 0956-053X. https://doi.org/10.1016/j.wasman.2014.09.030
11. Sanchez, F. A. C., Boudaoud, H., Camargo, M., & Pearce, J. M. (2020). Plastic recycling in additive manufacturing: A systematic literature review and opportunities for the circular economy. *Journal of Cleaner Production, 2020*(264), 121602.
12. Plastindia Foundation. (2018). Report on the Indian plastics industry. https://plastindia.org/pdf/Indian-Plastics-Industry-Report-2018-2.pdf
13. Hossain, R., Islam, Md. T., Shanker, R., Khan, D., Locock, K. E. S., Ghose, A., Schandl, H., Dhodapkar, R., & Sahajwalla, V. (2022). Plastic waste management in India: Challenges, opportunities, and roadmap for circular economy. *Sustainability, 14*(8), 4425. https://doi.org/10.3390/su14084425
14. Global Consumption of Plastic Materials by Region (1980–2015). https://www.plasticsinsight.com/global-consumption-plastic-materials-region-1980-2015/
15. Jayavelu, S., Muniamuthu, S., Rajkumar, C., Udaya Prakash, J., Hema Kumar, V. S., & Ganesan, S. (2023). Scope of waste plastic to reuse in India – A review. *Materials Today: Proceedings*. ISSN 2214-7853. https://doi.org/10.1016/j.matpr.2023.04.156
16. FICCI. (2018). Indian plastic industry: Challenges & opportunities. https://www.slideshare.net/TSMG-Chemicals/indian-plastic-industry-challenges-opportunities
17. Patil, A.; Patel, A.; Purohit, R. An overview of polymeric materials for automotive applications. *Materials Today: Proceedings, 4*, 3807–3815.
18. McKeen, L. Plastics used in medical devices. Retrieved March 29, 2021, from https://www.pentasil.eu/images/Plastics%20Used%20in%20Medical%20Devices.pdf
19. Piedmont Plastics. What is the best type of plastic for medical equipment? Retrieved March 29, 2021, from https://www.piedmontplastics.com/blog/best-plastics-for-medical-equipment
20. ELMER. What type of plastic is used in the body of an led bulb? Retrieved March 29, 2021, from https://elmerindia.com/knowledge/what-type-of-plastic-is-used-in-the-body-of-an-led-bulb-did-it-have-any-special-specification/
21. Ebbing, D., & Gammon, S. D. (2016). *General chemistry* (11th edn.). Cengage Learning. ISBN: 978-1-305-88729-9. See: https://chemistry.com.pk/books/general-chemistry-11e-ebbing-gammon/
22. Elias, H.-G. (1997). *An introduction to polymer science*. VCH.
23. Rasaq, S. A., Swarna, J., & Jaiswal, A. K. (2023). Green and sustainable pretreatment methods for cellulose extraction from lignocellulosic biomass and its applications: A Review. *Carbohydrate Polymer Technologies and Applications, 7*(8), 100396. https://doi.org/10.1016/j.carpta.2023.100396
24. Belgacem, M. N., & Gandini, A. (2008). The state of the art. In *Monomers, polymers and composites from renewable resources*. https://doi.org/10.1016/B978-0-08-045316-3.00001-6
25. Clark, J. H. (2006). Green chemistry: Today (and tomorrow). *Green Chemistry, 8*, 17–21.
26. Koczoń, P., Bartyzel, B., Luliano, A., Klensporf-Pawlik, D., Kowalska, D., Majewska, E., Tarnowska, K., Zieniuk, B., & Gruczyńska-Sękowska, E. (2022). Chemical structures, properties, and applications of selected crude oil-based and bio-based polymers. *Polymers, 14*(24), 5551. https://doi.org/10.3390/polym14245551

27. Scott, G. (2000). "Green" polymers. *Polymer Degradation and Stability, 68*, 1–7.
28. Billmeyer, F. W. (1971). *Textbook of polymer science*. Wiley–Interscience.
29. Shabudeen Syed, P. S. (2009). *Engineering chemistry for textile and fashion technology*. https://www.researchgate.net/publication/275028382_Polymers
30. Karak, N. (2009). *Fundamentals of polymers: Raw materials to finish products*. PHI Learning Pvt. Ltd.
31. Bhat, A. R., Kumar, R., & Mural, P. K. S. (2023). Natural fibre reinforced polymer composites: A comprehensive review of tribo-mechanical properties. *Tribology International, 189*, 108978. https://doi.org/10.1016/j.triboint.2023.108978
32. Li, Y., Xiao, Y., Yu, L., Ji, K., & Li, D. (2022). A review on the tooling technologies for composites manufacturing of aerospace structures: Materials, structures and processes. *Composites Part A: Applied Science and Manufacturing, 154*, 106762. https://doi.org/10.1016/j.compositesa.2021.106762
33. Hosny, M. M., Abdel-Aziz, M. H., Zoromba, M. Sh., & Al-Hossainy, A. F. (2024). Design and synthesis of high-performance polymer blend solar cell. *Journal of Materials Research and Technology, 28*, 4199–4211. https://doi.org/10.1016/j.jmrt.2024.01.047
34. Huang, H. D., Ren, P. G., Zhong, G. J., Olah, A., Li, Z. M., Baer, E., & Zhu, L. (2023). Promising strategies and new opportunities for high barrier polymer packaging films. *Progress in Polymer Science, 144*, 101722. https://doi.org/10.1016/j.progpolymsci.2023.101722
35. Nath, P. C., Sharma, R., Debnath, S., Sharma, M., Inbaraj, B. S., Dikkala, P. K., Nayak, P. K., & Sridhar, K. (2023). Recent trends in polysaccharide-based biodegradable polymers for smart food packaging industry. *International Journal of Biological Macromolecules, 253*(Part 8), 127524. https://doi.org/10.1016/j.ijbiomac.2023.127524
36. Hossain, M. D., Saha, S., Hassan, M. K., Yuen, A. C., He, Y., Saha, S., & Hittini, W. (2021). Flame behaviour, fire hazard and fire testing approach for lightweight composite claddings: A review. *Journal of Structural Fire Engineering, 12*(3), 257–292. https://doi.org/10.1108/JSFE-09-2020-0027
37. Seco, A., Echeverría, A. M., Marcelino, S., García, B., & Espuelas, S. (2020). Durability of polyester polymer concretes based on metallurgical wastes for the manufacture of construction and building products. *Construction and Building Materials, 240*, 117907. https://doi.org/10.1016/j.conbuildmat.2019.117907
38. Bahin, L., Tourlonias, M., Bueno, M. A., Sharma, K., & Rossi, R. M. (2023). Smart textiles with polymer optical fibre implementation for in-situ measurements of compression and bending. *Sensors and Actuators A: Physical, 350*, 114117. https://doi.org/10.1016/j.sna.2022.114117
39. Selvam, S., & Yim, J. H. (2023). Effective self-charge boosting sweat electrolyte textile supercapacitors array from bio-compatible polymer metal chelates. *Journal of Power Sources, 556*, 232511. https://doi.org/10.1016/j.jpowsour.2022.232511
40. Li, L., Han, L., Hu, H., & Zhang, R. (2023). A review on polymers and their composites for flexible electronics. *Materials Advances, 4*(3), 726–746. https://pubs.rsc.org/en/content/articlelanding/2023/ma/d2ma00940d
41. Salas, A., Pazniak, H., Gonzalez-Julian, J., Bianco, S., Amici, J., Ouisse, T., Roppolo, I., & Cocuzza, M. (2023). Development of polymeric/MXenes composites towards 3D printable electronics. *Composites, Part B: Engineering, 263*, 110854. https://doi.org/10.1016/j.compositesb.2023.110854
42. Moore, C. J. (2008). Synthetic polymers in the marine environment: A rapidly increasing, long-term threat. *Environmental Research, 108*(2), 131–139. https://doi.org/10.1016/j.envres.2008.07.025
43. Hazwani, M., Abdul Majid, M.S., Azaman, M. D., Ridzuan, M. J. M., & Cheng, E. M. (2023). Mechanical properties and flammability of pineapple leaf fibre (PALF) reinforced polymer composite with hybridized fire retardants. *Materials Today: Proceedings*. https://doi.org/10.1016/j.matpr.2023.09.198
44. Narita, C., & Yamada, K. (2024). Development of glossy and UV-resistant Urushi coatings by thermal polymerization. *Progress in Organic Coatings, 186*, 108032. https://doi.org/10.1016/j.porgcoat.2023.108032

45. Liu, M., Bauman, L., Nogueira, C. L., Aucoin, M. G., Anderson, W. A., & Zhao, B. (2022). Antimicrobial polymeric composites for high-touch surfaces in healthcare applications. *Current Opinion in Biomedical Engineering, 22*, 100395. https://doi.org/10.1016/j.cobme.2022.100395
46. Singh, R., Kaur, J., Gupta, K., Singh, M., Kanaoujiya, R., & Kaur, N. (2022). Recent advances and applications of polymeric materials in healthcare sector and COVID-19 management. *Materials Today: Proceedings, 62*(Part 6), 2878–2882. https://doi.org/10.1016/j.matpr.2022.02.472
47. He, Q., Sun, K., Shi, Z., Liu, Y., & Fan, R. (2023). Polymer dielectrics for capacitive energy storage: From theories, materials to industrial capacitors. *Materials Today, 68*, 298–333. https://doi.org/10.1016/j.mattod.2023.07.023
48. Balakrishnan, P., Thomas, M. S., Pothan, L. A., Thomas, S., & Sreekala, M. S. (2014). Polymer films for packaging. In *Encyclopedia of polymeric nanomaterials* (pp. 1–8). https://doi.org/10.1007/978-3-642-36199-9_406-1
49. Frank, A., & Biederbick, K. (1984). *Kunststoffkompendium*. Vogel-Buchverlag.
50. Kaufmann, H. (1988). Grundlagen Der Organischen Chemie, 8. Auflage, Birkh€auser Verlag, Basel.
51. Klein, R. (2011). *Laser welding of plastics*, 1st edn. Wiley-VCH Verlag GmbH & Co. KGaA. Published 2011 by Wiley-VCH Verlag GmbH & Co. KGaA.
52. Lange, J.-P. (2021). *ACS*. ACS Publications. https://pubs.acs.org/doi/10.1021/acssuschemeng.1c05013
53. Kumar, R. (2021). Tertiary and quaternary recycling of thermoplastics by additive manufacturing approach for thermal sustainability. *sciencsDirect*, https://www.sciencedirect.com/science/article/abs/pii/S2214785320360466#:~:text=The%20quaternary%20recycling%20is%20also,the%20energy%20and%20fuel%20production
54. Desidery, L., & Lanotte, M. (2022). Polymers and plastics: Types, properties, and manufacturing. In F. Giustozzi & S. Nizamuddin (Eds.), *Plastic waste for sustainable asphalt roads* (pp. 3–28). In Woodhead Publishing Series in Civil and Structural Engineering. Woodhead Publishing. ISBN 9780323857895. https://doi.org/10.1016/B978-0-323-85789-5.00001-0
55. Veerasingam, S., Ranjani, M., Venkatachalapathy, R., Bagaev, A., Mukhanov, V., Litvinyuk, D., Verzhevskaia, L., Guganathan, L., & Vethamony, P. (2020). Microplastics in different environmental compartments in India: Analytical methods, distribution, associated contaminants and research needs. *TrAC Trends in Analytical Chemistry, 133*, 116071.
56. Carney, A. B., & Eggert, H. (2019). Marine plastic pollution: Sources, impacts, and policy issues. *Review of Environmental Economics and Policy, 13*, 317–326.
57. Amrutha, K., & Warrier, A. K. (2020). The first report on the source-to-sink characterization of microplastic pollution from a riverine environment in tropical India. *Science of the Total Environment, 739*, 140377.
58. TERI. (2020). Plastic waste management: Turning challenges into opportunities. Retrieved January 15, 2021, from https://www.teriin.org/sites/default/files/2020-12/plastic-management_0.pdf
59. Santhosh, S., & Shrivastav, R. (2021). Plastic waste management: What can India learn from other countries. Retrieved January 28, 2021, from https://www.downtoearth.org.in/blog/waste/plastic-waste-management-what-can-india-learn-from-other-countries-67048
60. United Nations Development Programme (UNDP). Plastic Waste Management Programme (2018–2024). https://www.in.undp.org/content/india/en/home/projects/plastic-waste-management.html#:~:text=Plastic%20Waste%20Management%20Programme%20(2018%2D2024)&text=India%20generates%2015%20million%20tonnes,the%20waste%20pickers%2C%20mostly%20women
61. Bhattacharya, R., Chandrasekhar, K., Roy, P., & Khan, A. (2018, June). Challenges and opportunities: Plastic waste management in India. https://archive.nyu.edu/handle/2451/42242
62. Sikka, P. (2007). Plastic waste management in India. *Packages in India, 39*, 43.
63. Central Pollution Control Board. Annual Report for the Year 2018–2019 on Implementation of Plastic Waste Management Rules. https://cpcb.nic.in/uploads/plasticwaste/Annual_Report_2018-19_PWM.pdf

64. Centre for Science and Environment. Report on Managing Plastic Waste in India: Challenges and Agenda, 2020. https://www.cseindia.org/content/downloadreports/10352
65. Liang, Y., Tan, Q., Song, Q., & Li, J. (2021). An analysis of the plastic waste trade and management in Asia. *Waste Management, 119*, 242–253.
66. Andrady, A. L. (2017). The plastic in microplastics: A review. *Science Direct*. https://www.sciencedirect.com/science/article/abs/pii/S0025326X1730111X
67. Ashrafy, A., et al. (2023). Microplastics pollution: A brief review of its source and abundance in different aquatic ecosystems. *Science Direct*. https://www.sciencedirect.com/science/article/pii/S2772416622001711
68. Lee, Y., Cho, J., Sohn, J., & Kim, C. (2023). Health effects of microplastic exposures: Current issues and perspectives in South Korea. *64*(5), 301–308.
69. Li, Y., Tao, L., Wang, Q., Wang, F., Li, G., & Song, M. (2023). Potential health impact of microplastics: A review of environmental distribution, human exposure, and toxic effects. *Environment & Health, 1*(4), 249–257. https://doi.org/10.1021/envhealth.3c00052
70. Food and Agriculture Organization. Microplastics in food commodities. A food safety review on human exposure through dietary sources. Retrieved February 20, 2023, from https://www.fao.org/documents/card/en/c/cc2392en
71. Peng, L. C., Fu, D. D., Qi, H. Y., Lan, C. Q., Yu, H. M., & Ge, C. J. (2020). Micro- and nano-plastics in marine environment: Source, distribution and threats - A review. *Science of the Total Environment, 698*, 134254.
72. Kwon, O. Y., Kang, J. H., Shim, W. J., & Hong, S. (2023). Distribution characteristics of floating microplastics in the Eastern Coast of the South Sea in Korea. Retrieved February 20, 2023, from https://sciwatch.kiost.ac.kr/handle/2020.kiost/25159
73. Prata, J. C. (2018). Airborne microplastics: Consequences to human health? *Environmental Pollution, 234*, 115–126. [PubMed] [Google Scholar], 2017, pp. 170–186, ISSN 0956-053X. https://doi.org/10.1016/j.wasman.2017.04.014
74. Duru, C. E., Enedoh, M. C., & Duru, I. A. (2019). Physicochemical assessment of borehole water in a reclaimed section of Nekede mechanic village, Imo State, Nigeria. *Chemistry Africa*. https://doi.org/10.1007/s42250-019-00077-8
75. Mourshed, M., Masud, M. H., Rashid, F., & Joardder, M. U. H. (2017). Towards the effective plastic waste management in Bangladesh: A review. *Environmental Science and Pollution Research, 24*(35), 27021–27046.
76. Paletta Angelo, W. L.-L. (2019). Barriers and challenges to plastics valorisation in the context of a circular economy: Case studies from Italy. *Science Direct*. https://doi.org/10.1016/j.jclepro.2019.118149
77. Landon-Lane, M. (2018). Corporate social responsibility in marine plastic debris governance. *PubMed*. https://doi.org/10.1016/j.marpolbul.2017.11.054
78. Alhazmi, H., Almansour, F. H., & Aldhafeeri, Z. (2021). Plastic waste management: A review of existing life cycle assessment studies. https://www.mdpi.com/2071-1050/13/10/5340
79. Sathiyapoobalan, Sundaralingam, & Ramanathan, N. (2023). Retrieved from https://iopscience.iop.org/article/10.1088/2515-7620/acfecf/pdf
80. Oehlmann, J., Schulte-Oehlmann, U., & Werner Kloas, O. J. (2009). A critical analysis of the biological impacts of plasticizers on wildlife. *PubMed*. https://doi.org/10.1098/rstb.2008.0242
81. Teuten, E. L., et al. (2009). Transport and release of chemicals from plastics to the environment and to wildlife. https://doi.org/10.1098/rstb.2008.0284
82. World Economic Forum, Future of the Environment, How a 'trash-to-treasure' mentality can change plastic waste, Aug 25, 2021.
83. Klee, R. J., & Graedel, T. E. (2004). Elemental cycles: A status report on Human or natural dominance. *Annual Review of Environment and Resources, 29*, 69–107.
84. European Bioplastics. (2013). Bioplastics: Facts and figures. European Bioplastics.

85. Moldova Report. (2023). The GEF Small Grants Programme. Zero Plastic, Study of Plastic Waste in Republic Moldova, UNDP. https://www.bing.com/ck/a?!&&p=6b20f54e9de18ca7JmltdHM9MTcwNzUyMzIwMCZpZ3VpZD0wZTdiMDE2MS1kZGZhLTZiNmItM2Y0My0xMjAxZGM2MTZhN2EmaW5zaWQ9NTIwMQ&ptn=3&ver=2&hsh=3&fclid=0e7b0161-ddfa-6b6b-3f43-1201dc616a7a&psq=Study+on+Plastic+Waste+in+the+Republic+of+Moldova&u=a1aHR0cHM6Ly93d3cudW5kcC5vcmcvbW9sZG92YS9wdWJsaWNhdGlvbnMvc3R1ZHktcGxhc3RpYy13YXN0ZS1yZXB1YmxpYy1tb2xkb3Zh&ntb=1
86. *Plastic waste recycling: Existing Indian scenario and future opportunities*. (2022). Springer.
87. Song, J. H., & Song, R. J. (2009). *Biodegradable and compostable alternatives to conventional plastics*. Royal Society Publishing.
88. Shen, L., Nieuwlaat, E., Worrell, E., & Patel, M. K. (2011). Life cycle energy and GHG emissions of PET recycling: Change oriented effects. *International Journal of Life Cycle Analysis, 16*, 522–536.
89. Shen, L., Worrell, E., & Patel, M. K. (2010). Open loop recycling: A LCA case study of PET bottle-to-fibre recycling. *Resources, Conservation and Recycling, 55*, 34–52.
90. Garlapati, V. K. (2016). E-waste in India and developed countries: Management, recycling, business and biotechnological initiatives. *Renewable and Sustainable Energy Reviews, 54*, 874–881.
91. Cui, J., & Zhang, L. (2008). Metallurgical recovery of metals from electronic waste: A review. *Journal of Hazardous Materials, 158*(2–3), 228–256.
92. Gulliani, S., Volpe, M., Messineo, A., & Volpe, R. (2023, May). Recovery of metals and valuable chemicals from waste electric and electronic materials: A critical review of existing technologies. *RSC Sustainability, 1*, 1085–1108.
93. Kiddee, P., Naidu, R., & Wong, M. H. (2013). Electronic waste management approaches: An overview. *Waste Management, 2013*(33), 1237–1250.
94. Oguchi, M., Murakami, S., Sakankura, H., Kida, A., & Kameya, T. (2011). A preliminary categorization of end of life electrical and electronic equipment as secondary metal resources. *Waste Management, 31*, 2150–2160.
95. Pathaka, P., Srivastavab, R. R., & Ojasvic. (2017). Assessment of legislation and practices for the sustainable management of waste electrical and electronic equipment in India. *Renewable and Sustainable Energy Reviews, 78*, 220–232.
96. Gotro, J. (2011). Polymer innovation blog. The seven classifications of plastics: Are they all really recycled? February 23, 2011.
97. Harper, C. A. (n.d.). *Modern hand book*. McGraw Hill Professional, 24 March 2000. Technology & Engineering (1298 pp.).
98. MHUA, M. o. (2019). *Plastic waste managment*.
99. Bandarra, B. S., & Quina, M. J. (2021). Municipal solid waste incineration and sustainable development. In Y. Gao, W. Song, J. L. Liu, & S. Bashir (Eds.), *Advances in sustainable energy*. Springer. https://doi.org/10.1007/978-3-030-74406-9_23
100. OECD. (2022). *Global plastics outlook: Economic drivers, environmental impacts and policy options*. OECD Publishing.
101. UNEP. (2021). Drowning in plastics: Marine litter and plastic waste vital graphics. Nairobi.
102. OECD Global Plastics Outlook Database. https://doi.org/10.1787/c0821f81-en
103. Lu, J. W., Zhang, S., Hai, J., & Lei, M. (2017). Status and perspectives of municipal solid waste incineration in China: A comparison with developed regions. *Waste Management, 69*.
104. Central Pollution Control Board. (2017, September). *Consolidated guidelines for segregation, collection and disposal of plastic waste*.
105. Lamichhane, G., Acharya, A., Marahatha, R., Modi, B., Paudel, R., Adhikari, A., Raut, B. K., Aryal, S., & Parajuli, N. (2023). Microplastics in environment: Global concern, challenges, and controlling measures. *International Journal of Environmental Science and Technology, 20*, 4673–4694.
106. GAIA.·(2019). Plastic pollution and waste incineration. https://www.no-burn.org/wp-content/uploads/Plastic-x-Incineration-2019.pdf

107. OECD. (2022, February). Plastic pollution is growing relentlessly as waste management and recycling fall short, Says OECD (2022, February). https://www.oecd.org/environment/plastic-pollution-is-growing-relentlessly-as-waste-management-and-recycling-fallshort.htm#:~:text=Global%20plastic%20waste%20generation%20more,11%25%20from%20clothing%20and%20textiles
108. Kabeyi, M. J., & Olanrewaju, O. A. (2023). Review and design overview of plastic waste-to-pyrolysis oil conversion with implications on the energy transition. Hustad, J. E. (Ed.). https://doi.org/10.1155/2023/1821129
109. Sundaralingam, S., & Ramanathan, N. (2023). Efficient plastic categorization for recycling and real time annotated data collection with TensorFlow object detection model. https://iopscience.iop.org/article/10.1088/2515-7620/acfecf/pdf
110. Top 8 Recycling Technology Trends & Innovations in 2024. (n.d.). *Start us insight*. https://www.startus-insights.com/innovators-guide/recycling-technology-trends-innovation/
111. Gothai, E., Thamilselvan, R., Natesan, P., Keerthivasan, M., Kabinesh, K., et al. (2022). Plastic waste classification using CNN for supporting 3R's principle. *IEEE Explore*. https://ieeexplore.ieee.org/document/9740902
112. Patel, R. (2024). Advanced recycling technologies to efficiently recycle global waste (11 pp.). https://www.upperinc.com/blog/recycling-technologies/
113. Kalair, A. R., Seyedmahmoudian, M., Stojcevski, A., Abas, N., & Khan, N. (2021). Waste to energy conversion for a sustainable future. *Heliyon, 7*(10), e08155. ISSN 2405-8440. https://doi.org/10.1016/j.heliyon.2021.e08155
114. Ellen MacArthur Foundation. Plastics and the Circular Economy. https://www.ellenmacarthurfoundation.org/explore/plastics-and-the-circular-economy
115. Hindustan Unilever Ltd. Could-Based Digital Governance for Traceability, Accountability in Waste Supply Chain. https://www.undp.org/content/dam/india/docs/evaluation-reports/HUL%20Brochure.pdf
116. Hopewel, J., Dvorak, R., & Kosior, E. (n.d.). Plastics recycling: Challenges and opportunit. *PubMed Central*. https://www.ncbi.nlm.nih.gov/pmc/articles/PMC2873020/
117. https://www.aaapolymer.com/recyclable-plastic-materials/
118. *Life Cycle Initiative A Life Cycle Approach to Plastic Pollution*. Hosted by UNEP (2021). Retrieved from life cycle initiative. https://www.lifecycleinitiative.org/activities/life-cycle-assessment-in-high-impact-sectors/life-cycle-approach-to-plastic-pollution/
119. Shen, M., Hu, T., Huang, W., Song, B., Qin, M., Yi, H., Zeng, G., & Zhang, Y. (2021). Can incineration completely eliminate plastic wastes? An investigation of microplastics and heavy metals in the bottom ash and fly ash from an incineration plant. *Science of the Total Environment, 779*, 146528. ISSN 0048-9697. https://doi.org/10.1016/j.scitotenv.2021.146528
120. Alabi, O. A., Ologbonjaye, K. I., Awosolu, O., & Alalade, O. E. (2019). Public and environmental health effects of plastic wastes disposal: A review. *Journal of Toxicology and Risk Assessment, 5*(1). https://doi.org/10.23937/2572-4061.1510021

GPSR Compliance

The European Union's (EU) General Product Safety Regulation (GPSR) is a set of rules that requires consumer products to be safe and our obligations to ensure this.

If you have any concerns about our products, you can contact us on ProductSafety@springernature.com

In case Publisher is established outside the EU, the EU authorized representative is:

Springer Nature Customer Service Center GmbH
Europaplatz 3
69115 Heidelberg, Germany

The manufacturer's authorised representative in the EU is Springer Nature Customer Service Centre GmbH, Europaplatz 3, 69115 Heidelberg, Germany. If you have any concerns regarding our products, please contact ProductSafety@springernature.com

Printed and bound by CPI Group (UK) Ltd, Croydon, CR0 4YY
01/10/2026
02225862-0007